FISHERY MANAGEMENT

FISHERY MANAGEMENT

PROFESSOR ARVIND KUMAR

Environmental Biology Research Unit,

Post Graduate Department of Zoology,

S.K.M. University, Dumka - 814 101, India.

A.P.H. PUBLISHING CORPORATION

ANSARI ROAD, DARYA GANJ

NEW DELHI-110 002

Published by
S. B. Nangia
A.P.H. Publishing Corporation
4435-36/7, Ansari Road, Darya Ganj
New Delhi-110 002
☎ 23274050
email : aphbooks@vsnl.net

2018

Typesetting at
NEW APCON

Printed at
Balaji offset
Navin Shahdara, Delhi-32

PREFACE

Due to ever increasing demand for fish as source of white meat and the rapid overall development, the aquatic ecosystems are under constant threat. The destruction of fish habitat by alteration of river systems, increased water abstraction, over fishing of important fishes, pollution of fishery waters, land development and domestication of species have been going on indiscriminately. These man-made alterations have naturally been showing qualitative and quantitative changes in ecobiology and population structure of fish.

Though during the preceding years, India has made substantial progress in fish research and fish production, but it is not satisfactory in comparison to the western countries. So, the task involves intensive research in the field of Ichthyology and Fisheries Science for selection of ideal fish species for aquaculture and their genetic upgrading through selection, cross-breeding and lineage, resolving of ideal feeding conditions and optimal feeds, determination of ideal environment for rapid growth and weight gain as well as effective control of diseases. Equally important is suitably planned research for developing suitable management practices for lentic ecosystem in order to maintain equilibrium in the aquatic ecosystem for optimum Pisciculture.

Life has had an inevitable role in shaping of environmental evolution of this watery planet. During the last 50 years, the earth has experienced and recovered from many natural disasters. Due to population explosion, large scale of urbanization, mushroom growth of industries and unmindful disposal of waste materials, the ecobalance of the biosphere has been threatened. Ecodegradation has become a growing challenge to mankind all over the world. Various unplanned activities have not only altered the scenario of the environment but have also disrupted the delicate fabric of man-nature relation. Not only this, they have changed the economic and socio-cultural life of the people. Therefore, each and every fish biologist has to learn how to develop innovative research in order to preserve, conserve and promote the fish research for the days to come. Thus, keeping in view of highlighting some of the lesser-known facts, I have ventured to bring out such an innovative and comprehensive review volume with a hope that the fish biologists may undertake further investigation on their undeciphered facts, which are neglected hitherto.

The all 49 chapters are written by authors who are acknowledged experts in their respective fields with the intention to provide sufficient depth of the subjects to satisfy the needs at a level, which will be comprehensive and interesting. The book shall be useful to the students, teachers, scientists and researchers in the field of Environmental Sciences and other people with the similar interest.

My special thanks and appreciation go to the scientists whose contributions have enriched this. I wish to express my sincere gratitude to Professor (Dr.) Indu Dhan, Hon'ble Vice Chancellor, and Professor (Dr.) B. N. Jha, Pro Vice-chancellor, S. K. M. University, Dumka who have been the source of constant inspiration. I am especially thankful to Dr. Sikandar Prasad Yadav, Principal, S. P. College, Dumka for his encouragement. I am grateful to Professor S. P. Roy and Shri Tribhuvan Poddar of T. M. Bhagalpur University, Bhagalpur who have been a source of continuous encouragement and inspiration to me. I owe my special thanks to Professor M. C. Das, Hon'ble Vice Chancellor of

Sambalpur University, Professor H. R. Singh of Allahabad University, Dr. S. N. Kaul of NEERI, Nagpur, Professor N. C. Datta of Calcutta University, Professor Ajit Varma of JNU, New Delhi, Professor A. K. Mittal of Banaras Hindu University, Varanasi, Professor S. K. Konar of Kalyani University, Professor P. S. Murthy of Bangalore University, Professor P. Natrajan of Kerala University, Professor M. Vikram Reddy of Pondicherry University, Professor P. P. Sood of Saurashtra Univ., Rajkot, Professor B. Manihar Sharma of Manipur University, Professor G. C. Goswami of Gauhati University, Professor K. Kapoor of Jodhpur University, Professor S. P. Hosmani of Mysore University, Professor Tanmay Bhattacharya of Midnapore University, Professor P. C. Mishra of Sambalpur University, Professor B. D. Joshi of Hardwar University, Professor G. C. Pandey of Faizabad University, Professor K. C. Sharma of Ajmer University, Professor M. Raziuddin of Hazaribag University, Professor U. S. Bagde of Mumbai University, Dr. M. P. Sinha of Ranchi University, Professor Gurdeep Singh of I. S. M., Dhanbad, Dr. P. K. Goel of Karad and Dr. S. K. Maiti of ISM, Dhanbad. I also acknowledge the incentives provided by one of my research scholars, Dr. Chandan Bohra, Head of Zoology, B. S. K. College, Barharwa for helping me actively in bringing out this book.

I also express my deep sense of gratitude to my parents whose blessings have always prompted me to pursue academic activities deeply. I am also thankful to my sweet wife, Professor Kumari Bimla and my two lovely sons, Kumar Pallav Shivshankaran and Kumar Prasun Ramakrishnan whose natural smiles extended to me relief all through this tiresome endeavour.

Last but not the least, I am also thankful to Mr. S. B. Nangia, APH Publishing Corporation, New Delhi for publishing this book. Finally, I will always remain a debtor to all my well-wishers for their blessings, without which this book would not have come into existence.

Dumka, Professor ARVIND KUMAR
December 20, 2003.

Dedicated
to

My Father

Late Shri Brahmadeo Sinha

A Noted Freedom Fighter,

Educationist,

Gandhian,

Philosopher,

&

A Humanitarian.

CONTENTS

CHAPTER 1

Studies on Ecotaxonomy of Fishes in Jharkhand with Reference to Ecological Niche and Drainage Pattern

Arvind Kumar and Ajay Kumar

Environmental Biology Research Unit,
Post Graduate Department of Zoology,
S. K. M. University, Dumka - 814101, Jharkhand.

ABSTRACT

The present investigation was carried out to enumerate the qualitative distribution of Ichthyofauna throughout Jharkhand (21°58' 10"N - 25°19' 15"N and 83°20'E- 88°4' 44"E), India with reference to their ecological niche. 58 species of fishes have been recorded in different waterbodies (Ponds, ditches, rivers, canals, reservoirs, lakes, hillstreams, paddy fields, etc.) of Jharkhand. Fishes of order Cypriniformes predominated the Ichthyofauna with 23 species (18 species of family Cyprinidae and 5 of Cobitidae); followed by order Siluriformes with 16 species (5 of each family Bagridae and Schilbeidae, 3 of Sisoridae and one each of family Claridae, Heteropneustidae and Siluridae); 5 species each of Clupeiformes and Perciformes, 3 species of Mastacembeliformes and only one each of Beloniformes and Symbranchiformes. It is a record of the seasonal survey conducted between March 2000 to May 2002 site by site. Althogether 58 species of fishes belonging to 18 families of 8 orders have been identified. They have been described with their taxonomic position, local names, taxonomic description, place of common occurrence, consumer groups and methods used to catch them. Physio-Graphy and Drainage pattern of Jharkhand have also been discussed.

Keywords : *Ichthyofauna, Jharkhand, Taxonomic description, Consumer groups.*

INTRODUCTION

Jharkhand is a newly formed state of India and is endowed with not only dense forests and plently of mineral resources, it abodes rich faunal biodiversity also. Jharkhand has a large number of waterbodies like ponds, streams, lakes, swamps, tanks, reservoirs, hillstreams, rivers, etc. which provide the natural habitat for various types of aquatic biota.

But till now, nothing is known about faunal biodiversity of Jharkhand. except Molluscan fauna (Kumar and Kumar, 2002). Particularly, any information on Ichthyofauna of Jharkhand is not available; although, much informations are available from other parts of India (Hamilton, 1822; Day, 1869,1878; Chaudhuri, 1912,1916; Hora, 1921; Menon, 1947, 49; Mishra, 1959; Datta Gupta et.al. 1961; Anon, 1962; Srivastava, 1968; Verma, 1977; Srivastava, 1984; Datta Munshi and Srivastava, 1988 and Talwar & Jhingran, 1991). Presently, an attampt has been made to enumerate the different species of fishes and their distribution in the waterbodies of Jharkhand.

Physio-Graphy and Drainage pattern reflects not only the abundance and floral diversity but reflects the abundance and faunal diversity also.

PHYSIO-GRAPHY OF JHARKHAND

The story of the physiography of Jharkhand consists of several chapters. More than 90% of the plain are composed of the Archaen rocks. Block faulting has been responsible for the preservation of vegetations as coal fields where as uplifts, warps, tilts and even depressions of Tertiary period explain its polygentic landforms.

The Archaen rocks belong to two chief groups i.e.

(i) Sedimentary rocks associated with lavas and

(ii) Intrusive basic igneous rocks and wide spread granite.

The sedimentary and basic igneous groups are important in northen and southern Chota Nagpur, where as the granite is dominant in the east and west direction across the core of Chota Nagpur.

Structure plays a significant role in the evolution of landscape. The landscape of the higher western Basaltic plateau stands out as a great contrast to the neighbouring central tract of the Chota Nagpur highlands. The higher western plateau has been cut up by tributaries of the North Koel to the north and by sankh to the south into many isolated masses locally known as 'Pats'.

The Ranchi Peneplain extends to the east of the scraps of the high western plateau where rivers flow differently over hardx crystalline rocks and softer Dhawars with braided Channels.

The Ranchi Peneplain descends to Kodanna and Giridih on northen and eastern sides. To the west the spurs of the Peneplain plunge into broken country of Palamu, but the south-west the narrow ridge-merges into Ranchi surface. The broken country of Palamau shows evidence of diversification and occurs in piece meal north of the Auranga till the Vindhyan scarps are reached.

The lower Hazaribagh plateau shows a 1250 feet surface and extends from the Champaran and Kodarma towards Giridih. Damodar basin is the product of prolonged erosion over the Gondwana deposits.

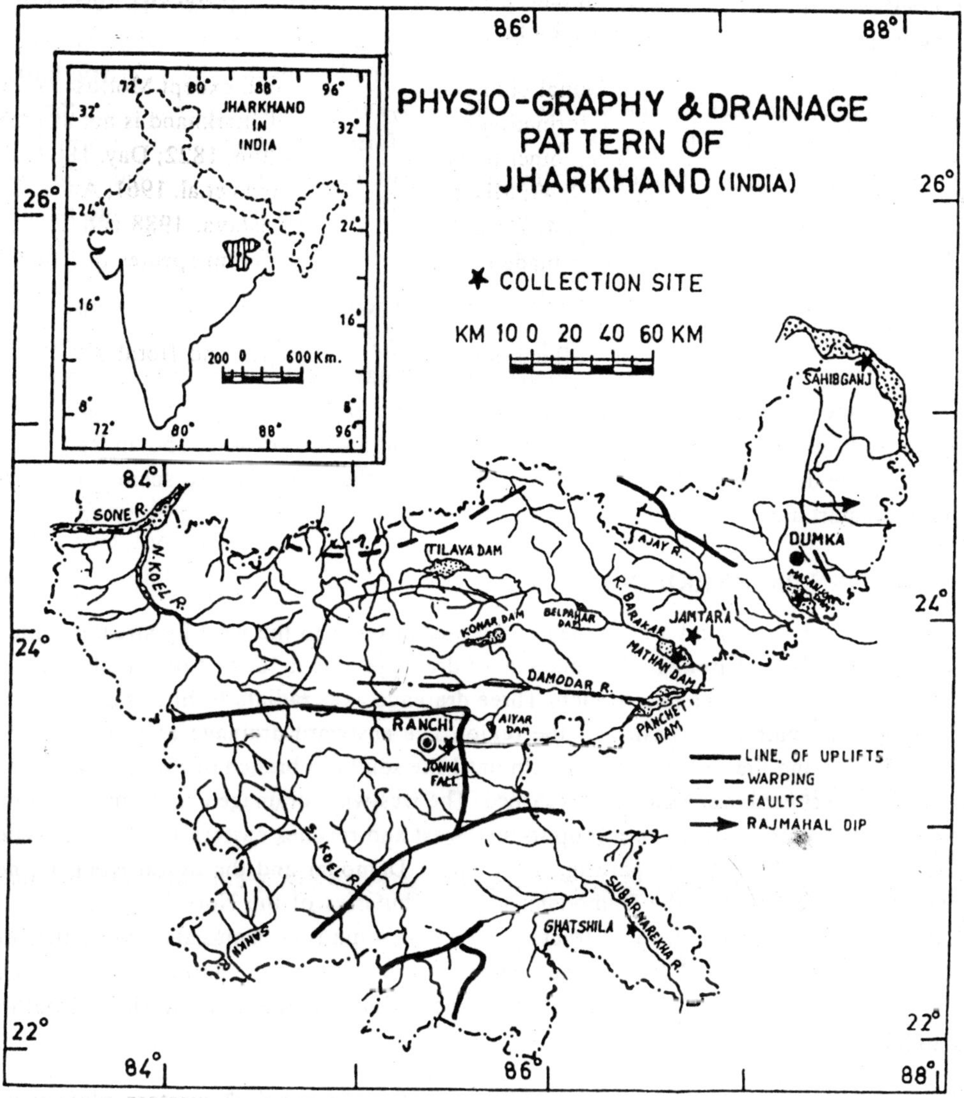

Fig 1.1 : Map showing the location of collection site of fishes in Jharkhand.

The plains of Singhbhum are composed either of Granite or of Schist or Shales. The gently undulating country is covered with a thick mantle of soil. The morphology of the Damodar basin is the product of prolonged erosion over the Gondwana deposits.

The Rajmahal highlands consist of a chain of level plateau of trap and of fresh Mesozoic beds. It is marked by scrap to the west and lava-capped plateau dips to the east, corresponding to the prevailing dip of the rocks.

The landscape in the vicinity of Jharia is marked by gentle undulations except towards Parasnath. Parasnath-Tundi ridge dominates the landscape rising to an elevation of 4,480 feet above sea-level. Tundi range is a mere eastward extension of Parasnath with the characteristic knife-edge crest. Cuboidal boulder are seen on the surface. Dolerite dykes, cut up by rains, cross the country. Knob-like summits of many hills locally known as 'Munda' form another type of landform. From the top of Mount Parasnath, the morphology of the central part of the basin is revealed in its sharp conical hill rocks clustering more to the south and west. The Jamunia meanders in the plain dotted with conical hills.

Water falls are also significantly paint the landscape where there are breaks in slope of varying degrees. The correspond to uplifts, warps and tilts of probably late Tertiary and Pleistocene periods. The Hundru fall (243'), on the Subarnarekha, Jonha fall or Gautamdhara fall on the Ganga Nadi, Dasam fall(130') on the Kanchi, Sadni fall (200') on the Sankh. Hirni fall on a small river, Tiger fall on the Barakar, Garha Nadi fall; Moti Jhara on a tributary of the Ganga with a drop of 150' in Rajmahal hills, Rajroppa falls on the discordant junction of Bhera Nadi and the Damodar are scenic spots of great geomorphologic importance.

DRAINAGE PATTERN OF JHARKHAND

The highlands of Jharkhand are drained on all sides except to the west by streams which feed the major rivers in the Bay of Bengal. The nucleus of drainage axis is located on an elevated tract of the plateau country south of Piska near Ranchi.Three drainage axes radiate from this elevated tract of the country which the main river basins of the region.The eastward drainage axis froms watershed between the Damodar and the Subarnarekha basin until the southern branch of this eastward drainage axis is reached, the westward drainage axis is reached. The westward drainage axis froms a semicircular water-parting till it takes a south eastern loop to turn west and reaches eastern limit of the state. The major drainage axis froms the water parting of the upper Damodar and the South Koel. Its nothern branch froms the divide between the Damodar and the tributaries of the North Koel.The southward drainage axis radiating from the elevated tract near Piska-Ladma road froms the divide between the subarnarekha and the south Koel basins.The drainage axis turns to the south-east in Singhbhum forming the watershed between the Karo, a tributary of the south Koel and the Karkari, a tributary of the subarnarekha.

A few minor drainage axes like the Vindhyan Scrap, Pat, Rajmahal, Dhanjori etc. are also occur.

The subarnarekha, South Koel, the Mor river etc. provide distinct examples of super imposition.

The initial drainage aligned with the former folded surface but their marks are not to be traced anywhere. The unhindered erosion depicts some of the chapters of intervention, deposition and creation of new drainage over Dalma lava surface, Dhanjori-Kolhan-Vidhyan surfaces. The relics of former streams may be accepted as Proto-Subararekha and Proto-Koel.

The rift vally of the Damodar initated many streams during the post-carboniferous Ice-age. The Damodar was formed in sympathy with the penman though, the former truncated streams might have been added with new drainage lines in relation to elevations during tertiary period. The Rajmahal and

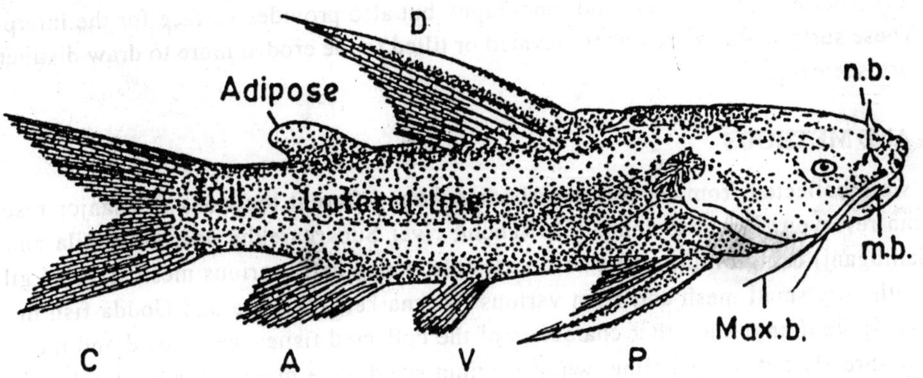

Fig. 1.2 a

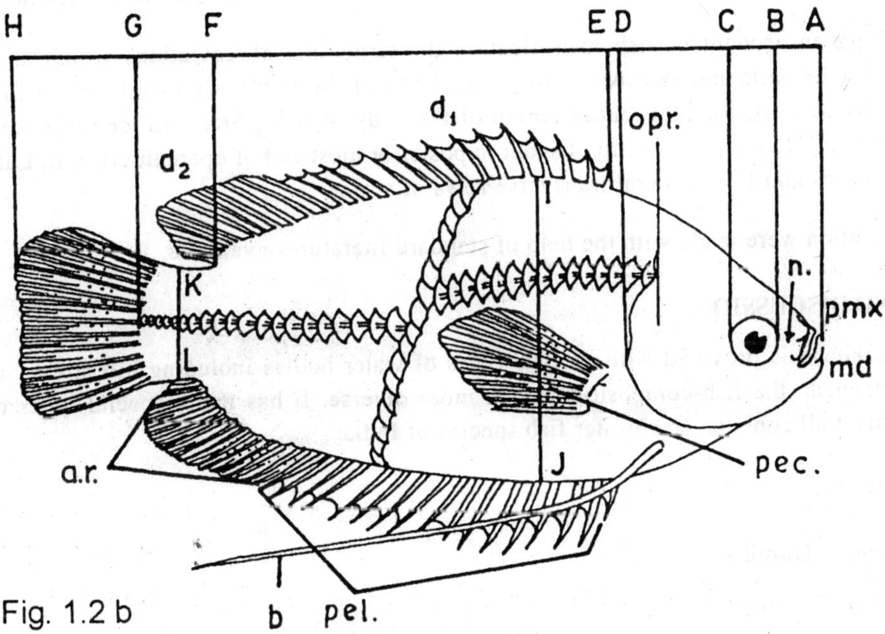

Fig. 1.2 b

Fig.1.2a: *Rita rita*(Ham.), labelled to show various parts of the body of a siluriod fish.
A. Anal fin, C. Caudal fin; D Dorsal fin; P Pectoral fin; mb. Maxillary barbel; n.b. Nasal barbel. V. Ventral fin.

Fig1.2b: *Colisa fasciatus* (Bloch & Schn.), labelled to show various parts of a scaly fish.
a.r. Anal rays; b. Barbel; dl. First dorsal fin; d2. Second dorsal fin; md. Mandible; n. Nostrils, Opr. Operculum; Pec. Pectoral fin; Pel. Pelvic/ventral fin; Pmx. Premaxilla; A-H. Total length; A-G. Standard length A.-D. Length of head. I.-.J. 1Height or depth of body; F.-G Length of caudal peduncle; K -L Height of caudal peduncle; B.-C. Diameter of eye; A.-B. Snout.

Deccan have flows buried former rivers and topography but also provides surface for the interplay of new streams. These surface also were either elevated or tilted to the eroded more to draw distinct river basins which are seen today.

MATERIALS AND METHODS

Fishes were collected from several waterbodies from different locations (Masanjor reservoir, Dumka; Mathan reservoir, Dhanbad;Jonhafall, Ranchi; river Subarnarekha near Ghatshila and river Ganga near Sahibganj) using divides drag nets, cast nets, scoop nets, various mesh sizes of gillnets, circular nets with very small mesh and from various fish markets (Jamtara and Godda fish markets) also. The colour spots if any; and other characters of the collected fishes were noted and the smaller specimens were directly put in formaline, while medium sized were given a longitudinal incision in abdominal region before to put in formaline and larger forms like *Pangasius pangasius, Channa striatus, wallago attu,* etc. were fixed by injecting 10% formaline into the muscles after making an incision in abdominal reason and finally kept in 10% formaline for identification and further studies.

In making an account of fish identification the terms and abbreviations which are used to describe the various parts and measurements of the body of the fishes are mentioned in Fig.1.2a & 1.2b and few are as : Total length (entire length of the body of fish); Standard length (total length - length of caudal fin); Lenth of head (tip of snout to posterior most end of operculum); L.I., Lateral-line of perforated scales and L.tr., Lateral transverse rows of scales.

Identification were made with the help of standard literatures available.

RESULTS AND DISCUSSION

As Jharkhand is provided with various types of water bodies including hill strems and river Ganges, which made the fish-composition much more diverse. It has many special types of fishes along with almost all common freshwater fish species of India.

ICHTHYO - FAUNA

1. *Gudusia chapra* (Hamilton)

Systematic Position

Order	-	Clupeiformes
Family	-	Clupeidae
Genus	-	Gudusia
Species	-	*Gudusia chapra* (Ham.)
Local Name	-	Suiya/Chapra/Chapri

D.14-16(2-3/12-13);P.13;V.8;A.21- 24(2/19-22);C.17;L.r.80-110; L.tr. 33-35.

PLATE I

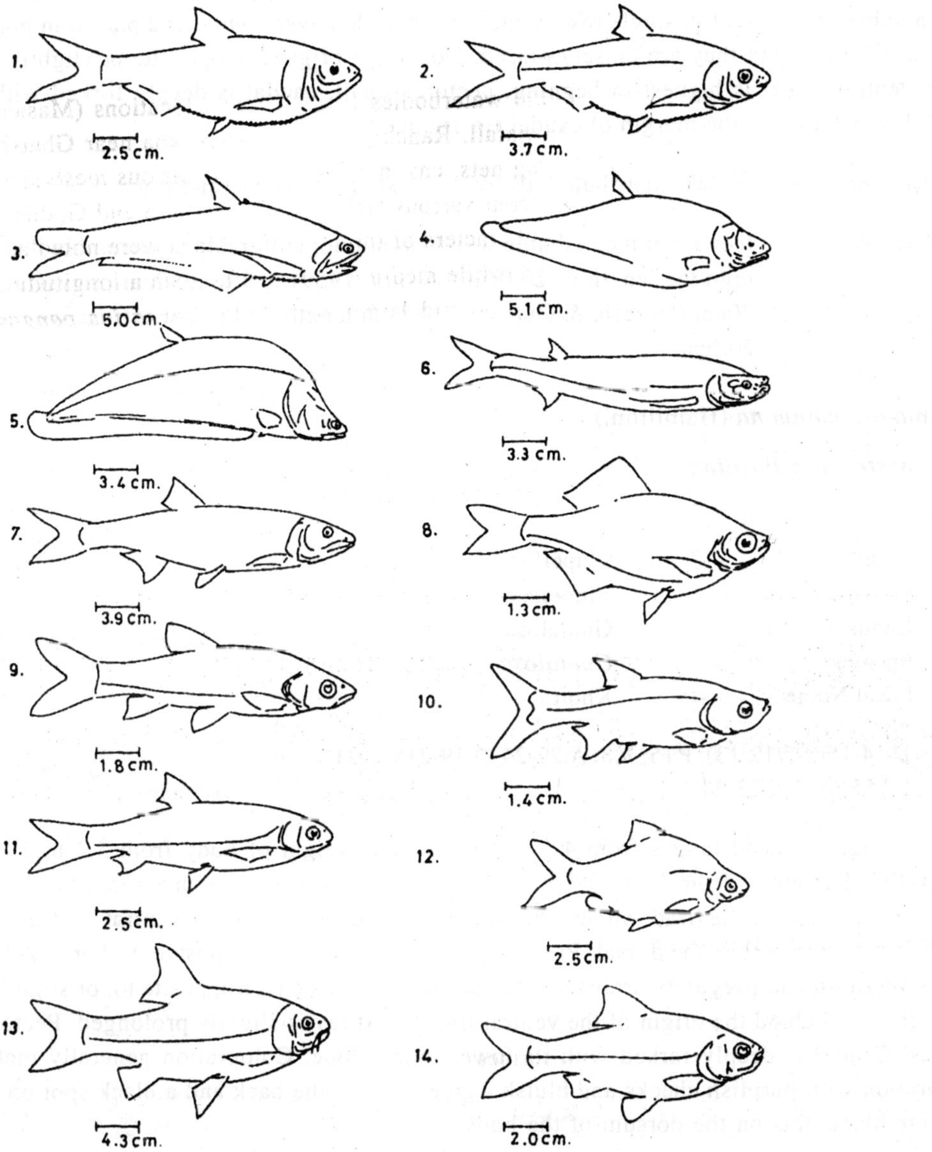

EXPLANATION OF PLATE -I

1. *Gudusia chapra* (Ham.)[Fig. 1]
2. *Gonialusa manmina* (Ham.)[Fig.2]
3. *Setipinna phasa* (Ham.)[Fig.3]
4. *Notopterus notopterus* (Pallas)[Fig.4]
5. *Notopterus chitala* (Ham.)[Fig.5]
6. *Oxygaster gora* (Ham.)[Fig.6]
7. *Barilus bola* (Ham.)[Fig.7]
8. *Daino devario* (Ham.)[Fig.8]
9. *Rasbora daniconius* (Ham.)[Fig.9]
10. *Amblypharyngodon mola* (Ham.)[Fig. 10]
11. *Aspidoparia morar* (Ham.)[Fig.11]
12. *Puntius conchonius* (Ham.)[Fig.12]
13. *Puntius sarana* (Ham.)[Fig 13]
14. *Puntius sophore* (Ham.)[Fig.14]

Length of head ranges from 3.9 to 4.5 and the height of body from 3.5 to 3.4 in the total length. Eye diameter is from 3.6 to 3.8 in the length of head.The eyes are with broad adipose eyelids.Teeth are absent. Abdominal profile is more convex than the dorsal. Preventral acutes 18-19 and postventral 9-10 in number and present in single row. Scales are smooth, silvery white and placed in horizontal rows but over the abdomen they are closely placed. Dorsal fin originates opposite or slightly before those of the ventrals, anal is highest in begning; pectoral small, caudal is deeply forked with its slightly longer lower lobe and the margin of caudal fin are black.

Distribution: Widely distributed all over Jharkhand in river sand ponds.

Remarks: It is a common ediable weedfish but not culturable at commercial scale. It grows upto the length of 20.0cm (Mishra, 1959); 6.7 to 20.0cm (Srivastava, 1968) and 20cm (Qureshi & Qureshi, 1983) in length. In our collections we recorded 8.8 to 20.1cm.

2. *Gonialosa manmina* (Hamilton.)

Systematic Position

Order	-	Clupeiformes
Family	-	Clupeidae
Genus	-	Gonialosa
Species	-	*Gonialosa manmina* (Ham.)
Local Name	-	Khoira

D.14-15(2-3/12-13); P.15; V.8; A.22-24 (3/19-21); C.21;
L.I.55-63; L.tr.22-24.

Length of head ranges from 4.7 to 5.0s and the height of body from 3.2 to 3.8 in the total length. Eye diameter is from 3.0 to 3.6 in the length of head. Eyes are with adipose eyelids and situated in the anterior half of the head. Snout is prominent and teeth are absent. Ventral profile of the body is a little more convex than the dorsal. Preventral acutes are 17-18 and postventral are 12-13 in number. Scales very thin and irregulary arranged. Dorsal fin originates either opposite to, or slightly in advance of, or slightly behind the origin of the ventral fin; its last ray is slightly prolonged. Pectoral reach the pelvics. Caudal is deeply forked with its lower longer lobe. Colouration generally metallic silvery colouration with purplish checks and bluish - green tint on the back and a black spot on the shoulder and fine black dots on the dorsum of the body.

Distribution: Found in almost all the major rivers of Jharkhand.

Remarks: **This fish is relished as good dish but not culturable at commercial scale and regarded as common weed fish. It grows upto the length of 27.9cm (Mishra, 1959); 9.7 to 12.2cm (Srivastava, 1968); and 28.0 cm (Qureshi &.Qureshi, 1983) in length. In our collections we recorded the length ranges from 10.1 to 28.3cm.**

3. *Setipinna phasa* (Hamilton)

Systematic Position

Order	-	Clupeiformes
Family	-	Engraulidae
Genus	-	Setipinna
Species	-	*Setipinna phasa* (Ham.)
Local Name	-	Bindi/phansa/phasba

D.1 + 14-15(2-3/12); P.14-15; V.7; A.70-80 (2/68-78); C.19; L.r.52;L.tr.14.

Length of head ranges from 6.1 to 7.2 and the height of body from 4.6 to 4.7 in the total length. Eye diameter is 4.4 in the length of head and eyes are covered with skin. Teeth are present on the jaws, palate and vomer bone. Preventral scutes are 15-16 and the postventral are 7 in number. Dorsal fin originates slightly posterior to the anal fin and it is much nearer to the snout than the base of the candal fin. The dorsal spine is short and strong. Superior ray of the pectoral fin is elongated to opposite the centre of anal fin. Anal fin is long and nonconfluent with the caudal. The caudal is forked with its longer lobes. Scales are decideous and some are found over the base of the anal and dorsal fins.Body is greenish along the back; sides are silvery and golden along the belly.

Distribution: It is found in Ganga at Sahibganj district and in other district which are associated with the boarder - line of Orissa (i.e. East Singhbhom, Saraykela, Kharsaban, West Singhbhiim, Simdega).

Remarks: It grows upto 30.4cm (Mishra, 1959); 13.0 to 22.8 cm (Srivastava, 1968) and 40.0cm (Qureshi & Qureshi,1983). In our collections we recorded the length ranges from 15.6 to 40.4cm. It is regarded as a good fish food and its landing is maximum during the end of monsoon on the bank of river Ganga in Sahibganj districts.

4. *Notopterus notopterus* (Pallas)

Systematic Position

Order	-	Clupeiformes
Family	-	Notopteridae
Genus	-	Notopterus
Species	-	*Notopterus notopterus* (Pallas)
Local Name	-	Pholi /Patola /Pholui /Pulli

D.7-8 (1-2/6-7); P.17; V.5-6; A. 100-110; C.19; L.I.225.

Length of head ranges from 4.5 to 5.6 and the height of body from 3.6 to 4.1 in the total length. Eye diameter is from 4.1 to 5.2 in the length of head. Teeth are present on the jaws, vomers, plate and also on the tongue. Dorsal profile is not so convex as that of abdomen. There is a double row of 28 to 33 serrations along the abdominal edge between the throat and the inseration of ventral fins. Dorsal fin is small and arises nearly midway between the snout and the end of the candal fin. Ventrals are minute. The anal originates just behind the ventrals and becomes confluent with the candal fin. Scales present on the checks are comparatively larger than the body scales. Lateral line is complete. Body is silvery white, with fine grey spots which are dark along the narrow back.

Distribution: It is distributed through out Jharkhand in rivers and ponds.

Remarks: It grows upto the length of 60.0 cm (Mishra, 1959); 15.0 to 17.0cm (Srivastava, 1968) and 25.0- 60.0cm (Qureshi & Qureshi, 1983). In our collections we recorded the length ranges from 16.7 to 51.3 cm. It is cultruable in all large lakes and reservoirs.

5. *Notopterus chitala* (Hamilton)

Systematic Position

Order	-	Clupeiformes
Family	-	Notopteridae
Genus	-	Notopterus
Species	-	*Notopterus chitala* (Ham.)
Local Name	-	Chital/Mohi/Pulli

D.9-10(1 -2/7-9);P.16;V.6;A.110-125;C.12-14;L.1.160-180.

Length of head ranges from 4.5 to 5.1 and the height of body from 3.5 to 4.1 in the total length. Eye diameter is from 6.9 - 8.1 in the length of head. Upper profile of head is deeply concave. Snout is prominent. Teeth are found on jaws, vomers, palatines and also on tongue. Dorsal profile of the body is more humped. About 50 serrations are arranged in double rows between throat and the insertion of the ventral fin. Dorsal fin originates nearer the tail than the snout. Ventrals minute. Anal fin is long and confluent with the caudal. Opercular scales are of normal size as present on whole body. Lateral-line is complete. The broad sides of body are silvery white, two or more black spots are present below lateral line near the caudal end.

Distribution: It is found in rivers, ponds and other water bodies throughout Jharkhand.

Remarks: It attains upto the length of 120.0cm (Mishra, 1959); 20.3 to 30.0cm (Srivastava, 1968) and 25.0cm (Qureshi & Qureshi, 1983). In our collections we recorded the length ranges from 19.9 to 34.7cm.

6. *Oxygaster gora* (Hamilton)

Systematic Position

Order	-	Cypriniformes
Family	-	Cyprinidae
Genus	-	Oxygaster
Species	-	*Oxygaster gora* (Ham.)
Local Name	-	Chel-hul/Ghora-chella.

D.9-10(2-3/7);P.12-15;V.8; A.I 5-16 (2/13-14); C. 19;
L.I. 140-160; L.tr. 18-20/16-18.

Length of head ranges from 5.0 to 5.1 and the height of body is 5.9 in the total length. Eye diameter is 5.0 to 6.3 in the length of head. Pharygeal teeth are present. Dorsal fin commences slightly in advance of the anal fin. Pectrol is long and the caudal is forked. Lateral-line is complete, but interrupted before the origin of the ventral fin. Scales are small, which extend forwards on the head to above the nostrils. Body is silvery white.

Distribution: Found all over Jharkhand but more in Garwaha, Palamu, Chatra, Hazaribag, Kodrama, Giridih, Deogher, Dumka and Godda districts.

Remarks: It has a delicate flavour and is also used as the biological tool for the controlling mosquito larvae. It attains the length of 22.7cm (Mishra, 1959); 14.5 to 25.5cm (Srivastava,1968) and 25.0cm (Qureshi & Qureshi, 1983). In our collections we recorded the length ranges from 16.6 to 25.3cm. 7.

7. *Barilus bola* (Hamilton)

Systematic Position

Order	-	Cypriniformes
Family	-	Cyprinidae
Genus	-	Barilus
Species	-	*Barilus bola* (Ham.)
Local Name	-	Buggarah/Bhola/Ghola/Bugguah.

D.10-11(3/7-8);P.13;V.9;A.13(3/10);C.19;L.1.88-94;
L.tr. 12-15/9-11.

Length of head ranges from 4.4 to 4.8 and the height of body from 5.1 to 6.0 in the total length. Eye diameter is from 4.6 to 7.1 in the length of head. Head is compressed and the snout is pointed. The cleft of mouth is deep. Barbels are absent. Dorsal fin commences midway between the angle of the preopercle and the base of the caudal. Anal fin is forked with slightly longer lower lobe. Lateral - line complete. Body is silvery with two or more rows of vertically arranged blue spots on sides.

Distribution: Found in rivers, ponds, reservoirs, including other water bodies throughout Jharkhand.

Remarks: This is the maingamefish of Jharkhand. It attains the length of 30.0cm (Mishra, 1959); 17.7 to 25.8 cm (Srivastava, 1968)and 30.0cm (Qureshi & Qureshi, 1983). In our collections we recored the length ranges from 16.1 to 30.0cm.

8. *Daino devario* (Hamilton)

Systematic Position

Order	-	Cypriniformes
Family	-	Cyprmidae
Genus	-	Danio

Species - *Danio devario* (Ham.)
Local Name - Debari/Bonkuaso/Patukari

D.I 8-19(3/15-16); P. 13-14; V.8; A.18-19(3/15-16); C. 19; L.I.41-48; L.tr. 11-11½/2½-5.

Length of head ranges from 5.1 to 5.2 and the height of body from 3.0 to 3.4 in the total length. Eye diameter is from 3.1 in the length of head. Lower jaw is longer than the upper. Barbles are absent. Dorsal fin commences midway between the anterior margin of the orbit and the base of the caudal fin. Caudal fin is lunate. Between the lateral-line and the base of the ventral fin 21½ rows of scales are present. Body is greenish above and dull white below and a bluish band is present along the mid lateral line which becomes broad and prominent posteriorly. A black spot is located at the posteior margin of the head.

Distribution: Found throughout Jharkhand in all types of water bodies.

Remarks: **This fish is regarded as the commom weed fish of Jharkhand. It grows upto the length of 10.0cm (Mishra, 1959); 7.3 to 7.7cm (Srivastava, 1968) and 10.0cm (Qureshi & Qureshi, 1983). In our collections we recorded the length ranges from 7.1 to 10.4cm.**

9. *Rasbora daniconius* (Hamilton)

Systematic Position

Order - Cyprinformes
Family - Cyprinidae
Genus - Rasbora
Species - *Rasbora daniconius* (Ham.)
Local Name - Dankoni/Dhera/Dendua

D.9(2/7); P.15; V.9; A.7(2/5); C.19; L.I.31-34; L.tr. 4½- 5½/4½-5.

Length of head ranges from 4.4 to 5.0 and the height of body from 4.4 to 5.1 in the total length. Eye diameter is from 3.5 to 4.1 in the length of head. Body is cylindrical and slightly compressed with rounded abdomen. Barbels are absent. Dorsal fin arises nearer to the origin of the ventrals than that of the anal. Caudal is forked. Lateral - line is concave and there are 21½ rows of scales between it and the base of the ventral fin. Pre-dorsal scales are 14 in number. A prominent blue-black band runs across the mid lateral-line extending from the posterior end of the orbit up to the tip of the tail. Lower half of the body is silvery white. The tips of the fins are brownish black or deep yellow.

Distribution: Found throughout Jharkhand in rivers, ponds, reservoirs etc.

Remarks: It grows upto the length of 20.0cm (Mishra, 1959); 8.0 to 8.8cm (Srivastava, 1968) and 14.0cm (Qureshi & Qureshi, 1983). In our collections we recovered the length ranges from 8.0 to 17.4cm. This fish is consumed generally by tribals and poor sections of Jharkhand.

10. *Amblypharyngodon mola* (Hamilton)

Systematic Position

Order	-	Cypnmformes
Family	-	Cyprinidae
Genus	-	Amblyphaeyngodon
Species	-	*Amblypharyngodon mola* (Ham.)
Local Name	-	Morara/Dhawai

D.2/7;P.15; V.9; A.7(2/5); C.19; L.I.65-75;L.tr. 12/12

Length of head ranges from 5.0 to 5.land the height of body ranges from 4.0 to 4.4 in the total length. Eye diameter is from 3.4 to 4.1 in the length of head. Dorsal profile is more convex than that of the abdomen. Mouth is anterior and lower jaw slightly more elongated. Dorsal fin arises midway between the front edge of the eye and the base of the caudal fin. Caudal is forked. Lateral-line is interrupted. A silvery lateral band is present on either side of the body and small black dots on the upper 3/4 surface are present.

Distribution: **It is found throughout Jharkhand.**

Remarks: **This fish is used as food by tribals and poor sections of Jharkhand only. It attains the length ranges from 8.0 to 8.5cm (Srivastava, 1968) and upto 9.0cm (Qureshi & Qureshi, 1983). In our collections we recorded the length ranges from 7.7 to 9.3cm.**

11. *Aspidoparia morar* (Hamilton)

Systematic Position

Order	-	Cypriniformes
Family	-	Cyprinidae
Genus	-	Aspidoparia
Species	-	*Aspidoparia morar* (Ham.)
Local Name	-	Kenwachi/Harda/Moran/Chippuah/Chelluah.

D.9-10(2-3/7-8); P.15;V.8-9;A.10-12(2/8-10);C.19;
L.I. 38-42;L.tr.5½-7½/3½-5.

Length of head ranges from 5.0 to 5.5 and the height of body is 4.1 to 5.2 in the total length. Eye diameter is 3.0 to 3.4 in the length of head. Mouth is small and inferior. Lower jaw is without lip and shorter than the upper jaw. Pharyngeal teeth are present. Dorsal fin commences midway between the posterior margin of the orbit and the base of the caudal fin, it is higher than long. Pectoral fin is longer than the head. Caudal is forked. Lateral-line is complete. 17 rows of scales are found before the base of the dorsal fin. Body is light yellow with the fins dark yellow.

Distribution: Distribution throughout Jharkhand but abundant in polluted zone of river subarnarekha near Ghatshila.

Remarks: It grows upto the length of 8.0 to 9.5cm (Sirvastava, 1968), and 18.0cm (Qureshi & Qureshi, 1983). In our collections we recorded the length ranges from 8.9 - 18.0cm.

12. *Puntius conchonius* (Hamilton)

Systematic Position

Order	-	Cypriniformes
Family	-	Cyprinidae
Genus	-	Puntius
Species	-	*Puntius conchonius* (Ham.)
Local Name	-	Kunchonpungti/Sidhri

D. 11(3/8); P.I 1-15; V.9; A.7-8(2-3/5); C.19; L.I.24-27;
L.tr. 5½- 6/6½

Length of head ranges from 4.9 to 5.1 and the height of body ranges from 2.5 to 2.6 in the total length. Eye diameter is from 3.0 to 3.5 in the length of head. Body is elevated and a slight concavity is there on the nape followed by a considerable rise to the base of the dorsal fin. The dorsal fin has its last undivided ray osseous, spinous and serrated. The dorsal fin commences midway between the anterior extremity of the orbit and the base of the caudal fin. The candal is forked all the fins are transparent lateral - line is in complete. All the scales with black bases. Body is silvery; dark along the back. A black spot is found on the lateral-line in the area between 18th to 20 scales.

Distribution: Distribution throughout Jharkhand and commonly found in rivers, ponds, ditches, reservoirs.

Rermarks: It grows upto the length of 12.5cm (Mishra, 1959); 8.0 to 8.4cm (Srivastava, 1968) and 13.0cm (Qureshi & Qureshi, 1983). In our collections we recorded the length ranges from 6.8 to 13.2 cm. This fish is consumed by the tribals and poor sections of the Jharkhand only.

13. *Puntis sarana* (Hamilton)

Systematic Position

Order	-	Cypriniformes
Family	-	Cyprinidae
Genus	-	Puntis
Species	-	*Puntis sarana* (Ham.)
Local Name	-	Giddi-kaoli/Durhie/Potah/Sarana.

D.ll-12(3/8-9);P.15-16;V.8-9;A.8(3/5);C.19;
L.I. 32-34; L.tr.5½-6½; Barbels 2 pairs.

Length of head ranges from 4.9 to 5.2 and the height of body from 3.6 to 3.7 in the total length. Eye diameter is from 4.1 to 4.8 in the length of head. Dorsal profile of the body is slightly elevated. Two pairs of barbics are present. Rostral pair is as long as the orbit, maxillary pair is slightly longer. Dorsal fin arises slightly nearer the snout than the base of the candal fin and opposite the insertion of the ventral fin. First last undivided ray is osseous, strong and finely serrated posteriorly. Caudal fin is forked. Lateral-line is complete and there are 3½ to 4 rows of scales between it and the base of the ventral fin. A row of 10 to 12 predorsal scales is present. Body colour silvery, darkest dorsally, opercles golden, fins dusky-brown to orange, some dark sports are found behind the opercle.

Distribution: It is a very common weed fish of all types of water bodies in all over Jharkhand.

Remarks: It attains the length of 30.0cm (Mishra, 1959); 11.2 to 37.5 (Srivastava, 1968) and 31.0cm (Qureshi & Qureshi, 1983). In our collections we recorded the length to ranges from 11.0 to 33.0cm.

14. *Puntius sophore* (Hamilton)

Systematic Position

Order	-	Cyprmiformes
Family	-	Cyprinidae
Genus	-	Puntius
Species	-	*Puntius sophore* (Ham.)
Local Name	-	Katcha-Karawa/Pattiah/Punti/Patia-kerundi.

D.11-12 (3/8-9); P. 15-17; V.9; A.8(3/5); C.I9; L.I. 23-26;
L.tr. 4½ -5½/5-½.

Length of head ranges from 5.0 to 5.1 and the height of body from 3.4 to 3.8 in the total length. Eye diameter is from 3.4 to 3.9 in the length of head. The dorsal profile is much more convex than that of the abdominal profile and a little concave over the ocipit. Upper jaw. is slightly longer. Barbel absent and the teeth are pharyngeal and crooked. Dorsal fin is inserted opposite the insertion of the ventrals. The candal fin is forked. Lateral line is complete and there are 3 to 3½ rows of scales between it and the base of the ventral fin. There are 8-9 rows of predorsal scales. Body is silvery with the darker dorsal surface and opercles are golden-red. There as two black spots; one at the base of the 3rd - 5th branched dorsal rays and the other at the base of the candal fin along the lateral-line over 22nd and 23rd and 24th scales.

Distribution: It is distributed throughout Jharkhand and found in all types of water bodies.

Remarks: **It attains upto the length of 6.3 to 10.4cm (Srivastava, 1968)and 13cm (Qureshi & Qureshi, 1983). In our collections we recorded the length ranges from 6.5 to 13.3cm. It is consumed by tribals and poor sections of Jharkhand.**

PLATE II

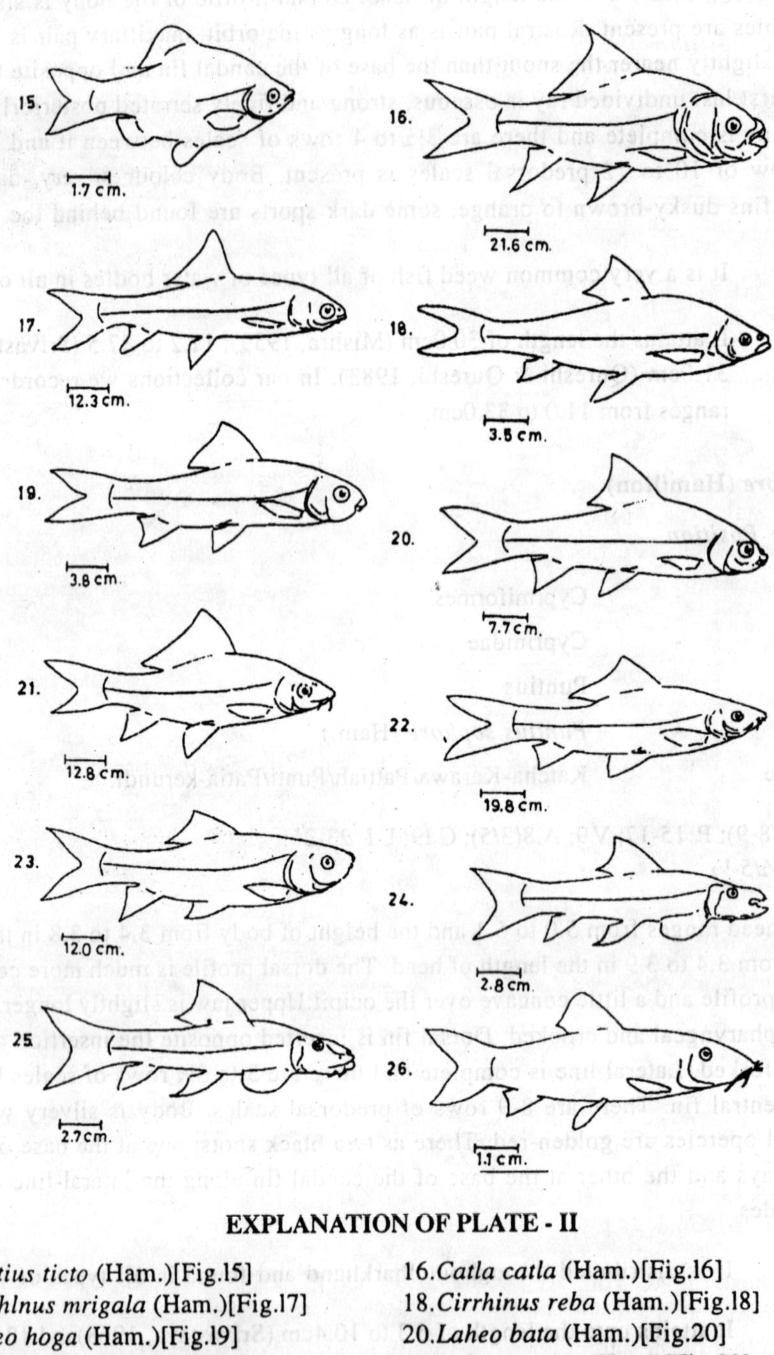

EXPLANATION OF PLATE - II

15.*Puntius ticto* (Ham.)[Fig.15] 16.*Catla catla* (Ham.)[Fig.16]
17.*Cirrhlnus mrigala* (Ham.)[Fig.17] 18.*Cirrhinus reba* (Ham.)[Fig.18]
19.*Labeo hoga* (Ham.)[Fig.19] 20.*Laheo bata* (Ham.)[Fig.20]
21.*Labeo calbasu* (Ham.)[Fig.21] 22.*Labeo gonius* (Ham.)[Fig.22]
23.*Labeo rohita* (Ham.)[Fig.23] 24.*Garra lamta* (Ham.)[Fig.24]
25.*Gara gotyla* (Gray)[Fig.25] 26.*Botia dario* (Ham.)[Fig.26]

15. *Puntius ticto* (Hamilton)

Systematic Position

Order	-	Cypriniformes
Family	-	Cyprinidae
Genus	-	Puntius
Species	-	*Puntius ticto* (Ham.)
Local Name	-	Kaoli/Kotree

D. 11; (3/8); P.13-15; V.9; A.7-8(2-3/5); C.19; L.I. 23-26;
L.tr. 5- 6/6- 6½.

Length of head ranges from 4.9 to 5.0 and the height of body from 2.9 to 3.3 in the total length. Eye diameter is 3.1 in the length of head. Body strongly compressed and elelated. Upper jaw is slightly longer than the lower jaw. Teeth are pharyngeal, crooked and pointed. Last undivided ray of the dorsal fin is osseous, strong and serrated. Caudal fin is forked. Lateral-line incomplete, ceasing after 14th scales. There are 4-5 rows of the scales between the lateral - line and the base of the ventral fins. A row of 10 to 11 predorsals is present. Body is silvery and fins are black. A black spot is located on the side of the tail before the base of the candal fin, and a spot near to the commencement of the lateral line.

Distribution: Found all over Jharkhand in all types of water bodies like ponds, lakes, rivers, resevoirs, ditches etc.

Remarks: It is consumed by tribals and poor sections of Jharkhand only. It attains upto the length of 10.0cm (Mishra, 1969); 6.2cm (Srivastava, 1968) and 10.0cm (Qureshi & Qureshi, 1983). In our collections the length ranges from 6.4 to 10.2cm.

16. *Catla catla* (Hamilton)

Systematic Position

Order	-	Cypriniformes
Family	-	Cyprinidae
Genus	-	Catla
Species	-	*Catla catla* (Ham.)
Local Name	-	Katla/Barkur/Barkura/Catla.

D.17-19(3-4/14-16); P.19-21; V.9; A.8(3/5); C.19;
L.I.40-43; L.tr.7½/6½-9.

Length of head ranges from 4.2 to 4.6 and the height of body from 3.0 to 3.4 in the total length. Eye diameter is from 6.1 to 6.9 in the length of head. Body is deep and stout. Dorsal profile of body is more convex than that of the abdominal one. Mouth is wide and the lower jaw is prominent. Teeth are pharyngeal and plough shaped. Dorsal fin commences slightly in advance of the ventral fin. Pectoral fin is located slightly behind the ventral fin. Caudal fin is forked. Scales are of moderate size. Lateral-line is complete and commences from the upper margin of operculum. The body is greyish above and silvery on the lateral and ventral sides. Fins are blackish in clolour. Scales have pink or coppery centre except those of the ventral side which are whitish.

Distribution: Found throughout Jharkhand in ponds, rivers, lakes, reservoirs, etc

Remarks: It is one of the fastest growing fish and considered as the most productive culturable fish of Jharkhand.It grows upto the length of 182.4cm (Mishra,1959) and 19.0cm to 28.0cm (Srivastava,1968) and 188.0cm (Qureshi & Qureshi, 1983). In our collections we recorded the length ranges from 17.6 cm to 67.9 cm.

17. *Cirrhinus mrigala* (Hamilton)

Systematic Position

Order - Cypriniformes
Family - Cyprinidae
Genus - Cirrhinus
Species - *Cirrhinus mrigala* (Ham.)
Local Name - Naim/Mrigala/Mirrgah/Mirca

D.15-16;(3/12-13);P.15-18;V.9;A.8(2/6);C.15;
L.I. 40-45; L.tr.6½/6½- 8½; Barbels 1 pair.

Length of head ranges from 5 to 5.3 and the height of body from 4.0 to 5.3 in the total length. Eye diameter is from 3.4 to 3.5 in the length of head. Eyes are golden in colour. Pharyngeal teeth are present. The dorsal fin is located nearer to the snout than to the base of the candal fin and arises opposite the 12th scales of the lateral line, it is less then the height of the body and its upper margin is slightly concave. Caudal fin is deeply forked and their lobes have inwardly convex edge. Body is silvery, dark grey along the back; sometimes with coppery tinge. Pectoral, ventral and the anal fins are orange and tinged with black.

Distribution: Found throughout Jharkhand in ponds, rivers, reservoirs etc.

Remarks: It is one of the fastest growing fish and considered as the most productive culturable fish of Jharkhand. It grows,upto the length of 90.0cm (Mishra, 1959); 16.0 to 26.0cm (Srivastava, 1968) and 56.0 to 90.0 cm (Qureshi & Qureshi, 1983). In our collections we recorded the length ranges from 22.0cm to 88.6 cm.

18. *Cirrhinus reba* (Hamilton)

Systematic Position

Order - Cypriniformes
Family - Cyprinidae
Genus - Cirrhinus
Species - *Cirrhinus reba* (Ham.)
Local Name - Rewah/Raicheng/Balta

D. 10-11; (2-3/8-9); P.16; V.9; A.8(2-3/5-6); C.19; L.I. 35-38;
L.tr. 7/5-9; Barbels 1 pair..

Length of head ranges from 5.0 to 6.0 and the height of body from 4.1 to 4.2 in the total length. Eye diameter is 4.1 to 5.6 in the length of head. Dorsal profile of the body is slightly more convex than the abdorminal one. Upper lip is fringed. Barbels are only two and rostral in position. Dorsal fin originates anterior to the ventral fin, its upper margin is concave. Caudal fin is deeply forked.

Lateral-line is complete. Scales are hexagonal and there are 5 rows between the lateral-line and the base of the ventral fin. Colour of the body is silvery. Scales are generally darkest at their edges. Sometimes bluish longitudinal bands are found on the body. Tip of pelvic and anal fins are orange.

Distribution: Found through out Jharkhand in all types of water bodies.

Remarks: It is not important as the cirrhinus mrigala in relation to their economic value and consumption. It attains upto the length of 13.8 to 22.8cm (Srivastava) and 30.0 cm (Qureshi & Qureshi, 1983). In our collections we recorded the length ranges from 11.5 to 30.1cm.

19. *Labeo boga* (Hamilton)

Systematic Position

Order	-	Cypriniformes
Family	-	Cyprinidae
Genus	-	Labeo
Species	-	*Labeo boga* (Ham.)
Local Name	-	Kalabattali / Jumuna fish.

D.11-13; (2-3/9-10); P.16; V.9; A.7(2/5);C. 19; L.I. 37-39;
L.tr. 6½-7/2; Barbels 1 pair.

Length of head ranges from 5.1 to 5.4 and the height of body from 5.3 to 5.6 in the total length. Eye diameter is 3.5 to 3.9 in the length of head. Dorsal profile more convex than the abdominal one. Lips are some what thick. Teeth are pharyngeal and arranged in three rows. Barbels are two, minute and maxillary in position. Dorsal fin commences much advance of the ventral fins. It is situated nearer the snout than the base of the candal fin, its upper margin is concave. Candal is deeply forked with both of its lobes equal. Lateral-line is complete and 5 rows of scales are present between it and the base of the ventral fin. Body is orange with the fins reddish tige, sometimes a dark spot is present on the shoulder.

Distribution: Found generally in the tributaries of Ganga river and also in ponds of the associated area.

Remarks: It grows upto the length of 30.0cm (Mishra,1959);10.5 to 12.3cm (Srivastava, 1968) and 30.0cm (Qureshi & Qureshi, 1983).In our collections we recorded the length ranges from 11.8 to 30.2 cm.

20. *Labeo bata* (Hamilton)

Systematic Position

Order	-	Cypriniformes
Family	-	Cyprmidae
Genus	-	Labco
Species	-	*Labeo bata* (Ham.)
Local Name	-	Gootellah/Bhanga/Bata.

D.11-12; (2-3/9-10); P.18; V.9; A.7(2/5); C.19; L.I. 37-40;
L.tr. 7/6-7; Barbels 1 pair.

Length of head ranges from 5.5 to 5.7 and the height of body from 4.4 to 4.8 in the total length.Eye diameter is 4.0 to 4.2 in the length of head. The dorsal profile of the body is more convex than the abdorminal portion.

21. *Labeo calbasu* (Hamilton)

Systematic Position

Order	-	Cypriniformes
Family	-	Cyprimdae
Genus	-	Labeo
Species	-	*Labeo calbasu* (Ham.)
Local Name	-	Kalbasu/Kala-beinse/Kursa/Kundna/Kalbus/Cuggera.

D.16-18; (3/13-15); P.19; V.9; A.7(2/5); C.19; L.I. 40-44;
L.tr. 7½/8; Barbels 2 pairs.

Length of head ranges from 5.1 to 5.9 and the height of body is 4.0 to 4.1 in the length of head. Lips are thick and fringed, teeth are pharyngeal, crooked and arranged in three rows. One pair of rostral and one pair of maxillary barbels are present. The rostral being slightly longer than the maxillary. Dorsal fin commences midway between the snout and the base of the candal fin, it arises slightly in advance of the ventral fins. Caudal fin is deeply forkede.Scales are large and 51/2 to 61/2 rows are present in between the lateral-line and the base of the ventral fin. Lateral - line is complete. Body is blackfsh with scarlet in the centre of the scales. Fins are black.

Distribution Distribution all over Jharkhand in rivers but more in tributries of Ganga river.

Remarks It is considerd as an excellent food fish. It attains the size of 91.2cm (Mishra,1959); 26.5cm (Srivastava, 1968) and 91.0cm (Qureshi & Qureshi,1983). In our collections we recorded the length ranges from 23.7 - 91.1cm.

22. *Labeo gonius* (Hamilton)

Systematic Position

Order	-	Cypriniformes
Family	-	Cyprinidae
Genus	-	Labeo
Species	-	*Labeo gonius* (Ham.)
Local Name	-	Cursa/ Collose/Kurchi/Goni/Cursua.

D.16-18; (2-3/13-15); P.17;V.7(2/5);C.19;
L.I. 74-84; L.tr. 16/17; Barbels 2 pairs.

Length of head ranges from 4.9 to 5.4 and the height of body is 4.1 to 4.5 in the total length. Eye diameter is from 4.5 to 4.9 in the length of head. Dorsal profile is more convex than the abdominal one. Month is wide, lips are thick and with a distinct inner fold in their entire circumference. Both the lips are fringed. A pair of short rostral and a pair of maxillary barbels are present. Teeth are pharyngeal, plough - shaped and arranged in 3 rows. Dorsal fin arises much nearer the snout than the base of the caudal. Cudal is deeply forked. Lateral - line complete and there are 10-13 rows of śmall scales between it and the base of the ventral fin. Body is greenish along the back becoming lighter on the sides upto the lateral-line, the rest of the body is silvery white. Black longitudinal bands extend from the anterior to the posterior end.

Distribution: Distribution throughout Jharkhand.

Remarks: It is extensively used for stocking the tanks as the common culturable fish. It attains the size of 152.0cm (Mishra,1959); 16.2 to 22.0cm (Srivastava, 1968) and 150.0 to 155.0cm (Qureshi & Qureshi,1983). In our collections we recorded the length ranges from 21.4 to 153.1cm.

23. *Labeo rohita* (Hamilton)

Systematic Position

Order	-	Cypriniformes
Family	-	Cyprinidae
Genus	-	Labeo
Species	-	*Labeo rohita* (Ham.)
Local Name		Rohu/Ruee

D.16(3/13); P.17; V.9; A.7(2/5); C.19;
L.I. 40-42; L.tr. 61/2; Barbels 1 pairs.

Length of head ranges from 4.4 to 5.0 and the height of body is 4.0 to 4.3 in the total length. Eye diameter is from 4.0 to 5.8 in the length of head, 3.0 in the interrorbital width, eye moderately large, not visible from below the ventral surface. Dorsal profile is more convex than abdomen. The snout is obtuse, depressed and lips very thick and fringed with a distinct innerfold covering both the jaws. Teeth are pharyngeal and arranged in three rows. Barbels are one pair, maxillary, short and thin. Dorsal fin arises about midway to the snout between the snout and the base of the caudal fin. Upper edge of the dorsal fin moderately concave; pectoral fin as long as head excluding the snout. Caudal fin deeply forked. Lateral line is complete. Scales large, 6½ rows between lateral - line and base of the ventral fin. Body bluish black along the back but reddish black along the sides and silvery beneath. There is a red mark on each scales surrounded by the dark margin. Fins are black and abdomen is silvery.

Distribution: Found through out Jharkhand in rivers, ponds, lakes, reservoir etc. as commonest culturable carp.

Remarks: Labeo rohita is one of the most popular and excelllent food fish of great economic important in Jharkhand. It attains the size of 91.2cm (Mishra, 1959); 24.2 to 30.0cm (Srivastava,1968) and 91.0cm (Qureshi & Qureshi, 1983). In our collections we recorded the length ranges from 24.9 to 91.0cm.

24. *Garra lamta* (Hamilton)

Systematic Position

Order	-	Cypriniformes
Family	-	Cyprinidae
Genus	-	Garra
Species	-	*Garra lamta* (Ham.)
Local Name	-	Pather-Chata

D.11(2-3/8-9); P.15; V.9;A.7(2/5);C. 17;
L.I.32-36; L.tr.4 - 4½/5; Barbels 2 pairs.

Length of head ranges from 5.0 to 5.4 and the height of body is 5.1 to 5.9 in the total length. Eye diameter is 5.2 to 7.9 in the length of head. Barbels are 2 pairs in number. Dorsal fin arises midway between the end of the snout and the base of the caudal fin and situated anterior to the origin of the ventrals. Caudal fin is slightly lobed. Lateral- line is complete. Body is greenish-brown or blue-green above and yellowish below. Generally a dark spot is present behind the gill opening. Fins are yellowish, stained darker at their margins.

Distribution: Throughout Jharkhand and mostly found in the fast flowing waters of mountains.

Remarks: A semi-cartilagenous adhesive disc is present on the ventral side of the lower lip. It attains atleast 20.0cm in length (Qureshi & Qureshi, 1983) But in our collections we recorded the length ranges from 6.8 to 20.7cm. It is generally consumed by tribals and poor sections of Jharkhand and fishing is done with the help of biocidal plant sap.

25. *Gara gotyla* (Gray)

Systematic Position

Order	-	Cypriniformes
Family	-	Cyprinidae
Genus	-	Garra
Species	-	*Garra gotyla* (Gray)
Local Name	-	Siltoka

D. 11 (3-2/8-9); P. 15; V.9; A.8(2/5); C. 17;
L.I. 32-36; L.tr.4½/5; Barbels 2 pairs.

Length of head ranges from 5.0 to 5.4 and the height of body is 5.1 to 5.9 in the total length. Eye diameter is 5.7 in the length of head. Eye directed slightly upwards and outwards. Dorsal fin arises midway between the end of snout and the base of the caudal, and anterior to the origin of the ventral. It is usually less in hight than the body. The pectoral is shorter than the head and caudal is slightly lobed. Lateral-line is complete. Body colour is greenish, with a bluesh green bands along the centre of body and extending along the middle of the caudal fin. Generally a dark spot behind the gill opening is present. Abdomen is yellowish-green. Fins are yellowish and stained darker at their margins.

Distribution: Throughout Jharkhand but abundant in mountain streams and in their associated waterbodies.

Remarks: Generally consumed by tribals and poor sections of Jharkhand and fishing is done with the help of biocidal sap by anaesthetizing the fish. It attains the length upto 20.0 to 22.0cm (Srivastava, 1959) but in our collections we recorded the length ranges from 16.7 to 22.3cm.

26. *Botia dario* (Hamilton)

Systematic Position

Order	-	Cypriniformes
Family	-	Cyprinidae
Genus	-	Botaia
Species	-	*Botia dario* (Ham.)
Local Name	-	Baghana

D. 12-13(3/9-10); P.14; V.8; A.7-8(2/5-6); C.19; Barbels 4 pairs.

Length of head ranges from 4.4 to 5.0 and the height of body 4.6 to 4.9 in the total length. Eye diameter is from 5.0 to 5.4 in the length of head. Barbels are eight, two pairs rostral, the anterior pair being longer than the posterior, one pair maxillary and one pair mandibular. The mandibular barbels are the shortest. The lower lip is fringed with 4 to 5 thick blunt lobes. The dorsal fin arises midway between the end of the snout and the base of the caudal fin. Scales are very indistinct. General colouration of the body is yellow. There are 7 or 8 oblique black bands over the body forming complete rings. 2 or 3 even more bands are present on either lobe of the caudal fin. There are 2 bands over the head, one in the nasal region covering the complete snout and the other in the region of the eye.

Distribution: Found in the districts associated with the boundary of Jharkhand.

Remarks: It is generally consumed by tribals and poor sections of Jharkhand only. It attains the length of 8.7cm (Mishra,1959); 5.8 to 6.3cm (Srivastava, 1968) and 9.0cm (Qureshi & Qureshi,1983). In our collections we recorded the length ranges from 5.3 to 9.2 cm.

PLATE III

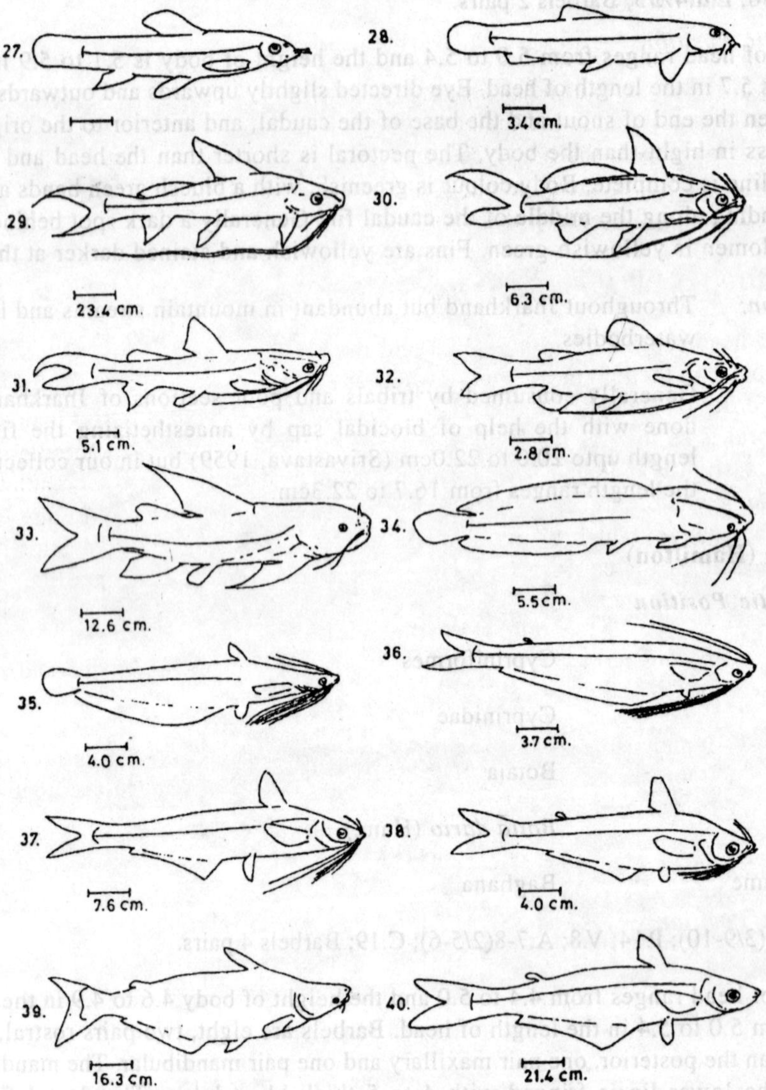

EXPLANATION OF PLATE-III

27. *Noemacheilus botia* (Hani.)[Fig.27]
28. *Lapidocephalichthys guntea* (Ham.)[Fig.28]
29. *Mystus aor* (Ham.)[Fig.29]
30. *Mystus cavasius* (Ham.)[Fig.30]
31. *Mystus seenghala* (Sykes) [Fig.31]
32. *Mystus vittatus* (Bloch)[Fig.32]
33. *Rita rita* (Ham.)[Fig.33]
34. *Clarias batrachus* (Linn.)[Fig.34]
35. *Hetropneustes fossilis* (Bloch)[Fig.35]
36. *Ailia coila* (Ham.)[Fig.36]
37. *Clupisoma garua* (Ham.)[Fig.37]
38. *Eutropiichthys vacha* (Ham.)[Fig.38]
39. *Pangasius pangusius* (Ham.)[Fig.39]
40. *Silonia silondia* (Ham.)[Fig.40]

27. *Noemacheilus botia* (Hamilton)

Systematic Position

Order	-	Cypriniformes
Family	-	Cobitidae
Genus	-	Noemacheilus
Species	-	*Noemacheilus botia* (Ham.)
Local Name	-	Natwa/Carri.

D.12-14(2/10-12); P.11; V.8; A.7(2/5); C.17; Barbels 3 pairs.

Length of head ranges from 4.4 to 5.3 and the height of body is 4.6 in the total length. Eye diameter is 3.6 to 4.1 in the length of head. Barbels are 3 pairs, long and the maxillary pair reaches to below the hind border of the eyes. Dorsal fin arises slightly nearer to the snout than to the base of the caudal fin. Pectoral is as long as the head. Ventral fin is inserted under the middle of the dorsal fin. The caudal fin is slightly notched. Scales are indistinct. Lateral-line is incomplete. Body almost cylindrical and somewhat depressed, greyish with 10-14 short, vertical bars on the lateral-line and a number of blotches above; which often form bands over the back. A black ocellus is present on the upper part of the base of the caudal fin which also bears seven irregular bars like '>'. Dorsal fin is orange with rows of black spots.

Distribution: Found throughout Jharkhand but abundant in hillstreams and other associated water bodies.

Remarks: It attains the size of 8.4cm (Mishra, 1959) and 8.0 cm (Srivastava, 1968) and 7.6cm (Qureshi & Qureshi,1983). But in our collections we recorded the size ranges from 7.1cm to 8.2cm.

28. *Lapidocephalichthys guntea* (Hamilton)

Systematic Position

Order	-	Cypriniformes
Family	-	Cobitidae
Genus	-	Lepidocephalichthys
Species	-	*Lapidocephalichthys guntea* (Ham.)
Local Name	-	Gunguch/Nakati.

D.8-9(2/6-7); P.8; V.7-8; A.7(2/5); C.16;
L.tr. 115; Barbels 4 pairs

Length of head ranges from 6.4 to 6.5 and the height of body is 5.4 to 6.2 in the total length. Eye diameter is 5.1 in the length of head. Body is compressed. There are two pairs of rostral and one pair of maxillary barbels, all are longer than the orbit. A fleshy flap from the lower surface of the mandible is given out on either side which joins the maxillary barbel and each has two or three barbels at its tip. Dorsal fin is situated opposite the ventrals and the caudal fin is entire. Scales are minute. Lateral-line is absent. General colouration of body is dirty yellowish-brown. A black band runs from the snout to the tail region. A black ocellus is present above the lateral band. The back is dotted black and the caudal and the dorsal fins are with numerous rows of dark spots.

Distribution: Throughout Jharkhand in the drainage river systems of hillstream.

Remarks: It attains 6.5 to 10.0 cm in length (Srivastava, 1968) and reaches a maximum length of 20.0 cm (Qureshi & Qureshi, 1983). In our collections we recorded the length ranges from 6.9 to 20.6 cm. It is generally consumed by tribals and poor sections of Jharkhand and fishing is done with the help of biocidal plant sap.

29. *Mystus aor* (Hamilton)

Systematic Position

Order	-	Siluriformes
Family	-	Bagridae
Genus	-	Mystus
Species	-	*Mystus aor* (Ham.)
Local Name	-	Ariathi/Alli/Darai tengar.

D.8(1/7) I 0; P.10-11(1/9-10); V.6;A.12-13 (3-4/9); C.17; Barbels 4 pairs.

Length of head ranges from 4.4 to 5.0 and the height of body from 6.0 to 8.5 in the total length. Eye diameter is from 5.1 to 7.8 in the head length of head. Width or gap of mouth is 4.0 in the length of head and the upper jaw is the longer. The median longitudinal furrow on the head, extends to the base of the occipital process. Barbels are 4 pairs, the maxillary barbels reach the base of the caudal fin or even beyond it. Teeth are present in a contineous semilunar band across the palate. Dorsal spine is finely serrated posteiorly with no distinct teeth, it is weak and shorter than the head. Pectoral spine is stronger than the dorsal. The base of the adipose dorsal fin equals that of rayed dorsal or even more. Body is bluish- leaden above, becoming white on abdomen. Fins are yellowish and tinted gray. A black spot on the posterior lower portion of the adipose fin is present.

Distribution: Found throughout Jharkhand in most of the rivers, tributtaries and other associated water bodies.

Remarks: It is mostly eaten by tribals and poor sections of the Jharkhand.It grows upto the length of 180.6cm (Mishra, 1959); 2.5 to 7.6cm (Srivastava, 1968) and average size of 45.0cm (Qureshi & Qureshi, 1983). In our collections we recorded the length ranges from 8.9 to 121.0cm.

30. *Mystus cavasius* (Hamilton)

Systematic Position

Order	-	Siluriformes
Family	-	Bagridae
Genus	-	Mystus
Species	-	*Mystus cavasius* (Ham.)
Local Name	-	Kavasi tengara/Guntea/Sutahawa-tengara.

D.8(1/7) I 0; P.9-10(1/8-9); V.6; A.11-13(3-4/7-9); C. 16; Barbels 4 pairs.

Length of head ranges from 5.4 to 6.0 and the height of body from 5.1 to 6.4 in the total length. Eye diameter is 3.1 to 3.9 in the lenght of head. Upper jaw is the longer. Width of month is less than half of the head length. The median longitudinal groove on the head is wide and extends up to the base of the occipital process. The occipital process has a shallow and narrow groove along its posterior half. Barbels are 4 pairs, maxillary extend beyond the base of the caudal fin. Both the jaws are with teeth. Palatine teeth are found in cotineous crescentic band. Dorsal spine higher than the body pointed. The fairly long adipose dorsal commences just behind the rayed one long, strong and serrated pectoral spine. Pelvic fin originates just behind the base of the dorsal fin. Caudal fin is forked, the uper bobe is longer and more pointed. Body is yellowish on the abdomen and checks, bluish band present along the lateral line, black spot covering the basal portion of the dorsal fin, pectoral, pelvic and anal fins are dull white.

Distribution: Distribution throughout the Jharkhand and found in almost all rivers, tributaries and other associated water bodies.

Remarks: It is considered as the popular eatable fish of the Jharkhand. It grows upto the size of 45.6 cm (Mishra,1959); 9.0 to 18.0 cm (Srivastava, 1968) and average size of 46.0 cm (Qureshi & Qureshi,1983). In our collections we recorded the length ranges from 10.2 to 46.3 cm.

31. *Mystus seenghala* (Sykes)

Systematic Position

Order	-	Siluriformes
Family	-	Bagridae
Genus	-	Mystus
Species	-	*Mystus seenghala* (Sykes)
Local Name	-	Ari/Aor/Singala/Tengara.

D.8(1/7) I 0; P.10(1/9);V.6;A.11-12(3/8-9);C.19-21; Barbels 4 pairs.

Length of head ranges from 4.0 to 4.2 and the height of body from 7.4 to 7.8 in the total length. Eye diameter is 7.0 to 7.7 in the length of head. Gape of mouth equals 1/3 of head length. Upper jaw is longer than the lower. Snout is spatulate. The median longitudinal groove on the head extends to the base of occipital process. Barbels are 4 pairs, maxillary reaching to middle or just beyond rayed dorsal. Teeth are present on the palate in a contineous crescentic band. Dorsal spine is weak and posteriorly serrated. Length of the base of the adipose dorsal is equal to the rayed dorsal. The interspace between the two fins is of the same length. Pectoral spine is stronger than the dorsal and serrated. Anal rays 11-12. Caudal fin is deeply forked and its upper lobe is the longer. Body is brownish-grey superiorly and silvery on sides and abdomen. A black spot is found on the hind end of the base of the adipose fin.

Distribution: Found throughout Jharkhand in rivers, tributaries, ponds etc.

Remarks: It is mostly consumed by tribals and poor sections of Jharkhand. It grows upto the length of 38.0 cm (Mishra, 1959); 17.8 to 26.5 cm (Srivastava, 1968) and upto 40.0 cm (Qureshi & Qureshi, 1983). In our collections we recorded the length ranges from 17.1 to 40.3 cm.

32. *Mystus vittatus* (Bloch)

Systematic Position

Order	-	Siluriformes
Family	-	Bagridae
Genus	-	Mystus
Species	-	*Mystus vittatus* (Bloch)
Local Name	-	Tengra/Kuntiah

D.8(1/7) I 0; P.9-10(1/8-9); V.6; A.9-12(2-3/8-9); C.17; Barbels 4 pairs.

Length of head ranges from 4.4 to 5. land the height of body from 4.7 to 5.3 in the total length. Eye diameter is from 4.4 to 5.8 in the head length of head. Median longitudinal groove on the head extends to midway behind post orbit. Opercle is with radiating lines. Barbels are 4 pairs, maxillary extend slightly behind the ventrals. Teeth are present in a semilunar band across the palate. Dorsal spine is finely serrated. Pectoral spine is also serrated and as long as head excluding snout. Base of the adipose fin is usually of the same length as the interspace between the two fins. Caudal fin is forked and its upper lobe is longer than the lower. Body is golden, with dark bluish shoulder spot. A broad black longitudinal band extends along and on either side of the lateral-line and another band is found along the dorsal side of the body.

Distribution: Distribution throughout Jharkhand and found in rivers, reservoirs, lakes etc.

Remarks: It is a very popular food fish for its pleasent smoky flavour among tribals and poor sections of the Jharkhand. It attains the size of 121.6cm (Mishra, 1959); 2.5 to 7.6cm (Srivastava, 1968) and average size of 45.0cm (Qureshi & Qureshi, 1983). In our collections we recorded the length ranges from 8.9 to 121.0cm.

33. *Rita rita* (Hamilton)

Systematic Position

Order	-	Siluriformes
Family	-	Bagridae
Genus	-	Rita
Species	-	*Rita rita* (Ham.)
Local Name	-	Rita/Chona/Hunna.

D.7(1/6) I 0; P.Il(I/10);V.8;A.13-14(4-5/9);C.19; Barbels 3 pairs.

Length of head ranges from 4.1 to 4.3 and the height of body from 5.4 to 5.9 in the total length. Eye diameter is from 8.2 to 10.0 in the head length of head upper surface of the head is covered with skin. Upper jaw is longer than the lower. Cubito-humeral process is about 3/4 of the head length. Barbels are 3 pairs, nasal very short, maxillary extending nearly to the end of the head and the mandibular a little behind. Teeth are villiform in upper jaw and in the anterior part of lower jaw. Internally there are 2-3 rows of rounded teeth on the lower jaw. Dorsal spine is very strong, serrated, and as long as or 1/3 longer than the head. Base of the adipose fin 3/5 of the rayed fin and of the interspace between the two fins. Pectoral spine is shorter than the dorsal and is denticulated on both the edges. Caudal is forked. Body is greenish-grey above and becoming lighter below.

Distribution:	Distribution throughout Jharkhand and found in rivers, reservoirs, lakes etc.
Remarks:	It attains the size of 121.6cm (Mishra, 1959); 2.5 to 7.6cm (Srivastava, 1968) and average size of 45.0cm (Qureshi & Qureshi, 1983). In our collections we recorded the length ranges from 8.9 to 121.0cm.

36. *Clarias batrachus* (Linnaeus)

Systematic Position

Order	-	Siluriformes
Family	-	Clariidae
Genus	-	Clarius
Species	-	*Clarias batrachus* (Lin.)
Local Name	-	Mangri/Magur/Mah-gur/Magurah.

D.62-67; P.9-12(1/8-11); V.6; A.48-58; C.I5-17; Barbels 4 pairs.

Length of head ranges from 5.4 to 5.6 and height of body from 6.4 to 7.3 in the total length. Eye diameter is from 7.9 to 9.8 in the length of head. The greatest width of head equals its length. Head is covered with fine granules superiorly. Two depressions are present on the head, one behind the eyes another near the nape. Upper jaw is the longer. Barbels 4 pairs, the maxillary reach the base or middle of the pectoral fin. Vomerine teeth are viliform and arranged in an uninterrupted band. Pectoral spine is finely serrated. Dorsal fin is very long, it originates a little behind the occipital process and ends a bit anterior to the base of the caudal fin. The caudal fin is free. All the fin are covered with thick skin. Body is brownish-black.

Distribution:	Found throughout Jharkhand in ponds, ditches Swamps etc.
Remarks:	It is used as most common culturable fish for domestic pools, ditches etc. and considered to be very nourishing and tasty fish. It grows up to the length of 45.6cm (Mishra,1959); 23.2 to 28.8cm (Srivastava, 1968 and 45.0cm (Qureshi & Qureshi, 1983). In our collections we recorded the length ranges from 21.7 to 45.3 cm.

35. *Heteropneustes fossilis* (Hamilton)

Systematic Position

Order	-	Silurifonmes
Family	-	Heteropneustidae
Genus	-	Heteropneustes
Species	-	*Heteropneustes fossilis* (Ham.)
Local Name	:	Pariasi/Pangsa/Pungas/Punwas/Jellum

D.6-7; P.8(1/7); V.6; A.60-79; C.19; Barbels 4 pairs.

Length of head ranges from 5.4 to 6.9 and height of body from 5.1 to 7.8 in the total length. Eye diameter is from 8.1 to 8.5 in the length of head. Barbels are 8, maxillaries reaching middle of the pectoral or even to the pelvic base. Vomerine teeth are found in a pyriform patch on either side, covering anteriorly and diverging posteriorly. Dorsal fin is short, spineless, with its origin before the anterior third of the body. Ventral fin originates below the origin of dorsal fin. Pectoral not united with caudal.Body is dark leaden-brown.

Distribution: Found throughout Jharkhand in ponds, ditches, Swamps etc.

Remarks: It is considered to be very nourishing and tasty fish. It grows upto the length of 30.4cm (Mishra, 1959); 8.4 to 23.5cm(Srivastava, 1968) and 30.0cm (Qureshi & Qureshi, 1983). In our collections we recorded the length ranges from 9.2 to 30.1 cm.

36. *Ailia coila* (Hamilton)

Systematic Position

Order	-	Siluriformes
Family	-	Schilbeidae
Genus	-	Ailia
Species	-	*Ailia coila* (Ham.)
Local Name	-	Kajoli/Bunsputta/Puttuli/Batausi

D.0;P.15(1/14);V.6;A.66;C.19; Barbels 4 pairs.

Length of head ranges from 6.8 to 6.9 of caudal 6.0 to 6.1 and height of body 5.9 to 6.0, in the total length. Eye diameter is 4.5 in the length of head, 1.74 from end of snout and 2.0 in the interorbital width. Body elongated, strongly compressed and abdomen rounded. Head short, greatly compressed, greatest width of the head is equal to its length behind the orbit. Width of mouth equal to 1/3 of the length of the head, upper jaw slightly longer.Barbels 4 pairs, one each of maxillary, nasal and two of mandibular generally longer than head. Rayed dorsal absent, adipose dorsal small and posteriorly free, pectoral spine slender and is as long as head; ventral fin short and is enclosed by pectoral, anal long but not confluent with caudal forked with lower lobe longer than upper lobe. Lateral-line complete and simple, body colour silvery, fins are stained grey at the margins.

Distribution: Distribution throughout Jharkhand and found in ponds, reservoirs, lakes, rivers etc.

Remarks: It is generally consumed by tribals and poor sections of Jharkhand only. It attains the length of 17.7cm (Mishra, 1959); 9.2 to 14.0 cm (Srivastava, 1968) and 30.0cm (Qureshi & Qureshi, 1983).In our collections we recorded the size ranges from 10.3 to 30.5 cm.

37. *Clupisoma garua* (Hamilton)

Systematic Position

Order	-	Siluriformes
Family	-	Schilbeidae
Genus	-	Clupisoma
Species	-	*Clupisoma garua* (Ham.)
Local Name	-	Buchua/Garua/Puttosi/Punia-buchua

D.8(1/7)/0; P.12(1/12); V.6; A.29-36(3/26-33); C.17; Barbels 4 pairs.

Length of head ranges from 5.4 to 6.3 and the height of body is 5.5 to 6.8 in the total length. Eye diameter is from 3.4 to 3.9 in the length of head. The eyes are with broad, circular adipose lid and situated partly on the lower surface of the head. Width or gap of the mouth is 2/5 in the length of head. Upper jaw is the longer. Barbels are 4 pairs, maxillaries reaching the middle or the end of the pelvic fins. Palatine and vomerine teeth on either side are contineous. Dorsal fin arises in advance of the ventral. Dorsal spine is slender, serrated and as long as or a little longer than dorsal spine. Anal fin is with 29-36 rays. Caudal fin is deeply forked, its lower lobe is longer than the upper.Body is greyish along the back and silvery along the sides and abdomen. Fins are tined gray.

Distribution: Found all over Jharkhand in rivers, ponds, lakes, reservoirs etc.

Remarks: It is considerd as the common catfish consumed generally by tribals and poor sections of Jharkhand. It grows upto the length of 60.8 cm (Mishra, 1959); 15.4 to 23.8 cm (Srivastava, 1968) and 60.0cm (Qureshi & Qureshi, 1983). In our collections we recorded the length ranges from 18.8 to 60.3 cm.

38. *Eutropiichthys vacha* (Hamilton)

Systematic Position

Order	-	Siluriformes
Family	-	Schilbeidae
Genus	-	Eutropiichthys
Species	-	*Eutropiichthys vacha* (Ham.)
Local Name	-	Bachwa/Vacha/Bacha/Bhacha/Butchwa

D.8(1/7) I 0; P. 14-18(1/13-17); V.6;A.44-51(3-4/41-47);
C.17; Barbels 4 pairs.

Length of head ranges from 5.4 to 5.6 and height of body from 5.0 to 5.4 in the total length. Eye diameter is 3.4 to 3.5 in the length of head. Eyes are with broad adipose lids. Upper jaw slightly longer. Barbels are 8; maxillary reaching nearly the base of the pectoral fin, nasal reaching just behind the head. Sharp teeth are present on the jaws; palatine and vomarine join to from an uninterrupted band. The dorsal spine is thin, serrated and as long as the head without snout. Pectoral spine is serrated and slightly longer than the dorsal spine. Anal fin is with 44 to 51 rays. Caudal fin is deeply forked. Body is silvey-greyish along the back. Edges of the pectoral and caudal fins are black.

Distribution: Found throughout Jharkhand in rivers, lakes etc.

Remarks: It grows up to the length of 30.4cm (Mishra, 1959); 9.0 to12.0 cm (Srivastava,1968) and 30.0cm (Qureshi & Qureshi, 1983). In our collections we recorded the size ranges from 10.2 to 30.1 cm.

39. *Pangasius pangasius* (Hamilton)

Systematic Position

Order	-	Siluriformes
Family	-	Schilbeidae
Genus	-	Pangasius
Species	-	*Pangasius pangasius* (Ham.)
Local Name	-	Pariasi/Pangsa/Pungas/Punwas/Jellum

D.8(1/7) I 0;P.13(1/12);V.6;A.31-34(4-5/27-29);
C.19; Barbels 2 pairs.

Length of head ranges from 5.4 to 6.0 and height of body from 4.1 to 5.0 in the total length. Eyes are situated in the anterior half of the head. Width of the mouth equals 3/7 of the length of the head. Upper jaw is longer than lower jaw. Teeth on the plate are arranged in a crescentic row. Barbels are 2 pairs, maxillary ones extending to the base of the pectoral fin. Dorsal spine is serrated, moderately strong and as long as the head behind the angle of mouth. Pectoral spine is serrated, strong and as long as the dorsal spine. Caudal is forked, lobes not sharply pointed. Body is silvery, darkest superiorly shot wth purple on sides, cheeks and the under surface of head is golden.

Distribution: Found throughout Jharkhand in rivers, lakes etc.

Remarks: It is useful in the control of molluse-borne diseases. It grows up to the length of 121.6cm (Mishra, 1959); 86.0 to 90.3cm (Srivastava,1968) and 122.0cm (Qureshi & Qureshi, 1983). In our collections we recorded the length ranges from 78.3 to 121.9 cm.

40. *Silonia silonclia* (Hamilton)

Systematic Position

Order	-	Siluriformes
Family	-	Schilbeidae
Genus	-	Silonia
Species		*Silonia silondia* (Ham.)
Local Name	-	Baikar/Banspati/Shilong/Silun/Dhain/Silondia-vacha.

D.8(1/7) 1 0;P.12-14(1/11-13);V.6;A.40-46(4/36-42);
C.17; Barbels 1 pair.

Length of head ranges from 5.1 to 5.4 and height of body from 5.1 to 5.9 in the total length. Eye diameter is 4.0 to 4.4 in the length of head. Eyes are with narrow adipose eyelids. Lower jaw is the longer and broadly pointed. Width or the gap of mouth is more than half of the head length. Only one pair of maxillary barbels are present which reach upto 1/3 of the orbit. Teeth are caniniform and arranged in a crescentic band across the plate. Dorsal spine is weak, finely serrated and as long as head behind midorbit. Pectoral spine is stronger, serrated and as long as head excluding snout. Caudal fin is deeply forked. Body is bluish along the back and silvery on the sides. Fins are stained grey and the lips are red.

Distribution: Found in rivers and reservoirs of the Jharkhand.

Remarks: It grows up to the length of 93.2cm (Mishra, 1959); 9.7 to 12.2 cm (Srivastava, 1968) and 182.0cm (Qureshi & Qureshi, 1983). In our collections we recorded the length ranges from 11.3 to 180.0 cm.

PLATE IV

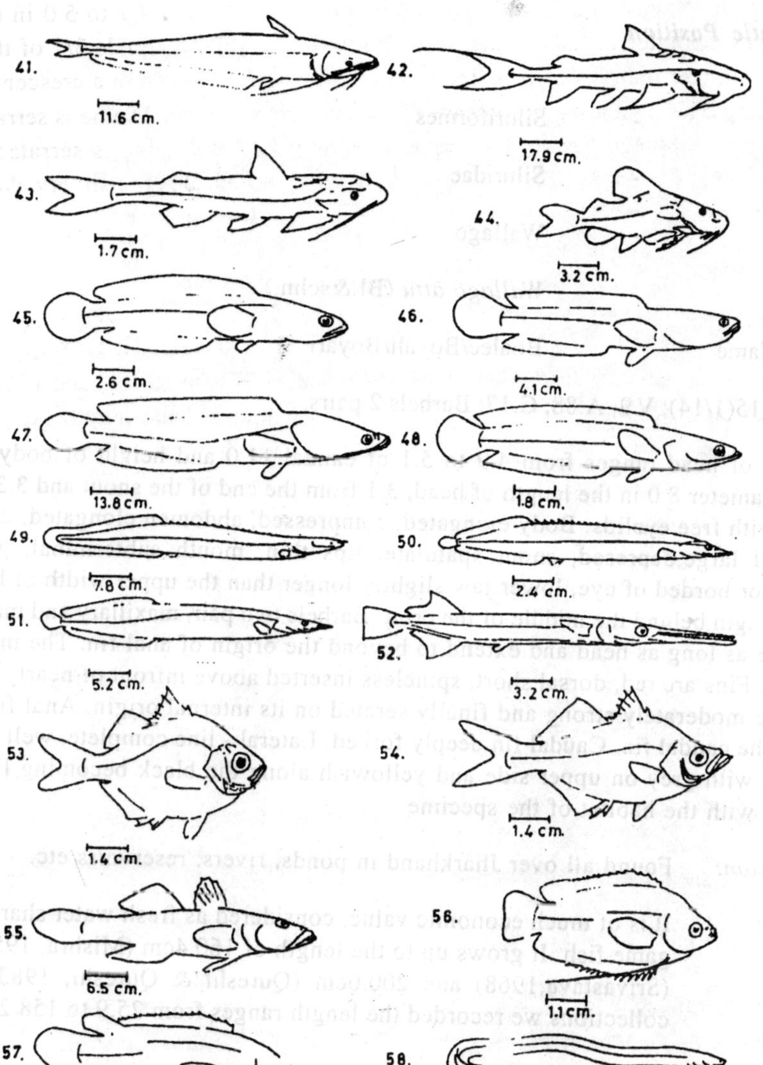

EXPLANATION OF PLATE - IV

41. *Wallago attu* [Fig.41]
42. *Bagarius bagarius* (Ham.)[Fig.42]
43. *Glyptothorax telchitta* (Ham.)[Fig.43]
44. *Hura hara* (Ham.)[Fig.44]
45. *Channa gachua* (Ham.)[Fig.45]
46. *Channa punctatus* (Bloch)[Fig.46]
47. *Channa marulius* (Ham.)[Fig.47]
48. *Channa striatus* (Bloch)[Fig.48]
49. *Mastacembelus armatus* (Lacepede)[Fig.49]
50. *Mastacembelus pancalus* (Ham.)[Fig.50]
51. *Macrognathus aculeatus* (Bloch)[Fig.51]
52. *Xenentodon cancila* (Ham.)[Fig.52]
53. *Chanda nama* (Ham.)[Fig.53]
54. *Chanda ranga* (Ham.)[Fig.54]
55. *Glossogohius giuris* (Ham.)[Fig.55]
56. *Anabas testudineus* (Bloch)[Fig.56]
57. *Colisa fasciatus* (Bloch & Schn.) [Fig.57]
58. *Monopterus cucliia* (Ham.)[Fig.58]

41. *Wallago attu* (Bloch and Schneider)

Systematic Position

Order	-	Siluriformes
Family	-	Siluridae
Genus	-	Wallago
Species	-	*Wallago attu* (Bl.&schn.)
Local Name	-	Boalee/Boyali/Boyari

D.5; P. 15(1/14); V.9; A.88; C.17; Barbels 2 pairs.

Length of head ranges from 4.9 to 5.1 of caudal 11.0 and height of body is 6.7 in the total length. Eye diameter 8.0 in the length of head, 3.1 from the end of the snout and 3.3 in the inter orbital width. Eyes with free eyelids. Body elongated, compressed, abdomen elongated, dorsal profile nearly straight. Head large,depresed; snout spatulate, lips thin, mouth subterminal, gap wide, reaching beyond anterior borded of eye, lower jaw slightly longer than the upper width of head approximately equal to the length behind the middle of the orbit. Barbels two pair, maxillary and mandibular, maxillary is about twice as long as head and extend to beyond the origin of anal fin. The mandibular pair is as long as snout. Fins are red, dorsal short, spineless inserted above infront of nearly as long as pectoral. Pectoral spine moderately strong and finally serated on its internal origin. Anal fin is long and is not confluent to the caudal fin. Caudal fin deeply forked. Lateral - line complete, well marked and simple. Colour white with grey on upper side and yellowish along the black becoming lighter below. Infact colour varies with the habitat of the specime

Distribution: Found all over Jharkhand in ponds, rivers, reservoirs etc.

Remarks: It is of much economic value, considered as fresh water shark and is also a good game fish. It grows up to the length of 160.4cm (Mishra, 1959); 25.2 to 28.4 cm (Srivastava,1968) and 200.0cm (Qureshi & Qureshi, 1983) in length. In our collections we recorded the length ranges from 25.9 to 158.2 cm.

42. *Bagarius bagarius* (Hamilton)

Systematic Position

Order	-	Siluriformes
Family	-	Sisoridae
Genus	-	Bagarius
Species	-	*Bagarius bagarius* (Ham.)
Local Name	-	Boonch/Baghar/Baghari/Shalun.

D.7(1/6)/0;P.13(1/12);V.6;A.13-15(3/10-12); C.17; Barbels 4 pairs.

Length of head ranges from 4.4 to 4.7 and the height of body is 8.0 in the total length including the caudal filament. Eye diameter is from 10.1 to 10.7 in the length of head. Upper jaw is longer and the gape of the mouth is about half of the head length. Occipital process does not reach the basal bone of the dorsal fin. Barbels are 4 pairs, maxillary with broad base and longer than the head reaching to anterior 1/3 of the pectoral fin. Teeth are sharp and of unequal size in both the jaws. Dorsal spine is smooth and its bony portion is as long as head without snout. Pectoral spine is stronger than the dorsal spine. Caudal fin is deeply forked and its upper lobe is prolonged. Skin is scabrous and the body is yellowish; with large black markings and cross bands in the region of rayed dorsal; adipose dorsal and tail. All fins are with a black base and often a dark band is present across each of them. Numerous round small, black markings are found over the entire body excluding the belly.

Distribution: Distribution throughout Jharkhand and found in reservoirs, lakes, rivers etc.

Remarks: It is considered as the freshwater shark. It grows upto the length of 150.0cm (Mishra, 1959); 9.8 to 11.4cm (Srivastava, 1968) and 152.0cm (Qureshi & Qureshi, 1983). But in our collections we recorded the length ranges from 10.3 to 152.2cm.

43. *Glyptothorax telchitta* (Hamilton)

Systematic Position

Order	-	Siluriformes
Family	-	Sisoridae
Genus	-	Glyptothorax
Species	-	*Glyptothorax telchitta* (Ham.)
Local Name	-	Telliah

D.7(1/6) I 0; P.9-10(1/8-9); V.6; A. 11-12 (2/9-10); C.17-18; Barbels 4 pairs.

Length of head ranges from 5.2 to 5.5 and the height of body is 7.1 to 8.0 in the total length. Eye diameter is from 12.0 to 13.0 in the length of head. Eyes are small and with circular pupils. Upper jaw is longer and upper lip is slightly fringed. Barbels are 4 pairs in number. Teeth are villiform and found on both the jaws. Vomero platine teeth are absent. Dorsal spine is covered with skin and is very finely serrated posteriorly. Dorsal fin is as high as the body. Pectoral spine is stronger. Ventral do not reach the anal. Caudal is deeply forked. Thoracic adhesive appratus is well developed. Body is brownish appearing to be yellowish due to the yellowish colour of the elevations. Black blotches are present on the surface of the body. Fins are yellow with black bands but the caudal is blackish with yellow edge.

Distribution: Commonly throughout Jharkhand and found in hillstreams generally but recorded in other waterbodies also.

Remarks: It attains 6.6 to 7.4 cm in length (Sirvastava, 1968), and upto 15 cm (Qureshi & Qureshi, 1983). In our collections we recorded the standrad length ranges from 13-17 cm. Generally consumed by tribals and poor sections of Jharkhand and fishing is done wish the help of biocidal plant sap.

44. *Hara hara* (Hamilton)

Systematic Position

Order	-	Siluriformes
Family	-	Sisoridae
Genus	-	Hara
Species	-	*Hara hara* (Ham.)
Local Name	-	Panahi

D.7-8(1-2/6) 0; P.7(1/6); V.6; A. 10-11(3/7-8); C.15; Barbels 4 pairs.

Length of head ranges from 4.1 to 4.4 and the height of body is 4.4 to 4.1 in the total length. Eye diameter is 10:0 in the length of head. Upper jaw is slightly longer. The nasal openings are situated close together and are divided by membranous flap bearing the short nasal barbels. Barbels are 4 pairs in number, the maxillaries reaching either upto the base of the pectoral or are produced a little beyond. Villiform teeth are present in the jaws but palate is without teeth. Dorsal spines are two, the first is short and second is strong and as long as the head. It is strongly serrated posterioly but weakly anteriorly. Adipose dorsal is present. Pectoral spine is strong and as long as the head. It is flattened slightly curved, denticulated internally and finely serrated externally. Caudal is forked. Skin is covered with small elevations, which in the posterior part of the body are arranged in parallel lines. Body is dirty yellowish-brown banded and blotched with black. Fins are with black bands and markings. All the barbels are annulated with black and yellowish - brown rings.

Distribution: Throughout Jharkhand.

Remarks: It atains 5.2 to 6.2cm (Srivastava, 1968); 15cm in length (Qureshi & Qureshi,1983). In our collections we recorded the length ranges from 5.7 to 15.2cm. It is generally consumed by tribals and poor sections of Jharkhand and fishing is done with the help of biocida plant sap.

45. *Channa gachua* (Hamilton)

Systematic Position

Order	-	Ophiocephaliformes
Family	-	Channidae
Genus	-	Channa
Species	-	*Channa gachua* (Ham.)
Local Name	-	Cheng/Chenga/Chamga

D.32-37;P.15-16;V.6;A.21-23;C.12;L.I.140-145;L.tr.3-4/6-7.

Length of head ranges from 3.4 to 4.1 and the height of body is 6.0 to 9.5 in the total length. Eye diameter is 6.0 in the length of head. Teeth are conical and present in the lower jaw as well as a few on the vomer and palatines. Pectoral fins are as long as the head behind the eye. Ventral is 2/5 of the length of pectoral fin. Head scales are broad and irregular. There are 4 to 5 rows of scales between the orbit and the angle of pre-opercle. Predorsals are 12 in numbers. Lateral - line curves downwards after 12th scales. Body is generally greenish becoming lighter beneath. Dorsal, anal and caudal fins are slaty grey with their margins orange. Pectoral has a deep blue base, transversely banded with orange and blue stripes.

Distribution: Distribution throughout Jharkhand and found in pools, ditches, ponds, lakes, river, reservoirs etc. but abundant in eutrophicated waterbodies.

Remarks: It is generally consumed by tribals and poor sections of Jharkhand. It attains the size of 32.5cm (Mishra,1959); 8.0 to 14.0cm (Srivastava, 1968) and 20.0cm (Qureshi & Qureshi, 1983). In our collections we recorded the length ranges from 6.5 to 21.8.0cm.

46. *Channa punctatus* (Bloch)

Systematic Position

Order	-	Ophiocephaliformes
Family	-	Channidae
Genus	-	Channa
Species	-	*Channa punctatus* (Bloch)
Local Name	-	Garai/Girai/Lata.

D.29-32; P.16-17; V.6; A.21-23; C.12; L.I.35-40; L.tr.4-5/6-9.

Length of head ranges from 3.3 to 3.5 and the height of body is 5.4 to 6.7 in the total length. Eye diameter is from 6.9 to 8.4 in the length of head. Lower jaw is longer. Teeth are conical and present in both the jaws,vomer and the palatine. Pelvic fins are 3/4 as long as pectorals. There are 5 rows of scales between the eye and the angle of preopercle. Predorsals are 12 in number. The scales are large and irregular on the summit of the head. Lateral-line straight. Body is generally greenish-grey becoming yellow below,several bands pass from the dorsum of the body down-wards to the middle of the sides. Fins are spotted.

Distribution: Distribution throughout Jharkhand and found in paddy fields, pools, ponds,ditches,canals, rivers, reservoirs, lakes and other eutrophic waterbodies.

Remarks: It is found abundantly in eutrophicated and muddy water bodies and generally consumed by tribals and poor sections of the Jharkhand. It grows upto the length of 30.4cm (Mishra, 1959) and 10.0cm to 15.3cm (Srivastava, 1968) and 30.0cm (Qureshi & Qureshi,1983). In our collections we recorded the length ranges from 12.3cm to 31.1cm.

47. *Channa marulius* (Hamilton)

Systematic Position

Order	-	Ophiocephaliformes
Family	-	Channidae
Genus	-	Channa
Species	-	*Channa marulius* (Ham.)
Local Name	-	Gajari/Bhor.

D.45-55; P.18; V.6; A.28-36; C.14;L.I.60-70; L.tr.41/2-61/2/11-13.

Length of head ranges from 4.1 to 5.0 and the height of body from 6.9 to 7.4 in the total length. Eye diameter is from 5.0 to 6.8 m the length of head.Teeth are villiform and present in several rows in the jaws, vomer and the palate. Pectoral fin is more than half the length of the head but not reaching the origin of the anal fin. Scales are of moderate size on the summit of head. There is a row of 10 scales between the orbit and the angle of preopercle. Predorsals are 16 in number. Lateral-line is straight upto the 17th scale, then it descends down 2 rows of scales and subsequently runs straight upto the centre of the caudal fin. Body is greyish-green becoming lighter below with a brilliant orange band passing from the tip of the snout, over the eyes, to the tip of the caudal fin. There are 4 to 5 round black blotches situated below the lateral-line and there is a large black prominent occelus with orange boundary at the upper part of the base of the caudal fin.

Distribution: Distribution throughout Jharkhand in ponds, pools, lakes, reservoirs, and other eutrophicated water bodies.

Remarks: It is considered as the favourate food fish. It grows upto the length of 120.0cm (Mishra, 1959) and 10.0cm to 25.0cm (Srivastava, 1968) and 122.0cm (Qureshi & Qureshi,1983). in our collections we recorded the length ranges from 14.8cm to 123.1cm.

48. *Channa striatus* (Bloch)

Systematic Position

Order	-	Ophiocephaliformes
Family	-	Channidae
Genus	-	Channa
Species	-	*Channa striatus* (Bloch)
Local Name	-	Morrul/Shol/Saur/Sowra.

D.37-45; P. 16-18; V.6; A.23-26; C.I3;
L.I.50-60; L.tr. 41/2-8/7-10.

Length of head ranges from 3.3 to 3.9 and the height of body is 5.9 to 7.9 in the total length. Eye diameter is from 6.0 to 7.4 in the length of head. Lower jaw is slightly longer, teeth are found in the jaws and the cardiform on the palatines. The pactoral does not quite reach to above the origin of the anal. There are 9 rows of scales between the eye and the angle of preopercle. Predorsal scales are 18-20 in number. I ateral - line curves downwards below 12th dorsal ray. Body is greyish to back above; dirty white below. Bands of grey or black colour pass down from the sides to the abdomen. Fins are greyish. A large black ocellus at the end of the base of dorsal fin is present.

Distribution: Distribution throughout Jharkhand and found in ponds, rivers, reservoirs, lakes and other eutrophic waterbodies.

Remarks: It prefers stagnment muddy water bodies. It attains the size of 90.0cm (Mishra, 1959) and 8.5cm to 28.5cm (Srivastava, 1968) and 91.0cm (Qureshi & Qureshi,1983). In our collections we recorded the length ranges from 16.4cm to 91.2cm.

49. *Mastacembelus armatus* (Lacepede)

Systematic Position

Order	-	Mastacembeliformes
Family	-	Mastacembelidae
Genus	-	Mastacembelus
Species	-	Mastacembelus armatus (Lacepede)
Local Name	-	Bamm/Bami/Bummi

D.32-39/74-90; P.23-25; A.78-91 (3/75-88).

Length of head ranges from 5.8 to 7.3 and the height of body from 10.4 to 11.8 in the total length. Eye diameter is from 11.1 to 11.6 in the length of head. Body elongated, cylindrical and eel-like. Snout long, tri-lobed and with a fleshy appendage. Cleft of mouth is narrow. The preorbital is with a spine which points backwards and reaches 1/3 distance of the orbit. Angle of preopercle is with 2-3 conspicuous denticulations. The dorsal spines commence over the middle of the pectoral fin. They increases in size from anterior to posterior direction, but last one is the longest. Origin of the soft part of the dorsal fin is behind the origin of anal. The anal has 3 spines. Pectoral fins are small and the ventral fins are absent. Scales are very minute and are similar on head and body. Body is brownish above and lighter below. There is a row of rounded black spots along the base of the dorsal fin.

Distribution: Distribution through Jharkhand and found in rivers, tributaries, lakes, reservoirs, ponds etc.

Remarks: It is generally consumed by tribabls and poor sections of Jharkhand. It grows upto the length of 60.9cm (Srivastava,1 968) and 60.0cm (Qureshi & Quresli, 1983). In our collections we recorded the length ranges from 17.6cm to 60.3cm.

50. *Mastacembelus / puncalus* (Hamilton)

Systematic Position

Order	-	Mastacembeliformes
Family	-	Mastacembelidae
Genus	-	Mastacembelus
Species	-	*Mastacembelus puncalus* (Ham.)
Local Name	-	Gaincha/Turi

D.24-26/30-42; P. 19; A.34-49(3/31 -46); C. 12.

Length of head ranges from 4.9 to 5.8 and the height of body is 7.1 to 9.8 in the total length. Eye diameter is 0.8 to 1.0 in the inerorbital width. Body is elongated, cylindrical and eel-lik is the longest and strongest. Origin of soft dorsal is behind the anal origin. Candal is not united with the dorsal and anal. Anal fin is with three spines, of which second is the longest and strongest, pectorals are small and the ventrals are absent. Caudal fin is rounded. Body is greenish-olive along the back and yellowish beneath. Yellowish-white spots are present over the side of body. Posterior part of the body is often vertically striped, soft dorsal, pectoral, caudal and anal fins are yellow with numerous black spots which are arranged in transverse bars on the anal fin.

Distribution: Distribution throughout Jharkhand and recorded in rivers, canals, ponds, lakes, reservoirs etc.

Remarks: It is generally consumed by tribals and poor sections of Jharkhand. It grows upto the length of 17.7cm (Mishra, 1959);18.0 to 49.0cm (Srivastava, 1968) and 18.0cm (Qureshi & Qureshi, 1983). In our collections we recorded the length ranges from 16.2 to 30.9.0cm.

51. *Macrognathus aculeatus* (Bloch)

Systematic Position

Order	-	Mastacembeliformes
Family	-	Mastacembelidae
Genus	-	Macrognathus
Species	-	*Macrognathus aculeatus* (Bloch)
Local Name	-	Goichi/Gaichee/Spmed-eel

D.16-22/44-54; P.19-23;A.47-55(3/44-52);C.15.

Length of head ranges from 5.1 to 5.9 and the height of body from 7.9 to 9.8 in the total length. Eye diameter is from 13.1 in the length of head. Body is elongated, cylindrical and eel-like. The snout is long, fleshy, trilobed, concave and transversely straited ventrally. Upper jaw is longer than the lower jaw. The cleft of mouth is narrow. Minute teeth are present in the jaws. First dorsal fin is composed the anal are separated by a notch above and below. Anal spines are three of which second is the longest and strongest. Pectorals are greatly reduced and the ventrals are absent. Scales of the head are longer than those of the body. Body is brownish or greenish yellow above and becoming yellowish or lighter below. A light coloured band is present along the body just above the lateral-line. 3-6 large black ocelli with white boundary are present along the base of soft dorsal on either side. Caudal fin has vertical bars.

Distribution: Distribution throughout Jharkhand in rivers,canals, ponds, lakes, reservoirs, etc.

Remarks: It is very commonly consumed by poor sections and tribals. It grows uptothe length of 8.0cm (Srivastava, 1968) and 38.0cm (Qureshi & Qureshi, 1983). In our collections we recorded the length ranges from 9.9cm to 38.7cm.

52. *Xenentoden cancila* (Hamilton)

Systematic Position

Order	-	Beloniformes
Family	-	Belonidae
Genus	-	Xenentodon
Species	-	*Xenentodon cancila* (Ham.)
Local Name	-	Kawa/Kakhya/Kauwa

D.15-18(2-3/13-15); P.11; V.6; A.16-18(2-3/14-16);C.15.

Length of head ranges from 2.4 to 2.7 and the height of body from 8.1 to 11.8 in the total length. Eye diameter is from 3.0 to 3.2 in the length of head, from the hind edge of opercle. Lower jaw is slightly longer than the upper jaw but both are proloned as beak. A deep longitudinal groove is present along the upper surface of the head. A row of large, sharp and widely separated teeth is present in both the jaws and an external row of numerous fine teeth is also present. Dorsal fin originates opposite to the anal. Pelvic commences nearer to the caudal base than to the hind broder of the eye. Lateral-line is not keeled. Scales are small and irregulary arranged. Body is subcylindrical, greenish-grey above and whitish below. Silvery streak with dark margin extends along the body from opposite the orbit and runs upto the middle of the base of the caudal fin. Upper portion of the body is with densely set black dots.

Distribution: Distribution throughout Jharkhand and found in rivers, canals, ponds, lakes, reservoirs, etc.

Remarks: It is generally consumed by poor sections and tribals of the Jharkhand. It attains the size of 30.4cm (Mishra, 1959); 21.7 to 24.5 cm (Srivastava, 1968) and about 30.0cm (Qureshi & Qureshi,1983). In our collections we re.orded the length ranges from 18.7 to 30.2 cm.

53. *Chanda nama* (Hamilton)

Systematic Position

Order	-	Perciformes
Family	-	Ambassidae
Genus	-	Chanda
Species	-	*Chanda nama* (Ham.)
Local Name	-	Chanda/Channe/Nama-Chanda.

D.1+7/14-18(1/13-17); P.12-13; V.6(1/5); A.17-21(3/14-18); C.17.

Length of head ranges from 4.1 to 4.5 and the height of body from 2.8 to 3.4 in the total length. Eye diameter is from 2.9 to 3.1 in the length of head. Body is greatly compressed and both the profiles of the body are equally convex. Lower jaw is much longer than the upper jaw. Preobital is slightly denticulated. The double horizontal border of the preopercle is also slightly denticulated at its lower edge, 2 large and curved canines are present on either side of the lower jaw. Dorsal fins are two and both are united at their bases. First dorsal is with 7 spines, a recumbent spine is also present before it. The third dorsal spine is the longest. Anal fin is with 3 spines. Scales are minute. Lateral-line may be broken, absent or even indistinct. Body is whiteish-yellow with black dots all over the body. The dots on the shoulder from an oblong vertical patch. Summit of head and the top of eyes are black.

Distribution: Distribution throughout Jharkhand and found in ponds, pools, ditches, rivers, lakes, reservoirs as a weed fishes.

Remarks: It is considered as the main aquarium fish (Glass fish or X-ray fish) of local variety. It is consumed only by tribal and poor sections of Jharkhand. It attains the size of about 7.6cm (Mishra, 1959) and 3.4cm to 6.5cm (Srivastava, 1968) and 8.0cm (Qureshi & Qureshi,1983). In our collections we recorded the length ranges from 3.1cm to 8.2cm.

54. *Chanda ranga* (Hamilton)

Systematic Position

Order	-	Perciformes
Family	-	Ambassidae
Genus	-	Chanda
Species	-	*Chanda ranga* (Ham.)
Local Name	-	Chanda/ranga-Chanda/Channa.

D.1+7/14-15(1/13-15); P.12-13; V.6(l/5); A.17-19(3/14-16);C.17.

Length of head ranges from 3.2 to 3.9 and the height of body from 2.4 to 2.7 in the total length. Eye diameter is from 2.4 to 2.6 in the length of head. Both the profiles of the body are very convex but the profile over the eyes is slightly concave. Preobital is with denticulations on its inferior edge, a strong tooth on its anterosuperior angle is directed towards the eye and 5 more teeth along the upper edge of the same are also present. There is another spine at the middle of the posterior margin of the orbit and its upper half has 5 more spines. Teeth are small, villiform and present in the jaws,vomer and palate. Dorsal fins are two, both are united at their base. The first dorsal fin is with 3 spines. Caudal fin is forked. Body is white with black spots. Closely set dots form a dark mark on the shoulder of each side. Margins of ventral fins are dark or grey.

Distribution: Distribution throughout Jharkhand and found in ponds, pools, ditches, rivers, lakes,-reservoirs, canals, etc.

Remarks: It is considered as the main aquarium fish (Indian Glass fish) of local variety and consumed only by tribal and poor sections of Jharkhand. It attains the size of about 7.2cm to 10.2cm (Mishra, 1959) and 3.2cm to 7.2cm (Srivastava, 1968) and 10.0cm (Qureshi & Qureshi, 1983). In our collections we recorded the length ranges from 4.1cm to 10.2cm.

55. *Glossogobius giuris* (Hamilton)

Systematic Position

Order	-	Perciformes
Family	-	Gobiidae
Genus	-	Glossogobius
Species	-	*Glossogobius giuris* (Ham.)
Local Name	-	Bhalia/Bhula/Bulla.

D.6 | 1/8-9; P.20;V.6(l/5);A.9-10(l/8-9);C. 17; L.I.30-35;L.tr.8-12.

Length of head ranges from 3.4 to 3.9 and the height of body is 4.9 to 5.3 in the total length. Eye diameter is from 6.1 to 7.9 in the head length. Head is depressed, pointed and the lower jaw is longer. Teeth are found in the jaws of which front teeth are caninoid. Scales extend over the head,body scales are caninoid while the head scales are cycloid. Dorsal fins are two in number, the first dorsal with 6 weak spines. Anal is pointed posteriorly and with one weak spine. The pelvic fins become united at their bases to from a sucking disc. Caudal fin is oblong. Lateral-line scales are 30-35. Body is with variable colours, generally olive to dusky-green above, lighter below with light black markings on the head. Irregular marking on the body are also present. Fins are yellow green with spots and dark markings.

Distribution: Distribution throughout Jharkhand and found mainly in rivers, canals, streams etc.

Remarks: It is consumed as the most popular fish. It attains the size of 30.4cm (Mishra,1959); 6.0 to 14.2cm (Srivastava, 1968) and 46.0cm (Qureshi & Qureshi, 1983). In our collections we recorded the length ranges from 7.3 to 46.3cm. It is generally consumed by the tribals and poor sections of Jharkhand.

56. *Anabas testudineus* (Bloch)

Systematic Position

Order	-	Perciformes
Family	-	Anabantidae
Genus	-	Anabas
Species	-	*Anabas testudineus* (Bloch)
Local Name	-	Coi/Kobhai/Kawai

D.17-18/8-10; P.15; V.6(1/5);A.9-10/9-11; C.17; L.I.28-32;L.tr.3-4/9/10.

Length of head ranges from 3.4 to 3.7 and the height of body is 3.0 to 3.9 in the total length. Eye diameter is from 4.4 to 5.1 in the length of head. Lower jaw is slightly longer than upper jaw. Preobital is strcngly denticulated but the posterrior edge of the opercle, subopercle and interopercle are strongly spinous. Teeth are present in the jaws,vormers and palatines. Dorsal fin is single and longer than the anal fin, it is with 17-18 spines. Its spinous portion is much longer than the soft one. Anal fin is with 9-10 spines. Scales are ctenoid and also found over the bases of dorsal, pectoral, anal and caudal fins. Lateral - line is interrupted. Body is light to dark green above and greenish-yellow to orange below. There are 4 wide cross bands on the body. A black spot is present on each side of the caudal peduncle and a black spot is also found at the posterior end of the opercle on both sides.

Distribution: Found throughout Jharkhand in ponds, pools, ditches, lakes, reservoirs, paddyfields and other eutrophic waterbodies.

Remarks: It has a very good flavour and is popular as food. It attains the size of 20.3cm (Mishra, 1959) and 13.0cm to 13.8cm (Srivastava, 1968) and 26.0cm (Qureshi & Qureshi, 1983). In our collections we recorded the length ranges from 12.4cm to 27.1cm.

57. *Colisafasciatus* (Bloch & Schneider)

Systematic Position

Order	-	Perciformes
Family	-	Anabantidae
Genus	-	Colisa
Species	-	*Colisa fasciatus* (Bloch & Sch.)
Local Name	-	Khosti/Khalisha

D.15-17/9/13; P.10; V.I; A.15-18/14-19; C.I5; L.I.29-3 l;L.tr.41/2-51/2/ 11-12.

Length of head ranges from 3.8 to 4.0 and the height of body from 3.0 to 3.4 in the total length. Eye diameter is from 3.2 to 3.4 in the length of head. Preorbitals, preopercles, subopercles and interopercles are serrated. Dorsal fin is single, its spinous portion is longer than the soft one. It is slightly shorter than the anal fin. Last dorsal fin is truncated. Sales are strongly ctenoid and some are present at the bases of the anal and the ventral fins. Body is greenish or bluish above and whitish below, 14 or more orange oblique bands descending dowards and backwards from the back to the abdomen, are found. Ventral fins are variegated with white and orange. Dorsal, caudal and anal fins are with orange spots. The border of the anal fin is deep red.

Distribution: Found throughout Jharkhand in ponds, pools, rivers, lakes, reservoirs, lakes, etc.

Remarks: It is considered as the favourate aquarium fish and generally consumed by tribal and poor sections of Jharkhand. It attains the size of about 8.4cm to 9.5cm (Srivastava,1968) and 6.0cm (Qureshi & Qureshi 1983). In our collections we recorded the length ranges from 5.3 cm to 10.2 cm.

58. *Monopteras cuchia* (Hamilton)

Systematic Position

Order	-	Symbranchiformes
Family	-	Amphipnoidae
Genus	-	Monopterus
Species	-	*Monopterus cuchia* (Ham.)
Local Name	-	Cuchia/Anhaya/Anhaya-baam.

D.Extremely rudimentry; P., V., A. and C.absent.

General appearance of the body is eel-like. Length of head ranges from 10.4 to 12.3 and the height of body is 22.3 in the total length. Eye diameter is from 10.1 to 12.8 in the length of head. Upper jaw is slightly longer than the lower and lips are fleshy. A band of large curved and backwardly directed teeth is present on the palatines. A single row of small teeth is present in the premaxillaries and a single row on the lower jaw. Rudimentary dorsal fin commences slightly before the anus. Paired fins are absent. Small but distinct scales are arranged in longitudinal rows. Gill openings are inferior and confluent to each other. Body is dark brown above with numerous black spots. The lower portion of the body is orange brown and without any spots.

Distribution: Distribution throughout Jharkhand and found in eutrophicated ponds, paddyfields, etc.

Remarks: It is consumed only by tribabls and poor sections of Jharkhand. It grows upto the length of 60.9cm (Mishra, 1959) and 35.0cm to 45.0cm (Srivastava, 1968) and 61.0 cm (Qureshi & Qureshi, 1983). In our collections we recorded the length ranges from 25.9cm to 57.3cm.

Although, it may be concluded that there are several fish species of much economical, moderately economical and less economical value are occurring in lotic and lentic types of waterbodies of Jharkhand. Hence, there is a great need of further detailed studies on the taxonomy of fishes of Jharkhand, which finally able to make the basis for the development of fisheries for academic as well as for economical purposes.

ACKNOWLEDGEMENTS

The authors are thankful to Sri Mohan Lal Mandal, Sri Ramanand Pandey and local fishermen for their support in carrying out the survey. I also thank to Sri Chandra Shekher and Sri Raj Madhavan for their excellent technical support.

REFERENCES

Anon (1962) : The wealth of India-Raw Materials,Vol. IV supplement, Fish and Fisheries, C.S.I.R.,New Delhi.

Beavan, R.(1877) : Handbook of the freshwater fishes of India, L, Reeve & Co., 5, Henrietta St., Convent Garden, London.

Berg, L.S. (1947) : Classification of fishes both recent and fossil, (English and Russian).

Datta Munsi, J.S. and Srivastava, M.P. (1988):Natural history of fishes and systematics of freshwater fishes of India, Narendra Pub.House, New Delhi.

Day, F.(1878) : The fishes of India, 4th ed, Jagmander Book Agency, New Delhi.

Hamilton, F.B.(1822) : An account of the fishes found in the river Ganges and its branches, Archibald constable and Co, Edinburg and London.

Jayram, K.C.(1999) : The freshwater fishes of the Indian Region, Narendra Publishing House, 1417, Kishan Dutt street, Maliwara, Delhi.

Menon, A.G.K.(1947) : A check-list of fishes of the Himalayan and the Indo-Gangetic Plains.Special Publication No.-l(Inland Fisheries Society of India).

Mishra, K.S.(1959) : An aid to the identification of the fishes of India and Pakistan,`ZSI, Calcutta.

Qureshi, T.A.and Qureshi, N.A.(1983) : Indian Fishes, Brij Brothers, Bhopal.

Singh, R.P. and Kumar, A.(1970) : Monograph of Bihar, 1st Ed, Bharti Bhawan, Patna.

Singh, Savindra (1993) : Geomorphology, Prayag Pustak Publication, Allahabad.

Srivastava, G.J.(1968) : Fishes of U.P.and Bihar, Vishwavidyalaya Prakashan, Varanasi.

Talwar, P.K. and Jhingran, A.G.(1991) : Inland fishes of India and adjacent countries, Oxford & IBH, New Delhi.

CHAPTER 2

RELATIVE CONDITION FACTOR (KN) OF THE BIGHEAD CARP, *ARISTICHTHYS NOBILIS* (RICHARDSON)

B.K. Mahapatra * and N.C. Datta

Fishery and Ecology Research Unit, Department of Zoology, University of Calcutta, 35 Ballygunge Circular Road, Kolkata-700 019

ABSTRACT

The analysis of length-weight relationship of *Aristichthys nobilis* reveals that cube law is not applicable to evaluate its well-being and hence, the relative condition factor (Kn) has been calculated from the data collected. The Kn value has been found to vary between 0.8492 - 1.3300.

Key words: *Relative condition factor, Bighead carp, Aristichthys nobilis*

INTRODUCTION

Assessment of mathematical relationship between length and weight of fish has wide practical applications in fishery biology. The relationship is also useful in evaluating the general condition or well-being of the fish species through the study of condition factor (K) or relative condition factor (Kn).

The condition factor or ponderal index and relative condition or relative condition factor or condition index or condition co-efficient of fish have been studied by a number of fishery biologists (Babu and Nair, 1983; De, 1986; Bandopadhyay and Datta, 1987). However, probably no such attempt has yet been made on bighead carp in a comprehensive manner. It is revealed from the study of length-weight relationship that this fish does not strictly follow the conventional Cube law (Mahapatra and Datta, 1998). Therefore, the present study of the relative condition factor of this carp in different size groups, sexes, seasons and total sample has been made since this species has immense cultural possibility.

* Present address : Division of Fisheries, ICAR Research Complex for NEH Region, Umroi Road, Barapani - 793 103, Meghalaya.

MATERIALS AND METHODS

The samples of *Aristichthys nobilis* were collected from the farm ponds of R.K.Ashram, KVK, Sunderbans (21°30'N to 23°N latitude and 88°E to 89°E longitude) during 1988 to 1993. A total of 358 fish were used for computation. By using the log length-weight relationship, the relative condition factor (Kn) was calculated as w/w^1 where is the observed weight and w^1 is the calculated weight for the observed length.

The relative condition factor (Kn) was calculated after categorising the total sample into following groups:

I. Different length groups

II. Fingerling, male, female and combined group

III. Different length groups of females

IV. Females of different seasons

V. Females of different months

RESULTS

Different length groups

The relative condition factor in different length groups is presented in Fig. (2.1). The Kn values in different length groups are found to vary between 0.8492 - 1.3300. The lowest and highest Kn values have been recorded 0.8492 in 50 mm and 1.3300 in 150 mm.

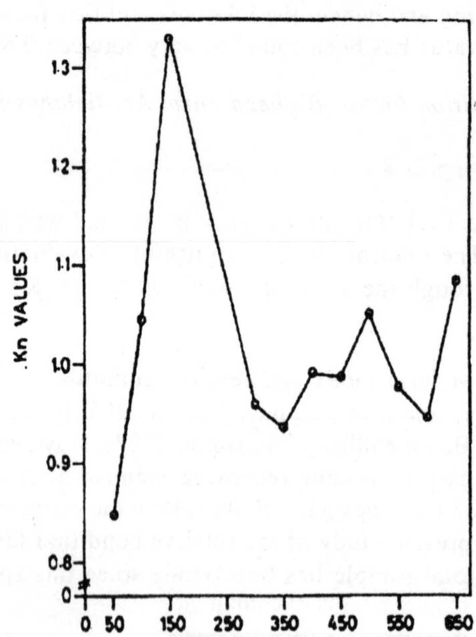

Fig. 2.1 Mean Kn values in different length of *A. nobilis*

Fingerling, male, female and combined group

The average Kn values of fingerlings, males, females, and combined group are shown in Fig. (2.2). The Kn values in sequence from higher to lower, are 1.0194, 1.0075, 1.0065 and 1.0020 in fingerlings, males, combined groups and females respectively.

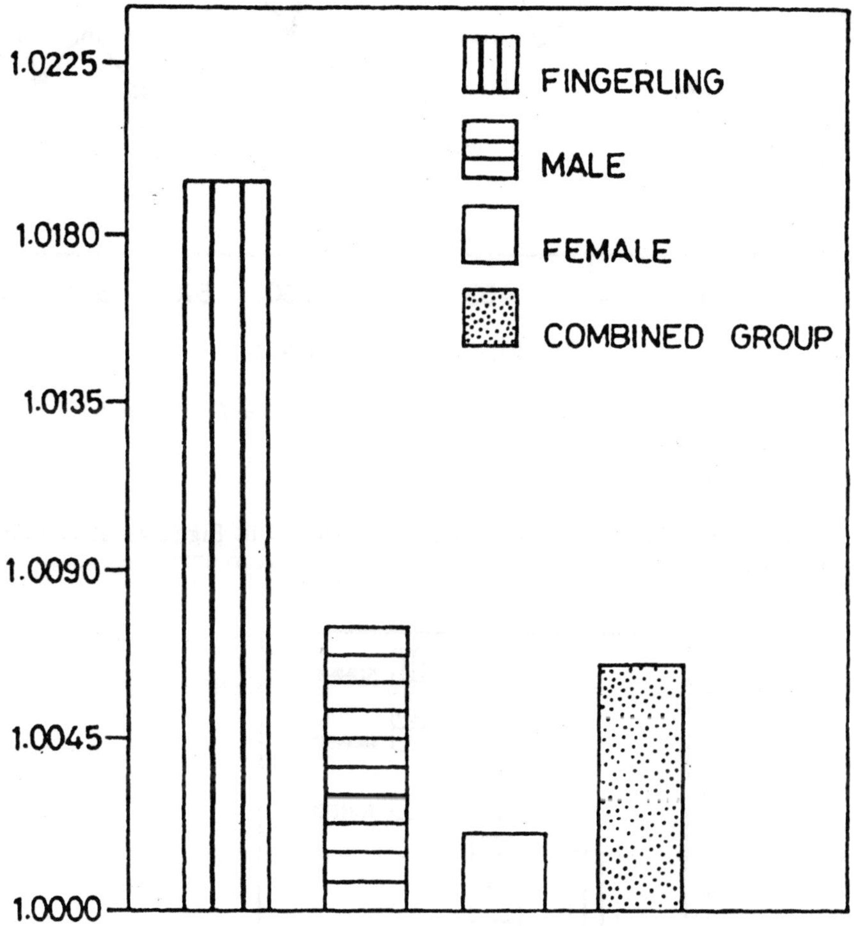

Fig. 2.2 Mean Kn values in fingerling, male female and combined group of *A. nobilis*

Females of different length groups

The Kn values of different length groups of female are shown in Fig. (2.3). The relative condition factor in different length groups is found to vary between 0.8874 and 1.1430. The highest value is in 510 mm and the lowest in 350 mm length groups.

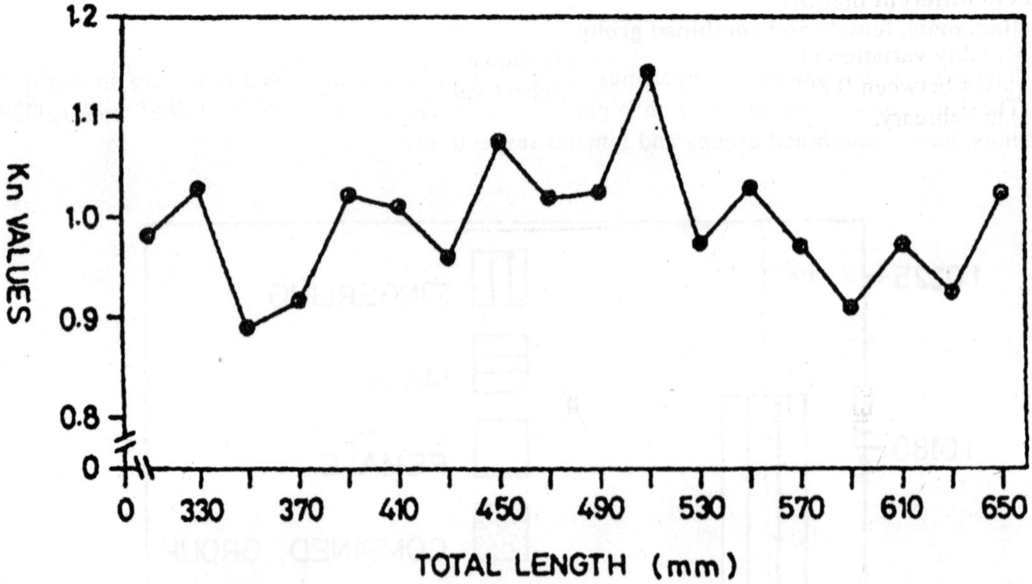

Fig. 2.3 Mean Kn values in different length of *A. nobilis* (female)

Females of different seasons

The variation of Kn values in different seasons is shown in Fig.(2.4). It is found that the highest value (1.0137) was recorded in monsoon whereas the lowest (0.9770) in winter.

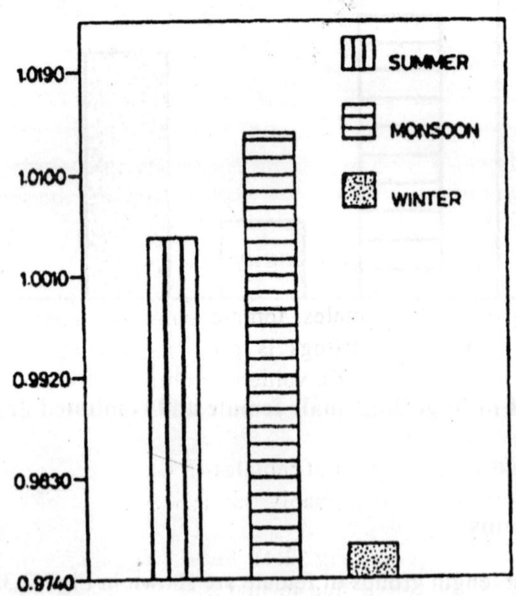

Fig. 2.4 Mean Kn values in different seasons of *A. nobilis* (female)

Females of different months

Monthly variation of Kn values of female is shown in Fig.(2.5). The values of relative condition factor varies between 0.8958 and 1.1037. The highest value (1.1037) is recorded in May and lowest (0.8958) in February.

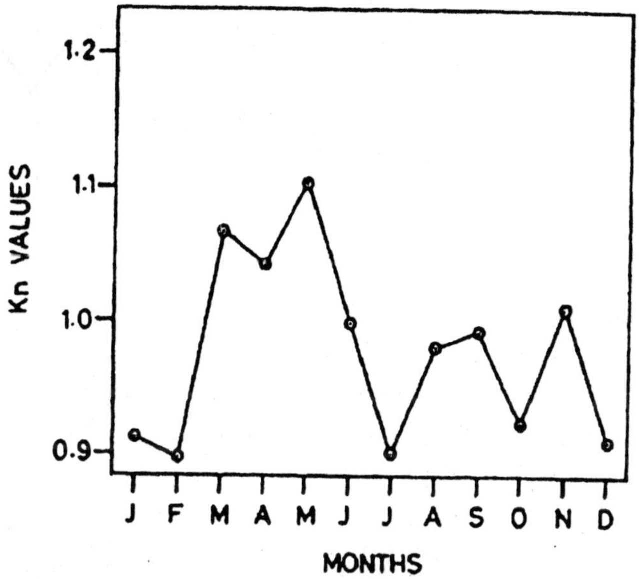

Fig. 2.5 Monthly fluctuations in mean Kn values of *A. nobilis* (female)

DISCUSSION

The gonad weight and feeding rate as influencing factor for the condition of fish have been emphasised by some previous workers (Le Cren, 1951; Bhatt, 1968; Babu and Nair, 1983; De, 1986). In *A. nobilis*, fluctuation of Kn values in between different length groups has been observed. In the Kn curve(Fig. 2.1) several peaks were observed including one at 150 mm and the second at 500 mm length groups. The first peak of the increase of the Kn value may be attributed to the breadth-wise growth. The second peak is due to the attainment of first maturity.

The Kn values among the fingerlings, males, females and combined groups (Fig.2.2) clearly show that higher Kn values in case of fingerlings is probably due to breadth-wise growth and voracious feeding habit in that stage. Low mean Kn value was noticed in females which may be due to spawning strain and spent condition.

The Kn curve (Fig. 2.3) in different length of females shows several peaks but the highest peak was noticed in 500 mm length group which is probably due to attainment of first maturity.

Higher Kn values in different seasons (Fig. 2.4) were noticed in monsoon whereas low Kn values in winter. Higher Kn value is probably due to maturity of gonads and low Kn value due to spawning strain, spent condition and low feeding rate during winter.

The Kn curve (Fig 2.5) in different months of female shows that highest Kn value in May is due to development of gonads as well as less strain and lowest in July possibly due to spent condition. Very likely metabolic strain due to spawning activities has not yet been regained.

Thus, it may be stated that feeding rate, physiological factors like maturity and spawning play an important role on the condition of *A. nobilis* which may be true in other fishes too.

ACKNOWLEDGEMENTS

The authors are greatly indebted to the authorities of Ramkrishna Ashram Krishi Vigyan Kendra, Nimpith for providing facilities to carry out the research work. The first author is also grateful to the Director, ICAR Research Complex for NEH Region for providing necessary facilities.

REFERENCES

Babu, N. and Nair, N.B. 1983. Condition factor of *Amblypharyngodon chackaiensis* Babu and Nair in Chackai boat channel (Trivandrum, Kerala, India). *Proc. Indian Acad. Sci., (Anim. Sci),* 92(1): 1-10.

Bal, J.W. and Jones, J.W.1960. On the growth of brown trout of Llyn Tegid. *Proc. Zool Soc. Lond.,* 134 : 1-41.

Bandopadhyay, B.K. and Datta, N.C. 1987. A study on the length-weight relationship and condition factor of *Scatophagus argus* (Linnaeus) from the brackish water impoundments of Hooghry-Matlah Estuary, West Bengal. Abstract, VIIth All India Seminar on Ichthyology, Visva-Bharati : 3.

Bhatt, V.S. 1968. Studies on the biology of some freshwater fishes. Part II *Heteropneustes fossilis* (Bloch). *Indian J.Fish.,* 15 : 99-115.

De, D.K. 1986. Studies on the food and feeding habit *of Hilsa ilisha* (Hamilton) of the Hooghly estuarine system and some aspects of its biology. Ph.D. Thesis, p. 28 5, University of Calcutta.

Le Cren, E.D. 1951. The length-weight relationship and seasonal cycle in gonad weight and condition factor in the perch *Perca fluviatilis. J. Anim. EcoL,* 20(2) : 201-219.

Mahapatra, B.K. and Datta, N.C. 1998. Length-weight relationship of bighead carp, *Aristichthys nobilis* (Richardson). *J. Inland Fish. Soc. India,* 30(1) : 101-104.

Qasim, S.Z. 1957. The biology of *Blennius pholis L.* (Teleostei). *Proc. Zool. Soc. Lond,* 128(2): 161-208.

Qayyam, A. and Qasim, S.Z. 1964. Studies on the biology of some freshwater fishes. Part I. *Ophiocephalus puntatus* (Bloch). *J. Bombay nat. Hist. Soc.,* 61 : 330-347.

CHAPTER 3

EFFECTS OF MELATONIN ON THE MELANOPHORES OF FISH, *LABEO ROHITA* (HAM.)

Shashi Patil

School of Studies in Zoology, Jiwaji University, Gwalior - 474011, India

ABSTRACT

In fish, *Labeo rohita* adapted to black background intraperitoneal administration of melatonin (MEL) *(in vivo)* Oresulted in paling of the integument. In both, innervated and denervated melanophores *in vitro* MEL caused aggregation of the pigment. *In vitro,* studies revealed ineffectiveness of alpha (phenoxybenzamine and yohimbine) and beta (propranolol) adrenolytic agents to inhibit MEL induced responses. Adenosine (ADE), theophylline (THE) and alpha-melanocyte stimulating hormone (α-MSH) caused rapid dispersion of the pigment aggregated by MEL. It is suggested from the study that MEL induced rapid and reversible melanosome aggregation within melanophores of *Labeo rohita* and the receptors for its action are different from adrenoceptors. The action of MEL is assumed to be linked to inhibition of adenylate cyclase.

Key words : *Melatonin, Labeo rohita, Aggregation, Dispersion, Receptors, Melanophores*

INTRODUCTION

Lerner and his colleagues (1958) first discovered melatonin from mammalian pineal organ and named it after its potent concentrating action on amphibian melanophores. In many fish species melatonin has been reported to exert potent melanosome aggregation (Fujii and Oshima, 1986), while in others it does not influence melanosome translocation within melanophores (Fujii, 1993; Visconti and Castrucci, 1993). In restricted area of the skin the differential responsiveness to MEL exists (Fujii and Oshima, 1986; Masagaki and Fujii, 1999)

In fish, *Labeo rohita,* alpha adrenoceptors (Jain and Patil, 1992), beta adrenoceptors (Patil and Jain, 1991) and melanin concentrating hormone (MCH) receptors (Jain and Patil, 1990) exist. The present study was conducted to study the effects of melatonm on the melanophores of *Labeo rohita.*

MATERIALS AND METHODS

Specimen of fish. *Labeo rohita* of either sex weighing average 7 g and measuring average 9 cm were acclimatized to laboratory conditions with normal day and light cycle at room temperature for few days before experimentation Specimen were collected from Govt. Fish Farm, Gwalior (M.P.). For *in vitro* study the isolated scales were removed from antero-dorsal trunk region and immediately immersed in physiological saline (PS) for about 20 minutes for equilibration before treatment with drugs under study (Jain and Patil, 1992). Denervated melanophores were obtained at 2-6 hrs from fish injected with reserpine (10 μg/g.i.p.).

The responses of control and drug treated melanophores were recorded using melanophore index (M.I.) prepared for *Labeo rohita* (Patil and Jain, 1993) based on Hogben and Slome (1931) melanophore index. As many as 5 melanophores were observed from 5 to 7 fishes each for each drug. All values are expressed as mean ± S.D.M. The experiments were carried at room temperature - 20-22°C.

The macroscopic colour changes were recorded utilizing a method based on a 9 point scale derived from Munsell Grey series (Sharma et al., 1996).

The drugs/chemicals used are : Melatonin (Sigma, USA), α-melanocyte stimulating hormone (Peninsula Labs. Europe Ltd., U.K.), adenosine (Sigma, USA), theophylline (Sigma, USA), propranolol hydrochloride (Cipla, Bombay), phenoxy benzamine hydrochloride (Sigma, USA), yohimbine hydrochloride (Paul Neuendorf, Germany), reserpine (Hindustan Ciba Geigy Ltd., India). Working solutions of these drugs were prepared by diluting the stock solutions with PS immediately before use.

RESULTS AND DISCUSSION

A black adapted fish. *Labeo rohita,* with general tint of the body equivalent to index number (I.N. 2.5) paled notably within 5 min (I.N. 4.0) after receiving an intraperitoneal injection of MEL 10 mg/kg (Fig 3.1). The maximum paling of skin was equivalent to I.N. 4.4 at 10 mm that lasted for about an hour from the treatment. The influence of hormone declined (I.N. = 4.1 - 2 hr) but appeared to be a long lasting one, as even after 5 hr of its administration the fish were found to be appreciably pale (I.N. =3.7). At 24 hr, on following day the pre-experimental shade was restored. This *in vivo* skin paling due to MEL is comparable with responses in other fishes (Bhargava and Jain. 1978; Prabhakar, 1988).

In vitro, melanophores were treated with MEL at concentrations ranging from 10^{-9}M to 10^{-5}M (Fig. 3.2). Minimal pigment aggregation was recorded at 10^{-8}M (M.I. 3.58, 10 mm) and maximum at 10^{-5}M (M.I. 2.1. 10 min). The melanophores showed no further aggregation even at higher concentration. The aggregation responses in *Labeo rohita* are consistent with earlier reports for dermal melanophores in amphibians (Sugden. 1991: Rollag and Lynch. 1993) and many fish species (Fujii and Miyashita. 1978; Iwakiri, 1983 and Fujii and Oshima, 1986). In denervated melanophores also the MEL produced melanosome aggregation as in innervated melanophores Iwakiri (1983) reported these responses due to MEL in both innervated and denervated melanophores in fish, *Oryzias latipes*

In the present study melanophores pretreated with alpha adrenergic blockers (Yohimbine and phenoxybenzamine) and beta adrenergic blocker (propranolol) at 10^{-4}M and 10^{-5}M were treated with MEL. These adrenergic blockers failed to inhibit the responses due to MEL. The study provides evidence for the existence of cellular receptors for the MEL that are different from adrenergic receptors

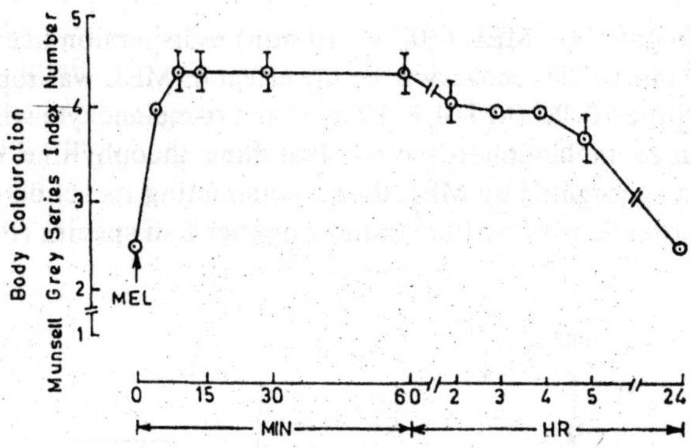

Fig. 3.1

In black-adapted fish, *Labeo rohita* placed over an illuminated black-background, a single injection of MEL (10 mg/kg) produced paling of the skin. Each point and the vertical bar represent the mean and ± S.D. (N=5) respectively.

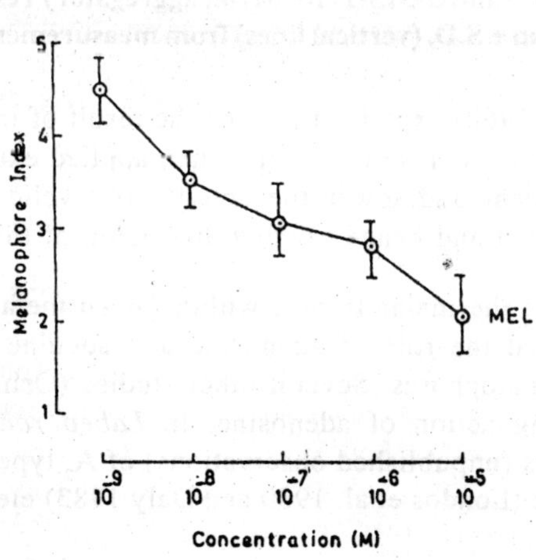

Fig. 3.2

Concentration-response curve of melatonin on isolated scale melanophores of *Labeo rohita*. Values are mean ± S.D. (vertical lines) of 25 measurement from 5 different fishes.

as reported in our earlier studies (Patil and Jam. 1989 Jain and Patil, 1992). Such possibility of existence of specific MEL receptors was also reported by Fujii and Miyashita (1978).

In melanophore aggregated by MEL (10^{-5}M; 10 min) redispersion occurred in PS (control) but with slow rate (M.I. 2.1, 10 min). The recovery from the effect of MEL was rapid with adenosine 10^{-2}M (M.I. 4.3; 10 min); theophylline 10^{-3}M (M.I. 4.1; 10 min) and α-melanocyte stimulating hormone 10^{-8}M (M.I. 4,6: 10 min). In *Labeo* melanophores methylxanthine theophylline (THE) induced pigment dispersion in melanophores aggregated by MEL thus counteracting its action (Fig. 3.3). The effects of methylxanthine are in accordance with earlier studies on other fish species (Fujii and Miyashita, 1976 and Nagahama et al 1980)

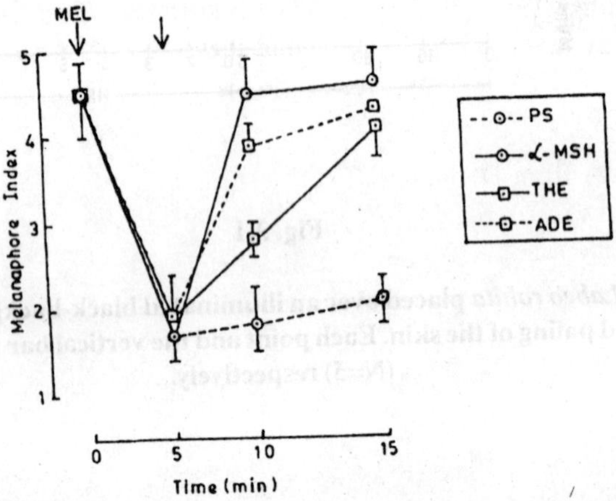

Fig. 3.3

Effects of ADE (10^{-2}M), THE (10^{-3}) and α-MSH (10^{-8}M) on aggregatory responses of *Labeo* melanophores to MEL. Values are mean ± S.D. (vertical lines) from measurement on 5 different fishes.

Butcher and Sutherland (1962) explained it to be the result of increased level of cAMP due to methylxanthines. These phosphodiesterase inhibitors when applied externally they penetrate the cell along the large chemical gradient and lower the activity of cyclic nucleotide phosphodiesterase resulting in increased cAMP level and hence causing dispersion of the pigment

Adenosine also dispersed the melanosomes within *Labeo* melanophores (Fig. 3.3). Fujii and Miyashita (1976) first suggested the role of adenosine and adenine nucleotides in the control of pigment dispersal in fish chromatophores. Several other studies (Oshima, 1989 and Namoto, 1992) support this pigment dispersing action of adenosine. In *Labeo rohita* there is evidence for the existence of adenosine receptors (unpublished observations) of A_2 type. These A_2-receptors stimulate the activity of adenylate cyclase (Londos et al. 1980 and Daly 1983) elevating the cAMP level leading to dispersion of the pigment

In the present study α-MSH mimicked the action of adenosine. It caused centrifugal translocation of the pigment in melanophores aggregated by MEL. The responses due to α-MSH are consistent with observations reported in other fishes (Sharma et al., 1996: Visconti et al. 1999 and Fujii. 2000).

In amphibian melanophores Seldenrijk et al. (1979) showed that the agents which elevate intracellular cAMP such as alpha-melanocyte stimulating hormone (α-MSH), forskolin and cell permeable analogues of cAMP countered the aggregation response of MEL.

In *Labeo,* too. it can be suggested that increased cAMP level may be responsible for pigment dispersion responses due to ADE, THE and α-MSH and MEL induced aggregation responses may be linked to inhibition of adenylate cyclase leading to decreased cAMP level.

It is concluded from the study that in fish, *Labeo rohita,* MEL induced concentration dependent rapid and reversible pigment aggregation and the cellular receptors mediating these responses appear to be different from the adrenergic receptors. As reported by many workers (Fujii and Miyashita, 1976,1978; Fujii and Oshima, 1986 and Fujii, 1993) it is speculated that in *Labeo* too signal transduction linked to MEL action may involve cAMP as second messenger.

ACKNOWLEDGEMENTS

Financial assistance from DST and UGC, New Delhi, is gratefully acknowledged. Thanks are due to Prof.R.Mathur and Prof A.K Jain, School of Studies in Zoology, Jiwaji University. Gwalior for guidance and encouragement.

REFERENCES

Bhargava, H.N and Jain, A K 1978 Some aspects of colour change mechanism in the Indian fresh-water siluroid *Heteropneustes fossilis (*Bloch.*).* Biochem. Expt. Biol XIV 4 : 359-373

Butcher, R.W. and Sutherland E. W. 1962 Adenosme 3'-5'-phosphate in biological materials. I. purification and properties of 3'-5' nucleotide phosphodiesterase and use of this enzyme to characterise adenosine 3'-5'-phosphate in human urine. J. Biol. Chem. 237 : 1244-250

Daly, J.W 1983. Role of ATP and adenosine receptors in physiological processes. In physiology and pharmacology of adenosine derivatives (Eds. Daly, J.W., Kuroda. Y, Phillis, J.W., Shimizu, H. and Ui. M.). Raven Press, New York. 275-290.

Fujji, R. 1993. Cytophysiology of fish chromatophores. Int. Rev. Cytol. 143 : 191-255.

Fujii, R. 2000. The regulation of motile activity in fish chromatophores. Pigment Cell Res. 13: 300 319.

Fujii, R. and Miyashita, Y. 1976. Beta-adrenoceptors, cyclic AMP and melanosome dispersion in guppy melanophores. Pigment Cell. 3 : 336-344.

Fujii, R. and Miyashita, Y. 1978. Receptor mechanisms in fish chromatophores-IV. Effects of melatonin and related substances on dermal and epidermal melanophores of the siluroid, *Parasilurus asotus.* Comp. Biochem. Physiol. 59C : 59-63.

Fujii, R. and Oshima, N. 1986. Control of chromatophore movements in teleost fishes. Zool Sci. 3 : 13-47

Hogben, L.T and Slome, D. 1931. The pigmentary effector system. VI. The dual character of the endocrine co-ordination in amphibian colour change. Proc. R. Soc. London (Biol.). 108 B : 10-53.

Iwakiri, M. 1983. Action of melatonin and its derivatives on melanophores in isolated scales of a teleost fish. J. Sci. Hiroshima Univ. Ser B, Div I. 31 85-98.

Jain, A.K. and Patil, S 1990. Evidence for melanin concentrating hormone (MCH) receptors **mediating** melanosome aggregation in *Labeo* melanophores Ind. J. Physiol. Pharmacol 34 : 187-190

Jain, A.K. and Patil, S. 1992 Alpha$_2$-adrenergic receptor activation induced melanophore response in a fresh-water teleost. *Labeo rohita* : an *in vitro* and *in vivo* study Proc. Nat. Acad Sci. India 62B 323-332

Londos, C. Cooper, D M.F and Wolf. J 1980 Subclasses of external adenosine receptors J Proc. Nat Acad Sci USA 77 2551-2554

Lerner, A.B.: Case. J.D.: Takahashi. Y: Lee T.H and Mori. W. 1958 Isolation of melatonin the pineal gland factor that lightens melanocytes. J. Am. Chem. Soc. 80 : 2587

Masagaki, A. and Fujii. R. 1999 Differential actions of melatonin on melanophores of the three line pencilfish. *Nannostomus trifasciatus.* Zool. Sci 16 : 35-42.

Nagahama, H.; Ohsaki. M and Muneoka. Y. 1980 Action of methylxanthines on melanophores in isolated scales of a fresh-water teleost. J. Sci. Hiroshima Univ. Ser B Div 1 28 : 27-49

Namoto, S. 1992 Effects of purine compounds and forskolin on melanophores of the medaka *Oryzias latipes.* Evidence for A$_2$-adenosine receptor. Comp. Biochem. Physiol. 103 C 2 : 391-398.

Oshima, N. 1989. Adenosine inhibits the release of neurotransmitters from melanosome aggregating nerves of fish. Comp. Biochem. Physiol. 93 C (2) : 207-211.

Patil, S. and Jain, A.K. 1989 The sympathetic neuro-melanophore transmission in a freshwater. Indian major carp, *Labeo rohita* (Ham.). Ind. J. Physiol. Pharmacol. 33 : 101-106.

Patil, S. and Jain, A.K. 1991. Evidence for the involvement of beta-adrenoceptors in darkening responses in the fish *Labeo rohita.* Proc. Nat. Acad. Sci. Letters, India. 14 : 191-193.

Patil, S. and Jain, A.K, 1993 Role of calcium in melanosome aggregation within *Labeo* melanophores. J. Bio Sci. 18 : 83-91.

Rollag, M.D and Lynch. G.R. 1993. Melatonin-induced desensitization in amphibian melanophores J. Exp. Zool. 265 : 488-495

Prabhakar, R.M. 1988 Photopigmentary responses in intact and eyeless fish *Labeo gonius* (Ham.). M.Phil. Dissertation Jiwaji University, Gwalior, M.P. India.

Seldenrijk, R ; Hup. D.R.W.: De Graan, P.N.E. and Van De Veerdonk, F.C.G. 1979 Morphological and physiological aspects of melanophores in primary culture from tadpole of *Xenopus laevis* Cell Tissue Res 198 397-409

Sharma, B.: Patil, S. and Jain A.K. 1996. Effect of melanocyte stimulating hormone (MSH) on melanophores in isolated scales and on the colour displayed by the skin *(in vivo)* in a fresh-water teleost. *Puntius sophore* (Ham) Proc Nat. Acad. Sci. India. 66B 139-150

Sugden, D 1991 Aggregation of pigment granules in single cultured *Xenopus laevis* melanophores by melatonin analogues. Br. J. Pharmacol. 104 : 922-927.

Visconti, M.A. and Castrucc, A.M.L. 1993 Melanotropin receptors in the cartilagenous fish *Petamotrygon reticulatus* and in the lung fish. *Lepidosiren paradoxa.* Comp. Biochem Physiol 106 C : 525-528.

Visconti, M.A.: Ramanzini G.C., Camargo. C.R. and Castrucci, A.M.L. 1999. Elasmobranch colour change A short review and novel data on hormone regulation. J. Exp. Zool. 284: 485-491.

CHAPTER 4

BIOACCUMULATION AND ELIMINATION OF CADMIUM IN FRESHWATER FINGERLINGS, *LABEO ROHITA* (HAM.)

PL.RM. Palaniappan, G. Jagadeesan+, P. Venkatachalam, N. Krishnakumar and S.Karthikeyan.

Department of Physics, Annamalai University, Annamalainagar, Tamil Nadu - 608 002, India.

+Department of Zoology, Annamalai University, Annamalainagar, Tamil Nadu - 608 002, India.

ABSTRACT

Bioaccumulation of cadmium is assessed in freshwater fish, *Labeo rohita* after induced acute and chronic exposures. Intake, accumulation and elimination of cadmium in different organs of fish have been discussed. The kidney was found to be the prime site of metal binding and brain accumulate least metal concentration. As the exposure concentration was increased, the biological magnification factor reduced and the cadmium concentration in tissues increased by increased exposure period. Also the treatment of chelating agent Calcium disodium ethylene diamine tetra acetic acid (CaNa2EDTA) reduced the metal toxicity significantly and follows the order, Kidney > Liver > Brain > Gill.

Keywords: *Bioaccumulation, Biomagnification, Cadmium, $CaNa_2EDTA$, tissues, Labeo rohita.*

INTRODUCTION

Cadmium is a ubiquitous, non-essential element which possesses high toxicity to both humans and aquatic organisms. According to the hypothesis proposed by Harris and Hohenemser (1978), cadmium is classified as the second most dangerous metal in our environment. In recent years, cadmium and cadmium compounds have been used extensively by various industries such as electroplating, battery manufacture, pigments and plastic production and this has produced sharp increases in contamination of air, water and soil. Population exposure to toxic trace metals are of great concern due to their non-biodegradable nature and long biological half-life for elimination from the body. There are many reports in the literature on metal accumulation in tissues. Usually accumulation differs between tissues and species. For example, in *Hamarus americanus* cadmium accumulated mainly in the digestive gland followed by the gills, wheras in *Callinectes sapidus*, cadmium accumulated mostly in the gills (Darmono, 1990). Cadmium toxicity to aquatic ecothermal animals depends on complex biochemical interactions and a balance between rates of absorption, detoxification and excretion. The main sites of cadmium deposit are the kidney and liver of vertebrate animals (Ferrari *et. al*, 1993). In the present investigation, an attempt has been made to study the accumulation pattern and the efficacy of an antidote $CaNa_2EDTA$ to induce the elimination of cadmium in Cd treated fingerling stage of freshwater fish, *Labeo rohita* (Ham.).

MATERIALS AND METHODS

Specimens of *Labeo rohita* were collected from the local fish farm and acclimatized to the laboratory conditions for fifteen days. Fishes were fed with oilless groundnut cake twice a day. Fishes ranging from 5-6 cm in length and weighing 8-10 gm were selected for experimental purpose. The physico-chemical parameters of water was estimated according to APHA (1989) and are follows: Dissolved oxygen: 7.2 ppm - 7.4 ppm; pH: 7.0 - 7.2; Temperature: 24.0 ±2.0^0 C; Salinity: 0.4 ppm - 0.5 ppm and Total hardness: 280 mg/l - 288 mg/l. Preliminary studies were carried out to find the median lethal concentration (LC_{50}) for 96 h by Probit analysis method (Finney, 1971).

The test specimens were divided into seven groups, each consisted of 20 fishes and stocked in 20 litre plastic aquaria, equipped with continuous air supply. The physico-chemical water parameters were measured systematically to maintain its optimum level. The water was changed at regular intervals along with waste feed and faecel materials. Group 1 was treated as control. Groups (2, 3), (4, 5) and (6, 7) were exposed to cadmium chloride of 2.5 ppm (acute, 1/3 of LC_{50}) for 14 days, 0.75 ppm (chronic, 1/10 of LC_{50}) for 30 days and 60 days respectively. Groups 3, 5 and 7 were again exposed to $CaNa_2EDTA$ of 1:3 ratio (Cd: $CaNa_2EDTA$) for another 7 days.

At the end of the experimental periods, the fishes were sacrificed, the organs were dissected and kept in frozen at -70^0 C until used. Before the sample preparation the tissues were dried in an oven at 80^0 C for 12 hours.

Estimation of cadmium

Following the method of Topping (1973), dried samples were processed for acid digestion by using perchloric acid and nitric acid in the ratio 1:3 respectively. Cadmium concentration in the samples were determined by Inductively Coupled Plasma-Atomic Emission Spectrometer (ISA JOBIN YVON 24 Model) installed at Centre for Advanced Study in Marine Biology, Portanova, Annamalai University.

RESULTS AND DISCUSSION

Table-4.1 summarize data on cadmium concentration (μg/g) in various tissues *of Labeo rohita*. It is evident from the table that cadmium is taken up by the animals from the surrounding medium and accumulated among various organs. The accumulation of metal in tissues of aquatic animals is dependent upon exposure concentration and period as well as some other factors such as salinity, temperature, interacting agents and metabolic activity of tissue in concern. It is also known that metal accumulation in tissues of fish is dependent upon the rate of uptake, storage and elimination. It is seen from the table that highest cadmium concentration is found in the kidney tissue and least in the brain tissue for both acute and chronic exposures. The accumulation of cadmium in the tissues follows

Table 4.1
Organ-wise average bioaccumulation (µg/g) of cadmium in *Labeo rohita* fingerlings during acute and chronic exposures.

Organ	Control	Acute-14 days-2.5 ppm			Chronic - 30 days 0.75 ppm			Chronic - 60 days 0.75 ppm		
		Cd	B.M.F.	Cd+EDTA	Cd	B.M.F.	Cd+EDTA	Cd	B.M.F.	Cd+EDTA
Kidney	6.142	81.146	32X	31.932	29.412	39X	12.020	75.067	100X	29.031
Liver	3.462	50.238	20X	23.311	22.611	30X	9.640	67.203	90X	37.442
Gill	2.280	32.481	13X	26.427	12.433	17X	9.000	45.612	61X	33.640
Muscle	2.674	30.832	12X	20.332	10.628	14X	6.642	29.524	31X	14.213
Brain	1.026	9.821	4X	5.121	4.514	6X	2.400	11.402	15X	7.631

B.M.F.: Biological Magnification Factor

the order: Kidney > Liver > Gill > Muscle > Brain. These results are consistent with the earlier workers (Mount and Stephen, 1967; Kumuda *et* al., 1972; Andres *et al.*, 2000). It has also been observed that the treatment of chelating agent $CaNa_2EDTA$ has reduced the accumulation of cadmium significantly.

Heavy metals primarily affect the kidney, which is involved in the cleaning processes of the body fluids and tissues. In the present study kidney tissue accumulates substantial amount of cadmium (81.146 µg/g, 29.412 µg/g & 75.067 µg/g) for acute and chronic exposures. Since kidney being a major target organ for metals, the perfusion rate is greater compared to other organs (Chris Kent, 1998). Hence the toxic substances present in the blood is delivered to kidneys in large quantities.

Next to kidney, liver accumulates a significant amount of cadmium (50.238 (µg/g, 22.611 (µg/g & 67.203 µg/g) for acute and chronic exposures. The metal binding protein metallothionein is of utmost importance in the accumulation of metals. The liver is the main organ for homeostasis in fish, and is in an advantageous position for clearing the blood substances entering the circulation from the gastro-intestinal tract passes through the liver before reaching the systemic circulation. Therefore, the liver removes the toxicants from the blood, bio-transform them into the bile and prevents distribution to other parts of the body (Coetzee, 1996). Hence, liver, which is a major producer of metal-binding proteins shows high concentrations of heavy metal, cadmium.

Gills are the predominant barriers that control the mechanism of absorption of bioavailable metals from the external medium (direct route of exposure). Hence the gills are one of the target of a direct contamination as they play a significant role in metal uptake, storage and eventually transfer to the internal compartments via blood transport (Andres *et al.*, 2000). In the present study, accumulation of cadmium in the gills are: 32.481 µg/g, 12.433 µg/g & 45.612 µg/g for acute and chronic toxicity respectively.

During acute and chronic exposures, fish accumulated maximum metal in kidney, followed by liver and gill. Muscle and brain were the least cadmium accumulated tissues. According to Bradley and Morris (1986), the muscle tissue is usually not a good indicator of environmental pollution in a freshwater ecosystem, and the fact that the muscle showed low metal concentration is therefore not surprising. Moreover the muscle tissue have larger surface area and the metals are evenly distributed, which leads to low metal significant 30.832 (μg/g, 10.628 μg/g & 29.524 μg/g). In the brain, the blood-brain barrier effectively decreases the amount of toxic substances that are transferred into the brain tissues (Chris Kent, 1998). Hence the least metal concentration (9.821 μg/g, 4.514 μg/g & 11.402 μg/g) observed in the brain is in the expected order and are consistent with the observation made by Vincent and Ambrose (1994) for *Catla catla.*

In the present study, the greatest biological magnification occurred at the lowest (0.75 ppm) cadmium exposure level. At this exposure level, the kidney accumulated approximately 39X and 100X the amount of cadmium present in the aquaria water for acute and chronic exposures. As the exposure concentration increases, the biological magnification reduced. At the highest exposure level (2.5 ppm), the kidney accumulated approximately 32X the amount of cadmium in the aquaria water.

Also, in the liver tissue, greatest biological magnification was found in the lowest (0.75 ppm) cadmium exposure level. At this exposure level, the liver accumulated 30X and 90X the amount of cadmium present in the aquaria water for acute and chronic exposures repectively. At the higher concentration, the biological magnification was reduced by 20X the amount of cadmium present in the aquaria water. Similar trend was observed in the gill, muscle and brain tissues. In the chronic exposure (60 days) the rate of accumulation increases due to the cadmium induced metallothionein synthesis in these tissues, which is an important transport and storage protein for cadmium (Friberg *et al.*, 1992)

Chelating agents are the most versatile and effective antidotes for metal intoxication. Sunda and Zamuda (1982) have demonstrated that organic chelation compounds can form stable complexes which can hardly be accumulated in organisms, thereby reducing the toxicity of the metals to organisms. Of the various chelating agents Calcium disodium ethylene diamine tetra acetic acid ($CaNa_2EDTA$) is an efficient chelator of many divalent and trivalent metals (Ellenhorn and Barceloux, 1988) for acute and chronic cadmium exposures. It penetrates cell membranes relatively poorly and chelates extracellular metal ions much more effectively than intracellular ions. The present results suggest that the $CaNa_2EDTA$ reduces the cadmium concentration in the tissues *of Labeo rohita* fingerlings significantly (19% - 62%). It is also evident from the present study, that cadmium is released from the organs at different rates (0.54 ppm to 6.57 ppm) and cadmium excretion is invariably found to be higher than the uptake and follows the order: Kidney > Liver > Brain > Gill.

ACKNOWLEDGEMENTS

The authors express their sincere thanks to Prof. S. Krishnan, Department of Physics, Annamalai University for his constant encouragement and to Prof. L.S. Ranganathan, Head, Department of Zoology, Annamalai University for providing the necessary laboratory facilities to carried out this study.

REFERENCES

Andres, S.; Ribeyre, P.; Tourencq, J.N. and Boudou, A. 2000. Interspecific comparison of cadmium and zinc contamination in the organs of four fish species along a polymetallic pollution gradient (Lot River. France). The Sci. of Total Environ. 248: 11-25.

APHA, 1989. Standard methods for the examination of water and waste water. 18th Ed. APHA Washington DC.

Bradley, R.W. and Morris, J.R. 1986. Heavy metals in fish from a series of metal contaminated lakes near Sudbury, Ontario, Water Air Soil Pollut. 27: 341-354.

Chris Kent. 1998. Basics of Toxicology. John Wiley & Sons, Inc. New York.

Coetzee, L. 1996. Bioaccumulation of metals in selected fish species and the effect of pH on aluminium toxicity in a cichlid *Oreochromis mossambicus.* M.Sc. Thesis, Rand Afrikaans University, South Africa.

Darmono, D.1990. Uptake of Cadmium and Nickel in Banana Prawn *(Penaeus merguiensis de man).* Bull. Environ. Contam. Toxicol. 45: 320-328.

Ellenhorn, M.J. and Barceloux, D.G. 1988. Medical Toxicology - Diagnosis and treatment of human poisoning. Elsevier (Eds.) New York.

Ferrari, L.; Alfredo Salibian; and Claudia Muino. V. 1993. Selective protection of temperature against cadmium acute toxicity to *Bufo arenarum Tadpoles.* Bull. Environ. Contam. Toxicol. 50: 212-218.

Finney, D.J. 1971. Probit analysis. Cambridge University Press.

Friberg, L.; linder, C.G. and T. Kjelltrom. 1992. Environmental Health Criteria 134-Cadmium, WHO, Geneva.

Harris, R.C. and Hohenemser, C. 1978. Mercury - Measuring and managing the risk. Environment. 20: 25-36.

Kumuda, H.; Kimura, S.; Yokata, M. and Matida, Y. 1972. Acute and chronic toxicity, uptake and retention of cadmium in freshwater organisms. Bull. Freshwat. Fish. Res. Lab. Tokyo. 22: 157-165.

Mount, D.I. and Stephan, C.E. 1967. A method of detecting cadmium poisoning in fish. J. Wild. Mgnst. 31: 168-172.

Sunda, WG. Zamuda, CD. 1982. Bioavailability of dissolved copper to the American oyster *Crassostrea virginica. I.* Importance of chemical speciation. Mar. Biol. 66: 77-82.

Topping, G. 1973. Heavy metals from Scottish water. Aquaculture. 1: 379-384.

Vincent, S. and Ambrose, T. 1994. Uptake of heavy metals. Cadmium and chromium in tissue of Indian major carp *Catla catla* (Ham.). Indian J. Environ. Hlth. 36: 200-204.

CHAPTER 5

EFFECT OF LIMNOLOGICAL PARAMETERS ON INDIGENOUS INLAND FISHERY DIVERSITY OF WESTERN BELT OF ORISSA

Mahapatra M.K.* & Mishra A**

**Deptartment of Zoology, Khariar College, Khariar, Nuapada - 766 107*

***Environmental Biology Unit, School of Life Sciences, Sambalpur University - 768019*

ABSTRACT

The experiment was conducted during Pre monsoon, Monsoon and Post Monsoon period of the year 2001 & 2002 at the School of Life Sciences, Sambalpur University, Orissa to study the effect of limnological Parameters on the inland fishery diversity of Western Orissa.

The limnological Parameters reveals that the condition of Western Orissa is not suitable for cultivation of carp species. The indigenous non-carp species have wide occurrence in Western Orissa. The total variety of 29 species have been reported in the study sites.

Key Words: *Monsoon, Limnology, Lentic, carp, Species.*

INTRODUCTION

The Western belt of Orissa occupy 30.78% of the total geographic area and 24.34% of the total population of Orissa. It comprises of 11 districts which lies between 19^0 - 22^0 N latitueds and 82^0 - 85^0 E longitudes (Mishra & Panda, 1992). The rain fall of Western Orissa is 65-70 cm / year due to poor monsoon. We can not change the monsoon, we can only equipped ourselves to adapt with the climate. Cultivation of carp species is impracticable due to harsh limnology; thus we should go for the widely available non carp-species in Western Orissa.

Poverty is the cause and effect of environmental degradation. For the overall development of Western Orissa, sustainable development of the available local resource is needed. The present study was conceived to look at the traditional water harvesting structures raised over vantage sides as life support systems providing protective irrigation to paddy fields and bases of fish production (either by culture or natural growth of popultion) with an assurance of availability of rice and curry (with proteins and minerals) for the people living closer to the earth.

METERIALS AND METHODS

The experiment was conducted at school of life sciences, Sambalpur University, Orissa in the year 2001 and 2002. The Subarnapur district of Western Orissa is the study site. Survey of India toposheet no. 73 (D/l), 64 (P/13), 64 (0/16), 73 (C/9) were taken for the study. Random sampling of villages were done after property analysing these toposheets. The samples were collected from foothill (FH). Central table land (CT) and river bank (RB) area. The sample ponds represent 5 types of traditional water harvesting structures such as *Sagar, Bundh, Kata, Muda* and *Chahla*. They are also in a proportion of dry (annual) and wet (perennial) Ponds.

The limnological paramters of different village ponds were analysed thrice i.e. Pre-monsoon, monsoon and post-monsoon in each year. The procedure followed from Saxena (1990) & Ghosh et.al, (1983). The fishes were captured by using lift net, cast net and drag net.

RESULTS & DISCUSSION

Water is one of the most important resources for the sustenance of life. Water temperature was considered as the most important single factor controlling mang profound influences in the medium (Welch, 1952, Ruttner, 1953). The temperature of a water body depends exclusively on the regional temperature pattern (Johnson, 1971). The water temperature is high in pre monsoon in comparison to post monsoon and low in monsoon. It is also observed that less fish diversity is reported in the water storing depression, where incoming water source is affected by sewage, agrochemicals, detergents etc. as it increase the water temperture.

Water bodies with slightly alkaline pH are consider to maintain good production and fish growth (Huet, 1971). The lowest pH was reported during monsoon (Dwivedi, et al. 1977, Jefferies et al. 1979, Kant & Anand, 1979). In FH, pH of water is more alkaline as compared to post monsoon. The pH of soil is slightly acidic while in CT & RB it is slightly alkaline.

Decrease in pH during winter may be attributed to reduction of photosynthesis activity and maximum pH during summer may be seen as a rusult of increased photosynthesis. We observed the pond having low pH or high acidity has low species diversity. Increase in alkalinity increase the diverstiy. It is may be due to runoff from near by lands. The free carbon dioxide is present is polluted water bodies. It appears generally when oxygen contact remain low or absent, In Pre monsoon the free CO_2 cone. increase due to rapid decompositon owing to high temperature.

Balkhi et al. (1987) reported that free carbon dioxide and total alkalinity are closely related. We observed these two are positively correlated. Dissolved oxygen is directly correlated with the productivity of water body. In summer an increase in depletion of dissolved oxygen content was correlated with the prevalance of high temperature (Hanan, 1979). C/N of soil is an important factor for the sustainability of the ecosystem. The C/N of soil increased in monsoon and post monsoon. It is also positively correlated with the fish diversity.

The diversity indices are governed by the physico-chemical region of water. We calculated the diversity indices by using Shannon-Weiver (1949) Method and Simpson (1949) method. The carp species were found absent in a major portion of the study sides while the indigenous non-carp species were reported a lot. The species *Chela untranchi* has wide occurence. The species diversity is more in post monsoon than monsoon and least is premonsoon. The *sagar* has more species diversity while the *chahla* has less. The trend is equable in the year 2001 & 2002. Changes is species diversity must be expected whether it be on a circadian, tidal, Seasonal or annual basis (Dills & Roger, 1979). We found more species diversity in perennial ponds. The diversity increases with sample size and effort, thus there is a species-area relationship (Williamson, 1981).

It clearly reveals that there is an all times increased fish distribution in RB. The occurrence of carp species is mostly observed near RB. The fish having less occurance is *Notopteres Chitala.*

Table - 5.1

Species diversity (Fish) at different sides (pooled data of two years)

Site/Season	Premonsoon	Monsoon	Post Monsoon
FH	11	19	17
CT	9	22	18
RB	12	18	23

Table - 5.2

Lentic Fishery diversity Observed in rural ponds of Western Orissa (Pooled data of two years)

Sl No.	Family	Name of the Species
1.	Notopteriae	*Notopteres notopteres*
2.	Notopteridae	*Notopteres chitala*
3.	Clupeidae	*Gadusia chapra*
4.	Cyprinidae	*Catla catla*
5.	Cyprinidae	*Labeo rohita*
6.	Cyprinidae	*Labeo bata*
7.	Cyprinidae	*Labeo calabasu*
8.	Cyprinidae	*Cirrihinus mrigala*
9.	Cyprinidae	*Cirrilunus reba*
10.	Cyprinidae	*Chela untrachi*
11.	Cyprinidae	*Punitus palehellus*
12.	Cyprinidae	*Punitus sarana*
13.	Cyprinidae	*Danio spinousus*
14.	Cyprinidae	*Labeo gonius*
15.	Claridae	*Rasbora daniconius*
16.	Claridae	*Clarius batrachus*
17.	Sagridae	*Mystus vittatus*
18.	Sagridae	*Mystus aor*
19.	Siluridae	*Wallago attu*
20.	Anabantidae	*Anabas testudieneus*
21.	Schilbedae	*Pseudotropis arthroriosides*
22.	Chennidae	*Channa punctatus*
23.	Mastucenbellidae	*Mastacembalus aculeate*
24.	Mugilidae	*Liza carsula*
25.	Chennidae	*Channa striatus*
26.	Sagridae	*Rita rita*
27.	Chennidae	*Tilapia Mosambicus*
28.	Cyprinidae	*Cyprinus carpio*
29.	Cyprinidae	*Ctenoporyngodon idella*

Table - 5.3
Frequency Abundance, density and diversity indices of inland fishes present in study sites of Western Orissa (Pooled data of two years).

Sl. No.	Frequency	Abundance	Density	Simpson's index	Shannon Weiver's index
1.	23.07	8.33	1.92	0.0008	0.0012
2.	7.69	4.02	0.53	0.0001	0.0002
3.	15.38	13.01	2.51	0.0009	0.0017
4.	23.07	9.00	2.17	0.0008	0.0016
5.	53.83	9.42	5.07	0.0059	0.0063
6.	46.14	7.21	3.53	0.0018	0.0029
7.	30.76	5.5	1.69	0.00001	0.00003
8.	38.45	6.2	2.57	0.0015	0.0019
9.	23.32	7.9	1.09	0.0007	0.0011
10.	61.52	5.12	3.15	0.0025	0.0031
11.	38.45	15.2	4.87	0.0098	0.0081
12.	23.09	9.13	2.12	0.001	0.0015
13.	30.06	5.02	1.05	0.0005	0.0024
14.	46.14	6.58	2.87	0.0019	0.0018
15.	38.45	4.61	1.76	0.0009	0.0027
16.	15.38	10	1.5	0.0011	0.0019
17.	55.83	4.19	2.23	0.0015	0.0017
18.	15.38	3.5	0.53	0.00006	0.00018
19.	38.45	4.0	1.5	0.005	0.0024
20.	40.42	5.9	2.5	0.0012	0.0019
21.	17.69	8.51	0.61	0.00008	0.0031
23.	31.23	7.8	3.3	0.0012	0.0017
24.	38.75	4.28	2.12	0.000001	0.000003
25	30.76	9.07	1.69	0.0019	0.0025
26.	31.59	6.12	2.87	0.000001	0.000005
27.	23.04	7.57	2.03	0.0682	0.0358
28.	48.52	16.09	5.92	0.0026	0.0032
29.	46.14	8.03	4.81	0.0025	0.0032

Table 5.4

Pooled data of two years (i.e. 2001 & 2002) showing $\overline{X}$ + SE of limnological Parameters from FH, CT & RB area of Western Orissa.

Parameters	Foot hill (FH) Ponds			Central table (CT) Ponds			River bank (RB) Ponds		
	Pre Monsoon	Monsoon	Post Monsoon	Pre Monsoon	Monsoon	Post Monsoon	Pre Monsoon	Monsoon	Post Monsoon
Temp. of Water in °C	38.57±3.18	34.56±3.89	30.28±2.33	37.59±3.33	36.07±1.57	29.72±1.57	35.43±3.43	33.21±5.01	30.82±2.89
pH of Water	9.82±1.28	7.59±0.23	8.57±0.31	9.21±0.59	8.29±0.84	7.82±0.81	8.12±0.29	7.19±0.12	6.89±0.23
pH of Soil	6.52±1.28	6.47±0.92	6.21±0.12	7.36±0.17	7.21±1.27	7.52±0.59	7.85±1.21	7.39±0.49	7.29±1.19
Free Carbon dioxide in PPM	138±18.3	98.57±10.21	93.21±8.17	113.57± 21.02	91.37±8.21	85.16± 11.48	92.55± 11.48	89.21±8.21	75.02±10.12
Total Alkalinty in PPM	3.04×10^2 ±1.08	2.59×10^2± 1.51	1.24×10^2± 0.33	4.8×10^2± 0.14	3.2×10^2± 0.51	5.2×10^2± 1.02	5.2×10^2± 1.02	3.92×10^2± 1.21	2.19×10^2± 0.89
Dissolved Oxygen in PPM	8.35±1.87	33.12±8.92	23.58±8.52	18.2±5.57	26.98±5.28	21.54±2.54	16.57±1.24	38.57±7.23	36.24±6.91
C/N of soil in mg/100g of soil	0.362± 0.0151	0.521± 0.0121	0.672± 0.0240	0.459± 0.0125	0.582± 0.0191	0.749± 0.0098	0.257± 0.0098	0.238± 0.0021	0.234± 0.0026

The Null hypothesis is rejected at 99% confidence levels in case of Temperature of Water pH of water, total alkalinity and dissolved oxygen content of water. Null hypothesis is rejected at 95% confidence level in case of free Carbon dioxide. It also reveals that there is no significant difference between pH of soil and C/N of soil. It was observed that there is no significant difference between Pre monsoon, monsoon and post monsoon of the year 2001 & 2002. The seasonal period has significant difference between themselves. There is no significant difference between FH, CT & RB area.

The change of diversity in different monsoonic period clearly reveals that a population can only be changed by birth (or hatching) death or migration. There are 12 families of fish species including 21 genera and 29 species present in the study sites. The growth of carp fishes is not adequate due to ill maintained limnological condition. As the monsoon flaunts western Orissa and prescribed limnological conditions are not seen, the non carp species should be taken care of appropriate limnological and microclimatic condition should be made to favour the carp fishes in the pre monsoon, monsoon & post monsoon. It needs awareness among the people about their teeming resources.

REFERENCES

Balkhi, M.H. Yusuf, A.K. & Qudri M.Y. (1987) - Hydrobiology of Arebar lake, Kashmir. Comp. Physiol. Ecol. 12 (3) 131 - 139.

Dills, G & Roger D.T (1974) - Micro invertebrate community as an indicator of acid/mine pollution Env. pollutant (6) 239 - 262.

Dwivedi, S.N., Bohra O.P, Reddy D.V., & Pillai, K.K. (1977) - Diurnal observation of abiotic factors at Kalkinada Bay, feeder channel and stocking pond of Brackish water fish farm, Kakinada (A.P) Comp. Physiol. Ecol. 2 (4) 248 - 250.

Ghosh, A.B., Bajaj, S.C., Hasan, R & Singh. D (1983) - Soil and water testing methods of IRRI.

Huet, M (1971) - Text book of culture, breeding and cultivation of fish. Fishing res. Ltd. Survey.

Jefferies, D.S, Cox C.M & Dillis, P.J. (1979). Depression of pH in lakes and strems in central ontarior Snowmelt, J. Fish. Res. Board Con 36 (6) PP. 640

Johnson, F. A. (1971) - Stream temperture in an Alpine area. J. Hydro biol. (Neither land) 14, 322 - 336

Kant,S & Ananda, V.K (1979) - Interelation of Phytoplankton and Physical factors in Manasar Lake, J & K Indian J. of Ecol. 5 (2) 134-147

Mahapatra, M.K. & Mishra, A (2001)- Limnological parameters flaunting lentic fishery diversity in Western Orissa. An M. Phil Dissertation of Sambalpur University.

Mahapatra, M. K. (2002) - Impact of monsoon on Hirakud command Area. Proc. of U.G.C. Seminar on growth of capital in Hirakud command Area, Problems & Possibilities pg 112 -115.

Mishra, A & Panda, P (1992) - Environment of western Orissa, Western Orissa : An Ethos, a Silver Jubliee publication of Sambalpur University.

Ruttner (1953) - Fundamentals of Limnology, University of Toronto Press

Saxena, M.M. (1990) - Environmetals analysis Agro Publ. India

Welch, P.S. (1952)- Limnology, Mc.Graw Hill Book Co. 2nd Ed. New Year.

Williamson, M (1981), Island Population,oxford University Press, London.

CHAPTER 6

INDUCED BREEDING OF *LABEO ROHITA* (HAMILTON-BUCHANAN) USING SINGLE APPLICATION OF OVAPRIM IN A CONTROLLED ENVIRONMENT CARP HATCHERY

Anita Jhajhria

Department of Zoology, Government P.G.College, Jalore- 334001, Rajasthan

ABSTRACT

Five induced breeding exercises of *Labeo rohita* (Hamilton-buchanan) were conducted in the modified CIFE-D81 Hatchery from 6.7.1999 to 25.7.2000 using the synthetic fish hormone ovaprim @0.2 ml (male) and 0.5 ml (female)/kg. body weight of brooder. The optimum water temperature (24.3-27.6°C) was observed during breeding of rohu. The eggs (1.00-2.59 lakhs) were produced by varying the injection time (7.15 a.m. - 3.30 p.m.). No Hatching occurred at 32.7°C temperature. The fertilization rate (50-85%) as well as hatching success (80-89%) indicates that ovaprim is a convenient ovultting agent for induced breeding.

Key words : *Ovaprim, Induced breeding, Labeo rohita (Hamilton-buchanan)*

INTRODUCTION

During the recent years, intensive aquaculture of carps has emerged as a thrust area and because of this reason, farmer's requirement for seed is increasing (Bhowmick *et.al.* 1997, Kesavanath and Radhaknshna, 1990; Tripathi, 1990; Bromage and Roberts, 1995). The seed production of cultivable carps is mainly based on the induced breeding and the inducing agents such as carp pituitary extracts (CPE) (Mahanta *et.al.* 1998) or synthetic hormonal preparations of the gonadotropin releasing hormone (GnRH) (Nandeesha *et.al.* 1990 b; Peter *et.al.* 1993; Lakra *et.dl.* 1996; Patino 1997; Barlinsky *et.al.* 1997 and Garg *et.al.* 2002.)

This paper is an attempt to work out the efficacy of ovaprim for reliable spawning and fingerling production as in Western Rajasthan it is rather difficult to meet the demands of seed supply through the natural breeding of carps.

MATERIALS AND METHODS

By operating cast and drag net in the local ponds and lakes the healthy brooders of *Labeo rohita* were collected in the happa. These were then brought to the modified CIFE-D81 Hatchery and were fed on oilcake + rice bran (1:1) for seven days.

The CIFE-D81 is a portable carp hatchery made up of low density translucent polyethylene material and consists of a breeding and a hatching unit. The breeding unit comprises of a cooling tower, large pools with spray shower and a water circulating system. The hatching unit comprises of vertical jars, each of 40 litres capacity.

After their acclimatization in the laboratory conditions ripe males with milt oozing freely and females with soft bulging rounded abdomen with reddish vent were selected for the breeding purposes. Male carp can be easily distinguished from the female during the breeding season by the roughness of the dorsal surface of its pectoral fins compared to the smooth surfaces of the female. The male and female brooders were administered ovaprim a synthetic fish hormone continuing calibrated amount of SGnRH (Salmon gonadotropin releasing hormone) and dopendone, a dopamine antagonist dissolved in an organic solvent. Ovaprim @ 0.5 ml/kg body weight of female brooder and @ 0.2 ml/kg body weight of male brooder was injected intramuscularly at an angle of 45^0 to the body of fish in the region of caudal peduncle by using a 2 ml hypodermic syringe fitted with B.D.H. no. 22 needle. Two male and one female brooder were exposed in the breeding unit of the modified CIFE- D81 Hatchery unit after giving the injection.

After successful breeding brooders were removed from the breeding tank. The brooders were weighed before and after induced spawning to evaluate the weight loss. The fertilization rate was assessed by counting the number of fertilized and unfertilized eggs. The eggs of rohu are of non-adhesive type and are red in colour having a mean diameter of 5 mm. These eggs were transferred to the hatching bucket of the hatching unit. Hatching success was evaluated by counting the number of spawn hatched out from the fertilized eggs. Some physico-chemical parameters were studied by following the methods of Welch (1948), APHA (1985), Strickland and Parson (1972), Murphy and Riley (1962) and Mullin and Riley (1956), in the breeding and the hatching units of the modified CIFE-D81 Hatchery.

RESULTS AND DISCUSSION

The results of five induced breeding exercises on *Labeo rohita* are summarised in the Table 6.1. The limnological parameters were also recorded in these breeding exercises (Table 6.2).

There are optimum temperature ranges for induced breeding of cultivated fishes and critical temperature limits, above and below which fish will not reproduce. In this study, the optimum water temperature range (24.3-27.6^{0}C) during breeding of rohu was observed. Water temperature plays a crucial role in the sexual maturation and breeding of carps.

Dwivedi and Reddy (1986) reported that in the controlled conditions, the temperature range (26-28.5^{0}C) is suitable for breeding Indian major carps. Chaudhuri and Singh (1984) observed the

Table 6.1 Result of induced fish breeding of rohu from 06.07.1999 to 25.07.2000.

Date of Breeding	Time of ovaprim injection	Broocer weight in kg.							Time & Date of egg laying	Spawning time in (Hrs.)	No. of eggs laid (lakhs)	Fertilization (%)	Hatching time (Hrs)	Hatching (%)	No. of spawn produced (Lakhs)
		Male				Female									
		Water temp. ⇩ Temp. decr.	Before breeding	After breeding	Weight loss	Before breeding	After breeding	Weight loss							
06.07.1999	9.00 am	27.6	2.700	2.398	0.302	2.500	1.960	0.540	7.10 pm 06.07.1999	10.10	2.63	72	42	80	1.51488
25.07.1999	7.15 am	29.2 ⇩ 24.3	2.100	1.996	0.104	1.900	1.700	0.200	7.00 pm 25.07.1999	12.15	1.10	50	39	89	0.48955
28.07.1999	10.30 am	32.7	2.400	2.151	0.249	2.000	1.700	0.300	11.50 pm 28.07.1999	13.20	1.00	12	-	-	-
16.07.2000	10.15 am	33.0 ⇩ 26.4	2.900	2.600	0.300	2.500	2.000	0.500	8.00 pm 16.07.2000	10.15	2.72	85	30	84	1.94208
25.07.2000	3.30 pm	26.2	3.200	3.000	0.200	3.000	2.670	0.300	5.40 am 26.07.2000	14.10	2.59	79	37	80	1.63688

optimum range of temperature for Indian and exotic carps to be between 24-31°C and they failed to spawn if the temperature is raised above 31°C. In the present work, also no spawn was produced at 32.7° C. Spawing time for carps was observed by Pandey and Singh (1997) between 14-20 hours.

Pandey *et.al.* (2002) recorded spawning time to be between 10-12 hours after administration of carp pituitary extract (CPE) and between 7-8 hours after injecting ovatide.

Nandeesha *et.al.* (1990b) observed after application of ovaprim the fertilization and hatching percent for rohu (98-99.6% and 90-95%). Lakra *et.al.* (1996) used ovaprim and reported fertilization and hatching percent (96.2% and 94.8%) for rohu. Pandey *et.al.* (2002) after injecting ovatide observed rate of fertilization (95-100%) and hatching success (90-98%) and by CPE recorded fertilization rate (75-85%) and hatching percent (70-78%) in *Labeo rohita.* In the present work, fertilization rate (50-85%) and hatching percent (80-89%) occurred after injecting ovaprim.

Thus, the results of the present study clearly demonstrate possibility of using the synthetic fish hormone preparation "ovaprim" for effective induced spawning and seed production in *Labeo rohita* (Hamilton-buchanan), in Western Rajasthan.

ACKNOWLEDGEMENTS

The author is grateful to Dr. Devendra Mohan, Associate Professor, Department of Zoology, Jai Narain Vyas University, Jodhpur for providing valuable suggestions and facilities to carry out the studies.

REFERENCES

APHA. 1985. American water works association and water environment federation, standard methods for the examination of water and waste water. 19th edition.

Barlinsky, D.L., William, K.V., Hodson, R.G. and Sullivan, C.V. 1997. Hormone induced spawning of summer flounder, *Paralichthys dentatus.* J. World Aquacult. Soc. 28: 79-86.

Bhowmick, R.M., Kowtal, G.V., Jana, R.K.and Gupta.S.D. 1997. Experiments on second spawning of Indian major carps in the same season by hypophysation. Aquaculture. 12: 149-155.

Bromage, N and Roberts, R.J. 1995. Broodstock management and egg and larval quality. Blackwell Science Publications, Oxford, U.K.

Chaudhuri, H and Singh, S.B.1984. Induced breeding of carps. ICAR Publication.

Dwivedi, S.N. and Reddy, A.K. 1986. Fish breeding in a controlled environment carp hatchery CIFE-D81. Aquaculture. 54: 27-36.

Garg, S.K., Bhatnagar, A., Kalla, A. and Johal, M.S.2002. Experimental Ichthyology. CBS Publishers and Distributors, New Delhi. pp. 140-141.

Kesavanath, P and Radhakrishnan, K.V. 1990. Carp seed production technology, special publication No. 2, Asian fisheries society, Indian branch, Mangalore.

Lakra, W.S., Mishra, A., Dayal, R. and Pandey, A.K. 1996. Breeding of Indian major carps with the synthetic hormone drug ovaprim in Uttar Pradesh. J. Adv. Zool. 17: 105-109.

Mahanta, P.C., Rao, K.G., Pandey, G.C.and Pandey, A.K. 1998. Induced double spawing of an Indian major carp *Labeo rohita,* in the agro climatic conditions of Assam. J. Adv. Zool. 19: 99-104.

Mullin, J.B. and Riley, J.P. 1956. The spectrophotometer determination of nitrate in natural water. Anal. Chem. Acta. pp. 464.

Murphy, J. and Riley, J.P. 1962. A modified single solution method for determination of phosphate in natural water. Anal. Chem. Acta. 27: 31-36.

Nandeesha, M.C., Rao, K.G., Jayanna, R., Parker, N.C., Varghese, T.J., Keshavanath, P and Shetty, H.P.C. 1990b. Induced spawning of Indian major carps through single application of ovaprim c. In : The second Asian Fisheries Forum Proceedings. Asian Fisheries Society, Manila, Philippines. (Eds. Hirano, R. and Hanyo, I.). pp. 581-585.

Pandey, A.C. and Singh. R.N. 1997. Breeding of *Catla catla, Labeo rohita* and *Cirrhinus mrigala* by ovaprim injection for quality seed production. J. Adv. Zool. 18: 38-41.

Pandey, A.K., Mahapatra, C.T., Kanungo, G., Sarkar, M., Sahoo, G.C. and Singh, B.N. 2002. Ovatide induced spawning in the Indian major carp, *Labeo rohita* (Hamilton-buchanan), Aquacult. 3: 1-4.

Patino, R. 1997. Manipulations of the reproductive system of fishes by means of exogenous chemicals. Prog. Fish. Cult. 59: 118-128.

Peter, R.E., Lin, H.R., Kraak, G.V. and Little, M. 1993. Releasing hormones, dopamine antagonists and induced spawning In: Recent advances in aquaculture (Eds. Muir, J.F. and Roberts, R.J.). Blackwell Science Publications, London, pp. 25-30.

Strickland, J.D.R. and Parson, T.R. 1972. A practical handbook of seawater analysis. Fish Research Board of Canada publications, Ottawa.

Triphati, S.D.1990. Aquaculture in India. In : Aquaculture in Asia. (Ed. Joseph, M.M.). Asian Fisheries Society, Indian Branch, College of Fisheries, Mangalore. pp. 191-222.

Welch, P.S. 1948. Limnological methods. McGraw Hill Book. Co.

CHAPTER 7

RANGE OF EXTENSION OF CAT FISH *BATASIO TRAVANCORIA* HORA & LAW TO RIVER THUNGA IN KARNATAKA PART OF WESTERN GHATS, INDIA.

Arunachalam, M*., A.Manimekalan, A.Johnson, R.Soranam, A.Sankaranarayanan & P.Sivakumar

SPK Centre for Environmental Sciences, Manonmaniam Sundaranar University, Alwarkurichi-627 412, Tamil Nadu.

During an intensive study in Kudremukh Wildlife Sanctuary as a part of National Agricultural Technology Programme on "Germplasm inventory, evaluation and gene banking of freshwater fishes of stream/rivers of peninsular India" two specimens of catfish *Batasio travancoria* Hora & Law was collected from Khuremuk Wildlife Sanctuary, Karnataka. This species is commonly called as Travancore batasio. Hora and Law (1941) originally described this species from Perunteraruvi, a tributary of Pamba river at Edakadathy, Kerala. Later Silas (1951) reported this species from Anamalai Hills; Jayaram *et al.* (1976) reported from Cardamom and Agastya hills of the Western Ghats and Raghunathan (1989) reported from Coorg district, Karnataka. Talwar and Jhingran (1991) reported the distribution of this species upto Kerala part of Ghats Western. Presently this species was collected from Vimalanathi at Munsar in Thunga river basin, Kudremukh National Park.

Many workers were studied the fish fauna of river Thunga and Bhadra (Chacko and Kuriyan, 1948; Chacko and Venkataraman, 1945; Chacko, 1948; Day, 1868; Hora and Misra, 1940; Hora, 1937; Rahimullah, 1943; Rao, 1920; Anuradha, 2000; Chandrashekhariah *et. al,* 2000). But this species was not reported by any of these workers from river Thunga and Bhadra. So the present collection shows the range of extension to the river Thunga river basin.

DIAGNOSIS

DI/7; A iii-iv 9-11; P I 7-9; Vi 5

Body elongate; head globular; occipital process not reaching the basal bone of dorsal fin; 8 barbels; dorsal spine weak;.caudal fin deeply forked; colour - greyish with a narrow lateral streak along the lateral sides. Morphometric characters are given in the table 7.1.

* Corresponding author (e-mail: arunaces@md5.vsnl.net.in)

Table 7.1
Morphometric characteristic of *Batasio travancoria* Hora & Law

	Morphometric Character	Measurements (in mm) 1	2	Proportion to SL Range	Mean	SD	Proportion to HL Range	Mean	SD
1	Total length		118.00	106.00					
2	Standard length	98.50	90.00						
3	Body depth	16.09	14.89	6.12 - 6.04	6.08	0.05	1.62- 1.53	1.58	0.07
4	Head length	26.11	22.78	3.77 - 3.95	3.86	0.13	1.00- 1.00	1.00	0.00
5	Snout length	12.65	11.44	7.79 - 7.87	7.83	0.06	2.06- 1.99	2.03	0.05
6	Mouth width	5.67	5.13	17.37 - 17.54	17.46	0.12	4.60- 4.44	4.52	0.12
7	Mouth height	9.21	8.85	10.69 - 10.17	10.43	0.37	2.83- 2.57	2.70	0.18
8	Eye diameter	5.10	4.10	19.31 - 21.95	20.63	1.86	5.12- 5.56	5.34	0.31
9	Inter orbital width	5.15	4.26	19.13 - 21.13	20.13	1.41	5.07- 5.35	5.21	0.20
10	Pre dorsal length	34.77	34.55	2.83 - 2.60	2.72	0.16	0.75- 0.66	0.71	0.06
11	Post dorsal length	56.23	53.62	1.75 - 1.68	1.72	0.05	0.46- 0.42	0.44	0.03
12	Pre pelvic length	46.54	42.89	2.12 - 2.10	2.11	0.01	0.56- 0.53	0.55	0.02
13	Distance between pectoral and pelvic	24.58	21.85	4.01 - 4.12	4.06	0.08	1.06- 1.04	1,05	0.01
14	Distance between pectoral and Vent	31.67	28.78	3.11 - 3.13	3.12	0.01	0.82- 0.79	0.81	0.02
15	Distance between pelvic and anal fin	19.10	17.91	5.16 - 5.03	5.09	0.09	1.37- 1.27	1.32	0.07
16	Distance between pelvic fin and vent	15.10	7.09	6.52 - 12.69	9.61	4.36	1.73- 3.21	2.47	1.05
17	Distance between anal fin and vent	15.15	13.26	6.50 - 6.79	6.64	0.20	1.72- 1.72	1.72	0.00
18	Dorsal fin height	13.58	9.66	7.25 - 9.32	8.29	1.46	1.92- 2.36	2.14	0.31
19	Pectoral fin length	16.23	13.32	6.07 - 6.76	6.41	0.49	1.61- 1.71	1.66	0.07
20	Pelvic fin length	15.90	11.92	6.19 - 7.55	6.87	0.96	1.64- 1.91	1.78	0.19
21	Anal fin height	11.84	6.30	8.32 - 14.29	11.30	4.22	2.21- 3.62	2.91	1.00
22	Anal fin length	22.48	17.82	4.38 - 5.05	4.72	0.47	1.16- 1.28	1.22	0.08
23	Caudal peduncle length	16.08	14.28	6.13 - 6.30	6.21	0.13	1.62- 1.60	1.61	0.02
24	Caudal peduncle depth	6.33	6.12	15.56 - 14.71	15.13	0.60	4.12- 3.72	3.92	0.28
25	Caudal fin length	22.00	16.00	4.48 - 5.63	5.05	0.81	1.19- 1.42	1.31	0.17

ECOLOGY

This species was collected from bank undercut with more tree roots, instream cover, with good canopy cover. Substrate type was muddy with detritus, with low flow. Habitat factors are given in the table 2.

Table 7.2
Habitat characteristics of Vimalanathi (Munsar)

Habitat factors	
Altitude (m)	628
Position	13^{0}19'21"N 75^{0}7'32.1"E
Stream order	2
Gradient (%)	2
Mean width (m)	14.2
Mean depth (cm)	62.75
Flow (m/sec)	0.434
Riparian cover (%)	60
Substrates (%)	
Bedrock	15
Boulder	30
Cobble	28.75
Gravel	13.75
Sand	7.5
Leaf litter	5
Fine sand	0

ACKNOWLEDGEMENTS

Senior author (M. Arunachalam) is grateful for the financial assistance from NATP under the mission mode programme of Germplasam Inventory and Gene banking of Freshwater Fishes (Sanction No. 27(281/98/NATP/MM-III 18 Dt 23.12.1999). We also thank the Mission Leader and Director Dr. D. Kapoor and Dr. S.P. Singh, Principal Investigator of the Lead Centre, National Bureau of Fish Genetic Resources, Lucknow for their leadership in this programme. We thank Sri. Chakrabarti, Principal Chief Conservator of Forests, Karnataka for the permission and moral support.

REFERENCES

Bhat Anuradha. 2000. Fish germplasm inventory of Sharavathi, Aghanashni, Bedti and Kali rivers, Uttara Kannada. Pp.148-151. *In:* Ponniah, A.G. and Gopalakrishnan, A. (Eds). Endemic Fish Diversity of Western Ghats. NBFGR-NATP Publicatopn-1, 347. p. National Bureau of Fish Genetic Resources, Lucknow, U.P., India.

Chacko. P.I. 1948. A survey of the fisheries of the Tungabhadra river. *Proc. Indian Acad. Sci.* 28:166-175.

Chacko, P.I. , Kuriyan, G.K. 1948. A survey of the fisheries of the Tungabhadra river. Proc. Indian Acad. Sci., 28(5):165-176.

Chandrashekhariah, H.N., Rahman, M.F. and Lakshmi Raghavan, S. 2000. Status of fish fauna in Karnataka. Pp.98-135 *In:* Ponniah, A.G. and Gopalakrishnan, A. (Eds). Endemic Fish Diversity of Western Ghats. NBFGR-NATP Publicatopn-1, 347. p. National Bureau of Fish Genetic Resources, Lucknow, U.P., India

Day F. 1868. Absernations on some Indian fishes. *Proc.2ool.Soc,* 580-585.

Hora, S.L. 1937. Notes on fishes in the Indian Museum. XXVIII. On three collections of fish from Mysore and Coorg, South India. *Rec. Indian Mus.* 9(l):5-28.

Hora, S.L. 1937. The game fishes of India. II. The Cachhwa or Butchwa, Eutropiichthys vacha (Hamilton). *J. Bombay Nat. Hist. Soc* 39(3):431-446.

Hora, S.L. and Law, N.C. 1941. The freshwater fish of Travancore. *Rec. Indian Mus.*, 43(2):233-256.

Hora, S.L. and Misra, K.S. 1940. On fishes of the genus Rohtee Sykes. *Rec. Indian Mus.* 42(1):155-172, 1 pl.

Jayaram, K.C., Indra, TJ. and Sunder Singh, M. 1976. On a collection of fish from the Cardamon Hills, South India, Madras J. Fish., 7:1-7.

Raghunathan, M.B. 1989. A study on the fish fauna of Coorg district, Karnataka. *Fishery Tech.*,

Rahimullah, M. 1943. Fish survey of Hyderabad state. Part 1. A preliminary report of fishes found in the Godavari, Purna, Kistna, Tungabhadra and Siddha rivers. *J. Bombay Nat. Hist. Soc.* 43:648-653

Rao, C.R.N. 1920. Some new species of Cyprinoid fish from Mysore. *Ann. Mag. Nat. Hist.*, (9) 6: 45 -64.

Silas, E.G. 1951. On a collection of fish from the Anamalai and Nelliampathi hill ranges (Western Ghats) with notes on its zoogeographical significances. *J. Bombay Nat. Hist. Soc.* 49(4):670-681.

CHAPTER 8

TOXICOLOGICAL EFFECTS OF CADMIUM CHLORIDE ON A FRESH WATER FISH, *TILAPIA MOSSAMBICA*, PETERS AND ITS ECOLOGICAL SIGNIFICANCE

Nilalohit Harichandan, Bibekananda Beura and A.K.Panigrahi

Laboratory of Environmental Toxicology, Environmental Science Research Centre, Berhampur University, BERHAMPUR-760 007, Orissa.

ABSTRACT

Fishes were exposed to graded series of concentrations of cadmium chloride for acute toxicity studies. The MAC value in 30 days was 1.25ppm of cadmium chloride. A safety concentration of 1. 2 ppm of Cadmium chloride I-^{50} was selected for this study. The LC_{10}, LC_{50}, LC_{90}, LC_{100}, determined for Tilapia fish was found to be 1.32,1.63,1.91 and 2.12ppm respectively after 30 days of exposure. Exposed fishes appeared lethargic showed erratic movements, loss of equilibrium, gradual onset of inactivity etc.when compared to the control fish. Inappetence and ataxia was observed in the exposed fish. The body weight of the $CdCI_2$ exposed fish decreased (6.17%) significantly, when compared to the control fish, where a significant increase in body weight was marked. When the exposed fish was transferred to toxicant free medium, no recovery was noted. A maximum of 22.32% decrease was noted in BSI of $CdCI_2$ exposed fish, where 17.94% recovery was noted after 28 days of recovery. The HSI decreased by 71.72% after 28 days of exposure and 27% recovery was noted after 28days of recovery. The KSI increased by 68% after 21 days of exposure and by 24% on 28d of exposure and 27% recovery was noted after 28days of recovery. The exposed fish could not recover to its pre-test activity.

INTRODUCTION

The purpose of acute toxicity test is to assess various abnormalities caused due to administration of a chemical to animals on occasion or other and to determine the order of lethality of the chemical. Considerable efforts have been directed towards the establishment of short-cut procedure. In aquatic toxicology, acute lethal toxicity tests with fish or invertebrates are usually intended to assess the numerical value of toxicity, to compare potencies of toxicants, to assess the effect of environmental variables on toxicity. Besides, statistical analysis of toxicity is essentially a tool in all these studies. Fathead minnows when exposed to low concentrations of methyl mercuric chloride (Olson *et al.*, 1973) showed no effect on survival at concentrations causing an acceptable residue. Panigrahi and Misra (1980) found severe poisoning symptoms in *Tilapia mossambica* exposed to inorganic mercury. Relative toxicity of Dichlorovos (DDVP) was studied by Rath and Misra (1979) on *Tilapia mossambica,* Peters belonging to 3 different age groups. Gouda *et* al. (1981) tested the toxicity of dimecron, sevin and lindax to *Anabas scandens* and *Heteropneustees fossilis.*

Cadmium toxicity of gold fish, *Carassius auratus* by McCarty *et al.*, (1978), lethal and sub-lethal effects of Binary mixtures of cyanide and hexavalent chromium, zinc or ammonia to the fathead minnow *(Pimephales promelas)* ; all these add to the knowledge of heavy metal toxicity on fishes. McCarty *et al.* (1978) studied the toxicity of cadmium to gold fish, *Carassius auratus* in hard and soft water. Panigrahi and Misra (1978,80) and Panigrahi (1980) reported the behavioral changes observed in inorganic mercury exposed fresh water fishes. Macleod and Passah (1973) reported the behavioral changes induced by mercury in Rainbow trout. Larson and Lewander (1973) reported metabolic effects of starvation in the eel, *Anguilla anguilla* L., Gibilin and Massaro (1973) observed behavioral changes induced by methyl mercury in Rainbow trout. Bhatia *et al.* (1973) reported behavioral changes induced by dieldrin in albino rats. Although adequate literature on acute toxicity of different pesticides and mercury based chemicals, mercury contained effluents and solid wastes to fish is available, no adeqaute information is available regarding this metal, cadmium and its acute toxicity. This study was designed to find out the lethal concentration values and behavioural changes induced by Cadmium chloride on *Oreochromis mossambicus sy. Tilapia mossambica,* Peters.

MATERIALS AND METHODS

Test fish, *Tilapia* were collected, acclimatized in the laboratory. A graded series of concentrations of Cadmium chloride ranging from 1 mg I^{-1} to 20 mg I^{-1} were prepared. 10 healthy fish were exposed to each concentration in 10 litre glass jars. The experiments were conducted in chlorine-free tap water.

The mortality rate of test-fish was studied following the method described by Finney(1971). Observation on the toxicity of Cadmium chloride was made at 24, 48, 72, 96 hours and 28 days after the experimental animals were first exposed. Individuals showing no respiratory movements, no opercular movements and no response to a tactile stimulus were recorded as dead, and were immediately removed. The test fish exposed to lower range of the heavy metal was exposed for a period of five weeks to find out the maximum allowable concentration (MAC), where no mortality was noticed and this was expressed as mg I^{-1}. LC_{50}. Different values such as LC_{10}, LC_{50}, LC_{100} were deduced from graphical interpolation (Finney, 1971).

Fishes were observed for behavioral changes influenced by Cadmium chloride. Experimental fishes were sacrificed; brain, liver, kidney and gill were dissected out carefully and washed thoroughly with distilled water. The sacrificed fish were weighed. Brain, liver and kidney were separated and weighed carefully in a single pan electric balance and the different somatic indices were calculated. All the parameters werenoted at 7-day interval and the percent change in parameters were computed.

RESULTS

The MAC value of Cadmium chloride was found to be 1.25ppm for 30 days and to be on the safe side 1.20ppm was considered for 28 days of the exposure for sub-lethal toxicological studies. With the increase in exposure period, the concentration of the chemical decreased showing the existence of a negative correlation. No mortality was observed in the control set.The following table indicate the lethal concentration values obtained from statistical interpolations.

Lethal concentration value	Cadmium chloride toxicant concentration (ppm)	
	24 h	48 h
LC_{10}	4.5ppm	4.2
LC_{50}	9.5ppm	9.2
LC_{100}	14.6ppm	14.4
Exposure period 30 days		
MAC	1.25	
Used concentration	1.20	
LC Value after 30 days of exposure		
LC_{10}	1.32	
LC_{50}	1.63	
LC_{90}	1.91	
LC_{100}	2.12	

All the exposed fish appeared lethargic after exposure to Cadmium chloride. The major clinical symptoms such as inappetance and ataxia appeared after 2 to 3 days exposure. At higher concentration of the Cadmium chloride, the exposed fish showed erratic movements. The other signs of toxicity such as loss of equilibrium, gradual onset of inactivity, erratic swimming with irregular collision to the inner glass walls of the aquarium were observed. Circular movement of the fish was recorded. The exposed fish shed fins and huge destruction of dermal mucous layer was marked. Scar and ulceration of the gill lamellae was observed in the cadmium chloride exposed fish. Some symptoms were similar to the preliminary symptoms of Epizootic Ulcerative Syndrome (EUS) and was marked in the exposed fish. At higher exposure periods, profuse bleeding was marked in the gills. Infection of eyes of test fish was not observed. In the present investigation, peculiar symptoms were marked. The cadmium chloride exposed fish showed initial symptoms like mercury poisoning but later on the exposed fish became over excited and periodic outbursts in irratic swimming was noted, ultimately the excited fish died of suffocation. When the excited fishes were heavily oxygenated and kept in undisturbed water, survived. The exposed fishes showed high level of sensitivity. Any type of external stimuli like a flash of light, some one coming closer to the aquarium, any type of movement, even any sound, the sluggish fish instantly starts vigorous movement and many a times colloid with the glass of the aquarium frequently and showed circular movements. After 1/2 minutes, the movements slow down and the fish becomes senseless, ventillation rate almost stops. If heavily oxygenated, the fish survives otherwise the exposed fish dies. During this later period of slowing down the movement, the colour of the fish becomes black due to dispersion of melanin pigment. This black colour disappears, when the system was heavily oxygenated. At times, the shocked fish automatically comes back to normal, and the black colour disappears. This type of self recovery was only in 30% of the cases. Heavy oxygenation can help to around 65% fishes to survive but the rest of the excited fish die instantly and never recover from the shock. The black colour persists for some time and after death the black colour again disappears. In cadmium exposed fish lesions develope on the fins and later the skin erodes and the bones were clearly visible and becomes naked and shedding of fins were noted in few exposed fish towards later period of exposure or at higher exposure period.

Body weight of the control fish increased by 3.96% after 28 days and by 7.86% after 56 days of exposure. Whereas, the body weight of the exposed fish gradually decreased during the exposure period and a maximum of 6.17% decrease was observed after 28 days of exposure (Fig. 8.1), when compared to control fish. No recovery in body weight was marked, when the exposed fish was transferred to toxicant free normal medium. Rather further depletion upto 8.92% over the 28 d exposure value was marked, on 28th day of recovery (Fig. 8.1). Recovery studies clearly indicated that transferring the exposed fish to normal water medium has no effect. Rather higher depletion in body weight was marked. The body weight showed a linear increase showing the existence of significant positive correlation ($r = 0.984$, $P \leq 0.01$) with the exposure period in the control set. Whereas, a significant negative correlation was observed between the exposure period and body weight. With the increase in exposure period, the body weight significantly ($r = -0.898$, $p \leq 0.0\ 05$) declined.

Autopsy studies revealed that the liver and brain of exposed fish were congested and tender. The brain somatic index initially increased from 1.13 to 1.22 after 7days of exposure.Than, the BSI decreased gradually with the increase in exposure period. On 28th day of exposure, the BSI decreased from 1.12 to 0.87 (Fig. 8.2). After 28d of exposure to cadmium chloride, the exposed fish was transferred to toxicant free medium for recovery studies. The BSI further decreased from 0.87 to 0.72 on 14th day of recovery. Interestingly, on 28th day of recovery, the BSI increased to 1.09 but the value was less than the control value (Fig. 8.2). The Hepato Somatic Index (HSI or LSI) remained almost at the same level in the control fish throughout the experimental period. The HSI gradually and significantly decreased with the increase in exposure period. The value depleted from 2.37 to 0.67 on 28th day of exposure (Fig. 8.2). Partial recovery from 0.67 to 1.09 on 14d and to 1.31 on 28d of recovery was noted, when compared to control value. A consistent Kidney Somatic Index (KSI) was recorded in the control fish in the entire period of experimentation. Interestingly and unlike BSI and HSI, the KSI value in the exposed fish when compared to control fish, increased with the increase in exposure period. The KSI value increased from 0.25 to 0.42 on 21st day of exposure and on 28th day of exposure the KSI value decreased from 0.25 to 0.31 (Fig. 8.3). The exposed fish was transferred to toxicant free medium for recovery studies. No recovery was noted. The KSI value increased in all the recovery periods.

The liver somatic index and brain somatic index decreased ($r = -0.986$, $p \leq 0.01$ and $r = -0.931$, $p \leq 0.01$) with the increase in exposure period, when compared to control fish ($r = 0.634$, p = NS and $r = 0.140$, p = NS). The percent inhibition of BSI intitially increased and after 7days of exposure the BSI and LSI decreased in exposed fish with the increase in exposure period. The control fish remained clinically healthy throughout the period of experiment. The percent increase in KSI increased with the increase in exposure period and a maximum of 68% increase was recorded on 21st day and 24%

increase was noted on 28th day of exposure.The percent decrease of BSI, increased with the increase in exposure period after 7days of exposure and a maximum decrease of 22.32% was recorded on 28th day of exposure. When the exposed fish was transferred to toxicant free medium 17.94% recovery was noted on 28d of recovery in BSI.In case of LSI, a similar trend was observed but here the index decreased maximum by 71.72%, when compared to control fish, after 28 days of exposure. When the fish was transferred to toxicant free medium, partial recovery by 17.91% and 27% was marked after 14 and 28d of recovery. In case of KSI, the value increased by 24% on 28d of exposure and after 28d of recovery, ther KSI value increased by 48%. This value indicated 24% further effect on the exposed fish. The values clearly indicated that during exposure, the kidney was drastically damaged and no recovery was marked. The exposed fish could not recover to its pre-exposure activity.

DISCUSSION

The mortality percentage for any fixed period increased with the increasing concentration of Cadmium chloride. There was an increase in mortality percentage with the increase in exposure period, also. The proportion of these animals that died in a fixed time increased with the increase in the toxicant concentration and the median tolerance limit value decreased with the exposure period (Panigrahi and Misra, 1978,80; and Panigrahi, 1984). The close response approach to assessment of pollutant toxicity involves several assumptions. The route of entry of the toxicant is generally agreed to be via the gills and thus enters directly into the circulatory system. This caused damage to tissues as a result of which there is depression in active metabolism (Macleod and Passah, 1973). Depression in active metabolism that directly reduced scope of activity was demonstrated by Fry (1957). Probably this has relevance with the impaired swimming ability, erratic movements, paralysis, anorexia and ataxia which were observed in all test fish at higher exposure periods.

High percentage mortality of fish due to the action of Cadmium chloride might be due to the pathological changes, as Mathur (1969) pointed out that fish subjected to insecticides such as lindane, dieldrin, DDT and BHC die due to pathological disorders. During the experimental period temperature, hardness, time of exposure size of the test-fish and all such factors were monitored since the interference of one of these factors will alter the TLM values and MAC values. At present, the sub-lethal concentration selected for further study is 1.20ppm of cadmium chloride for 28 days, where no death was recorded. The rapid absorption of Cadmium chloride through the gill, skin and gastro-intestinal tract of fish was well evident in the present investigation. The toxicity of Cadmium chloride becomes more apparent in a very shorter period in aquatic animals. Panigrahi and Misra (1978,79,80) and Panigrahi (1980) documented a detailed behavioral changes in relation to inorganic-mercury poisoning. The above authors had observed swelling of eyes and consequent blindness. Such type of symptom was not observed in Cadmium chloride exposed fish, probably such a symptom is unique in mercury contamination. Macleod and Passah (1973) reported that loss of appetite, loss of weight, nervousness, dizziness, loss of equilibrium, erratic swimming and gradual onset of inactivity were the sub-clinical effects of inorganic mercury intoxication. Similar symptoms were marked in cadmium exposure here.

The observed depression in active metabolism in cadmium exposed fish were indicative of damage to nervous tissues, inhibition of enzymes or a vital system, was totally in agreement with Panigrahi (1980). Fry (1957) demonstrated the depression in active metabolism directly reduced "scope for activity". This in turn, may result in decreased growth, impaired swimming ability with erratic out burst at times. Loss in weight due to starvation was noted by Larsen and Lewander (1973). But the loss due to starvation was very less. Panigrahi and Misra (1978, 1980) reported the loss of body

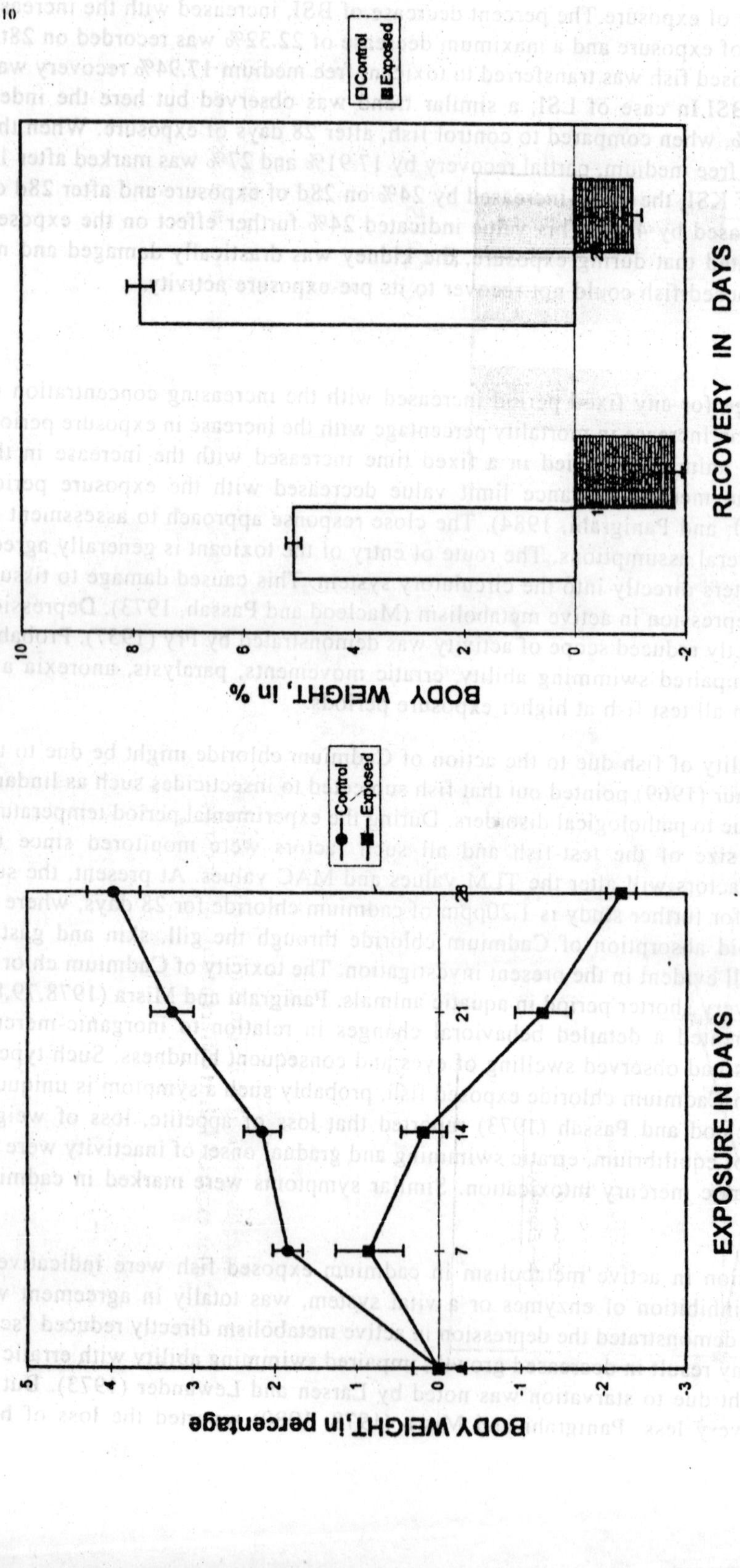

Showing change in body weight of the control and Cadmium chloride exposed fish at different days of exposure.

Fig. 8.1

Showing change in body weight of the control and Cadmium chloride exposed fish at different days of recovery.

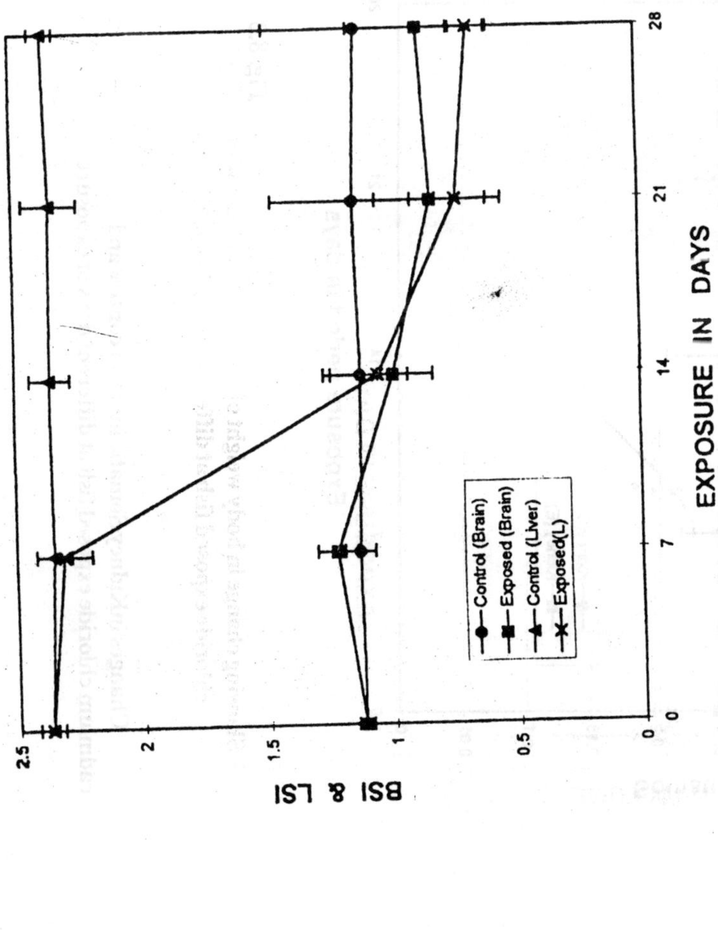

Showing change in LSI & BSI in control and Cadmium chloride exposed fish at different days of exposure

Fig. 8.2

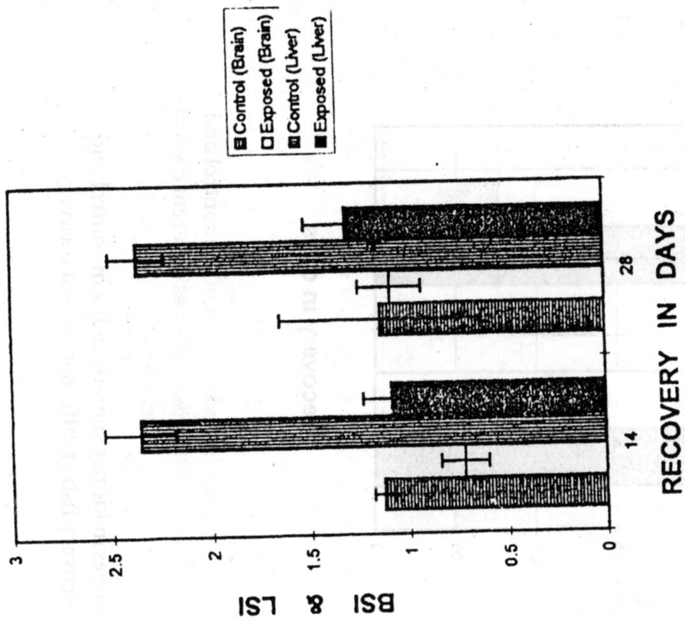

Showing change in BSI & LSI in control and cadmium chloride exposed fish at different days of recovery

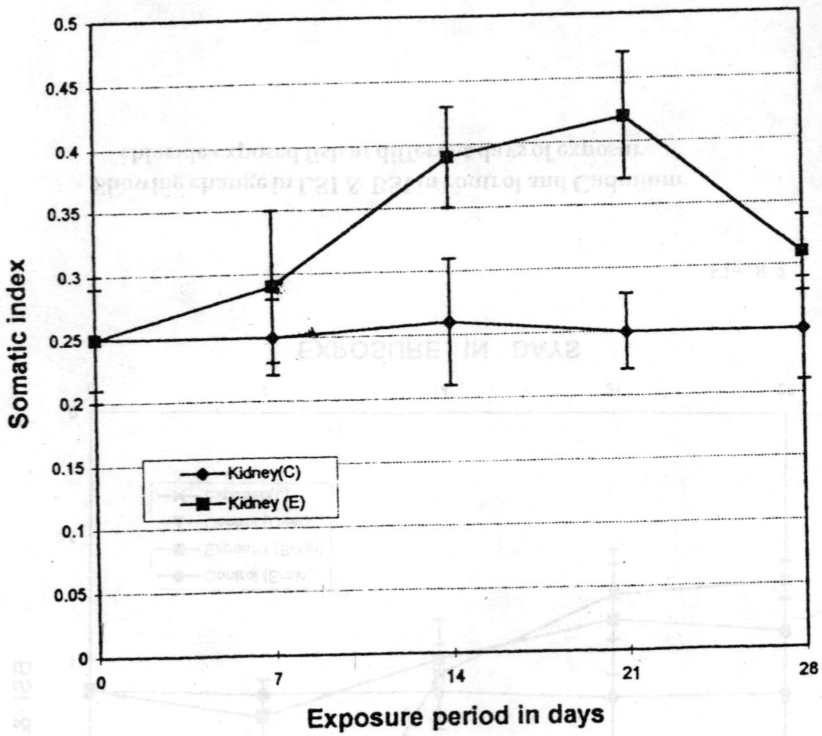

Changes inKidney Somatic index in control and cadmium chloride exposed fish at different days of exposure

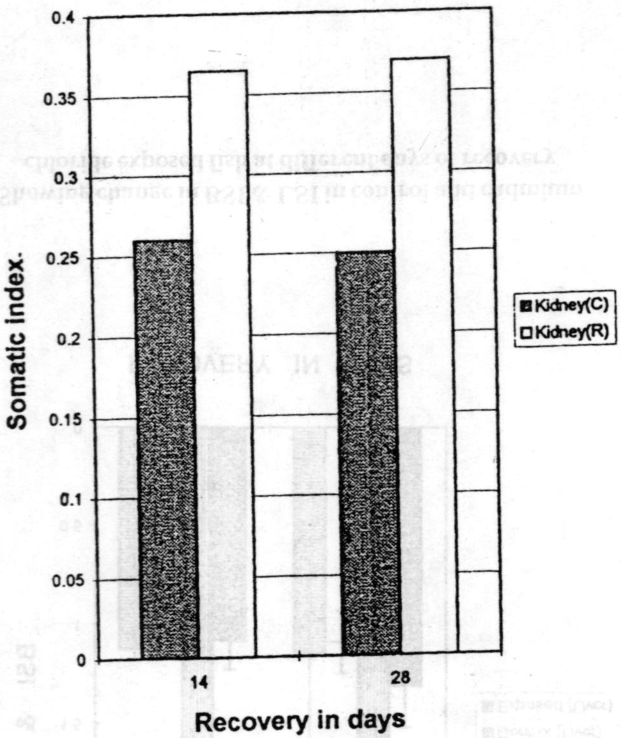

Changes in kidney somatic index in control and recovery fish at different days of recovery.

Fig. 8.3

weight due to mercury intoxication and confirmed that this loss in body weight was only due to mercury stress. In addition, exposed fish maintained their feeding habit only after a short span of time. Hence the weight loss cannot be correlated with starvation but can only be related with heavy metal intoxication. Whereas, the control fish showed increase in weight. The differences observed between control fish and exposed fish in terms of body weight is only due to the administration of Cadmium chloride in exposed aquarium.

Borg *et al.* (1970) reported inappetance, muscular weakness, ataxia and loss of body weight as the main clinical symptoms of mercury poisoning. On autopsy, he found muscular atrophy, which can be correlated with the weight loss in the goshawks. Hanko *et al.* (1970) reported loss of appetite, weakness of the extremities, excitation in the animal and loss of body weight due to mercury poisoning in chickens. A decline in liver somatic index an brain somatic index in exposed fish could be correlated with the degeneration of cells and decrease in other macromolecular variables. Concisely, at this stage it is not possible to establish the cause of loss in body weight and decrease in liver and brain somatic index, except a generalised comment on the behavior of the exposed fish to heavy metal stress that the observed symptoms were most probably be due to cadmium contamination. The significant increase in KSI in both the exposure period and recovery period indicate positively that cadmium has a significant effect on the kidney and the kidney of the exposed fish was badly damaged by cadmium chloride. The non recovery of kidney during recovery period also indicate positive permanent damage to the system under study. Hence, with the recognition that all heavy metals are potentially lethal to fish even at relatively low concentrations. It is now a normal practice to test all new chemicals for their toxicity to fish. The LC_{50} and MAC values can be helpful to understand about the amount of heavy metal, cadmium that can be discharged into the environment by any industry, which discharges cadmium in the effluent or in any other waste discharged by the industry.

ACKNOWLEDGEMENTS

The authors wish to thank the Head, Department of Botany, Berhampur University for library and laboratory facilities.

REFERENCES

Bhatia, S.C., Sharma,S.C. and Venkrtasubramanian,T.A.(1973): Effects of dieldrin on hepatic carbohydtrate metabolism and protein biosynthesis in vivo. Toxicol. Appl.Pharmacol.,24:216.

Borg, K., Wanntorp, H., Erne, K and Hanko, E. (1969): Alkyl mercury poisoning in terrestrial Swedish wild-life.Viltry, 6, 301-379.

Finney, D.J.(1971): Estimation of the median effective dose.In: Probit analysis. University Press, Cambridge.

Fry, F.E.J.(1957):The aquatic respiration of fish, P.1-63. In M.E. Brown[ed.] The physiology of fishes. Vol.1. Academic Press, Inc., NewYork, N.Y.

Gibilin.F.J. and Massaro. E.J. (1973): Pharmacodynamics of methyl mercury in the rainbow trout *(Salmo gairdnen):* Tissue uptake, distribution and excretion. Toxicol. Appl. Pharmacol., 24, 81-91.

Gouda, R.K., Tripathy, N.K. and Das, C.C. (1981): Toxicity of Dimecron Sevin and Lindex to *Anabas scandens* and *Heteropneustes fossilis*. Comp. Physiol. Ecol., 6(3), 170-172.

Hanko, E.,Erne, K.,Wanntorp, H. and Borg, K. (1970): Poisoning in ferrets by tissues of alkyl mercury fed chickens. Acta vet. Scand., 11, 268-282.

Larsson, A., and Lewander. K. (1973): Metabolic effects of starvation in the eel, *Anguilla anguilla,L.* Comp.Biochem.Physiol.,44,367-374.

MacLeod, J.C. and Pessah, E.(1973): Temperature effects on mercury accumulation, toxicity, and metabolic rate in Rainbow trout *(Samo gairdnen)*J.Fish.Res.Board Can.,30:485-492.

McCarty,L.S.,Henry,J.A.C.,and Houston,A.H. (1978): Toxicity of cadmium to goldfish, *Carassius auratus,* in hard and soft water. J.Fish. Res.Board Can.,35 (1):35-42.

Olson. K.R., Bergman, H.L and Fromm, P.O. (1973): Uptake of methyl mercuric chloride by Trout: A study of uptake pathways into the whole Animal and uptake by erythrocytes in vitro. J.Fish.Res.Board Can.,30,1293-1299.

Panigrahi, A.K. and Misra B.N. (1978); Toxicological effects of inorganic mercury on a fresh water fish, *Anabas scandens,* Cuv. & Val and its ecological implications. Environ.Pollut., 16,31-39. Panigrahi.A.K. and Misra.B.N. (1979): The effect of mercury on the morphology of erythrocytes in *Anabas scandens* Cuv. & Val. Bull.Environ. Contam. Toxicol., 23, 784-787.

Panigrahi. A.K. and Misra. B.N. (1980): Toxicological effects of inorganic mercury on a fresh water fish, *Tilapia mossambica,*Peters. Arch. Toxicol.,44,269-278.

Panigrahi, A.K. (1980): Toxicological effects of mercurial compounds on fresh water fishes and its ecological implications. Ph.D. thesis. Berhampur Universrty, Orissa.

Panigrahy, S.C.(1985): Toxcicological effects of Metacid-50 on a freshwater fish and its ecological implications Ph.D. thesis. Berhampur University, Orissa.

Rath, S. And Misra, B.N. (1979): Sub-lethal effects of dochlorovos on respiratory metabolism of *Tilapia mossambica* of 3 age groups. Exp. Geront., 14:37-41.

Rath, S. And Misra, B.N. (1980): Changes in nucleic acids and protein content of *Tilapia mossambica* exposed to dichlorovos. Indian J. Fish., 27 (1&2):76-81.

CHAPTER 9

Evaluation of Growth Patterns of Two Indian Major Carps - Rohu (Labeo Rohita) and Mrigal (Cirrhinus Mrigala) with Tilapia (Sarotherodon Mossambicus) Under Mixed Rearing Conditions

T. A. Sethuramalingam and P. Ramakrishnan

Centre for Aquafeed and Nutrition (CAFeN)
Research Department of Zoology
St. Xavier's College (Autonomous) Palayankottai - 627002. India.

ABSTRACT

The food utilization and growth pattern of rohu, mrigal with tilapia fingerlings under mixed rearing conditions were studied. When reared individually mrigal exhibited a reduction in growth but recovered with a marginal increase of weight when combined with rohu and tilapia. The growth rate of rohu decreased significantly ($p<0.05$) from 406.57. J/g fish/day in individual rearing to 132.39 J/g fish/day when reared with mrigal. But the growth rate increased significantly to 341.24 J/g fish/day when reared in combination with mrigal and tilapia. Rohu consumed 1981.47 J/g fish/day in individual rearing but exhibited a feeding rate of 1488.63 J/g fish/day in combination with tilapia. Despite of this feeding rate, rohu did not show reduction in growth rate because of a significant increase ($p<0.05$) in the food conversion efficiency rrom 16.52% in individual rearing to 24.14% in combination with tilapia. The interaction between mrigal and rohu positively influenced the energy budget of tilapia. Even though the feeding rate of tilapia decreased from 2793.91 J/g fish/day in individual rearing to 1807.22 J/g fish/day when reared with mrigal, the growth rate increased significantly from 388.97 J/g fish/day to 523.31 J/g fish/day. The increased growth rate even at low level of feeding is attributed to the decreased metabolic rate.

INTRODUCTION

In recent years aquaculture practice has been efficiently increased to enhance fish production by exploiting all the niches through composite and polyculture techniques. The pinnacle of composite fish culture is to reduced the competition among a cultured species to a minimum as they exploit all available area of the pond. Increase in growth of some fishes when reared with other species is often reported to have a great benefit for fish farmers (Jena et al., 1998). Enhancement in the number and biomass of brown trout *Salmo trutta* was influenced by the introduction of rainbow trout. Growth of catla and grass carp have shown enhanced growth in mixed culture than cultured individually. (Basavaraju et al.,1987). Williams et.al., (1987) have reported that the mean weight of channal cat fish, *Icta Lurus punctatus* with blue tilapia, *Oreocromis aurea* reared together was significently higher than that of channal cat fish in the cages. grown without blue tilapia. Raising Nile tilapia with pig or other fishes in integrated culture increases the fish biomass (Woynarovich., 1979; Schroeder and Hapher, 1979; Chiayvareesajja et.al., 1989; Nitithamyong et.al., 1991). Similar findings are also reported by Mork, (1982); Holm, (1989); Tripathi, (1990); Nortvedt and Holm, (1993); Elangoven and Lethi, (1994); Mohanty, (1995). The positive interaction between and among the fish species cultured has become one of the important features contributing to the success of poly-culture (Murty et.al.,1978; Sen and Chakraberty,1979). Moreover any alteration in food consumption and utilization in the species could alter the production of fishes. So it is important to study the cause of change in the growth of fish pattern during species. Interaction in mixed rearing conditions. Hence the present study was aimed to find out the growth pattern of carps (rohu and mrigal) under mixed rearing conditions along with tilapia. Tilapia was preferred in the study because of its proliferic breeding habit and also the combination implies a thorough understanding of the interaction of tilapia with carps in composite culture.

MATERIALS AND METHODS

Fingerlings of rohu and mrigal were procured from Fish Farmers Development Agency (FFDA), Manimuthar, Tirunelveli Dist. and transported to Centre for Aquafeed and Nutrition(CAFeN) laboratory, St.Xavier's College, Palayankottai and acclamatized in the round troughs (40 lit. capacity) for 15 days. Tilapia were collected from nearby ponds of Potal near Palayankottai. Fish of 2.4±0.27g were recruited from stock and divided into three series. In the first series, five fish were reared individually, in the second series combination of two fishes such as mrigal and rohu, rohu and tilapia and mrigal and tilapia were maintained. The third series contain all the three fishes (mrigal, rohu and tilapia). All the three series were maintained atleast with triplicate and were held in round plastic troughs (20 lit. capacity) interconnected with pipes supplied with aerated water with a flow rate of 4 lit/hr. The experimental fishes were fed *ad. libitum* with formulated pelleted diet which were specially prepared for carp with 40% crude protein level (Sethuramalingam, 2002). In the second series of experiments a wire mesh was placed inside each trough between the two types of fishes during the course of feeding to find out the individual food consumption. The collection of unfed and faeces were made prior to the feeding and were oven dried and packed. The experiments were run for 49 days which were carried out at temperature of 26± 1^{0}C; DO 6.2 0.23 ml O/lit and 12:12 LL:DD cycles. After the experimental period, fishes from each series were sacrificed (Maynard and Loosli, 1962) and wet and dry weight were taken. The energy content of the test fishes were also estimated using a semi microbomb calorimeter (Jauncey, 1982)

The scheme of energy budget was followed by IBP(Petrusewicz and MacFadyen, 1970)

C = P+R+U+F   C = Consumption   P = Growth   R = Metabolism,

F = Faeces   U = Urine

Absorption (A) was calculated by substracting F from C. (A = C - F);. P = Weight difference of animals at the begining and at the end of the experiments. Total metabolism was deducted as the difference between absorption and growth. (R = A - P). Urine is not considered in this work.

Rate of feeding (FR)/absorption (AR)/Conversion (CR) = Food consumed/ absorbed/converted (g)/wet weight of the fish(g) X time (days) Absorption efficiency(AE)% = Food absorbed (g)/food consumed(g) X 100 Gross conversion efficiency(GCE-K1%) = Growth(g)/food consumed(g) X 100 Net conversion efficiency (NCE-K2%) = Growth(g)/food absorbed(g) X 100 Absorption of individual fish in combination could not be calculated because the faeces collection was impossible in that series hence it was assumed that absorption efficiency of each fish in any combination was same as in the individual rearing.

One way ANOVA was performed to examine the level of significance among the various series and correlation coefficient was employed to find out the relationship between and among each series/ combinations (Snedecor and Cochran, 1967)

RESULT AND DISCUSSION

The food consumption and utilization of fishes have been greatly influenced by the interspecific interation of rohu, mrigal and tilapia. The bottom feeder mrigal consumed as low as 1219.91 J/g fish/ day when reared individually. However the increase in the body weight was evident significantly ($p<0.05$) amidst rohu and tilapia because of increase in food intake (Table 9.1). Elangoven and Lethi (1994) reported a weight gain of catla when combined with rohu or tilapia over monoculture of catla.

Upon individual rearing the feeding and growth rates of rohu declined significantly ($p<0.05$) from 1981.47 to 1387.86 J/g fish/ day and 406.57 to 132.39 J/g fish/day respectively than reared with mrigal (Table 9.1b). However in combination with tilapia and mrigal, the reduction in weight gain by this interaction between mrigal and rohu was counter balanced and the growth rate of rohu increased significantly ($p<0.05$) to 341.24 J/g fish/day. This report was well supported by Elizabeth(1986), in which Channa had a negative effect on rohu's growth; but when mrigal was introduced with rohu and Channa combinations, the negative growth effect on rohu was nullified. The same result was accepted by Beem and Gebhert, (1988) with a weight gain in *Ictalurus punctatus* when cultured with *Salmo graidneri* at different age groups. The reduction in the growth of rohu in the presence of mrigal may be attributed to the cumulative effect of reduced food consumption and increased metabolic rate. The interaction between rohu and tilapia was found to be positive and influenced the growth of rohu very well. Though the feeding rate of rohu declined during mixed rearing with tilapia combination, there was no decline in growth rate when reared individually. This may be due to lesser fraction of energy utilized by rohu for metabolism which in turn increased the growth rate and conversion efficiency.

Rearing of rohu mrigal with tilapia has a positive effect on the energy budget of tilapia. Eventhough the feeding rate of tilapia decreased drastically when reared with mrigal, the growth rate increased and the metabolic rate decreased. This was evident by the concomitant increase in conversion efficiency of tilapia as tllapia *(Oreochromis mossambicus)* was reared with Catla and rohu,

Table 9.1(a) **Energy budget of *Mrigal* reared individually and in combination with *Rohu* and *Tilapia*.**

Rearing conditions	Feeding rate (FR)	Absorption rate (AR)	Conversion rate (CR)	Metabolic rate (MR)	Absorption efficiency (AE) (%)	Gross conversion efficiency (GCE) (%)	Net conversion (NCE) (%)
Mrigal & combination							
Individual	1219.91± 81.28	1069.93± 24.66	121.73± 18.04	921.33± 18.62	85.34± 3.67	10.43± 1.83	11.43± 1.07
Mrigal + Rohu	1551.86± 33.68	1352.7± 18.03	220.9± 12.27	1142.9± 12.41	87.72± 2.91	14.28± 2.64	16.37± 1.34
Mrigal + Tilapia	1495.82± 53.22	903.3± 28.86	154.92 12.63	747.36± 32.18	61.64± 2.88	10.52± 2.07	17.05± 1.92
Mrigal + Rohu + Tilapa	-	-	169.88± 16.71	-	-	-	-

Table 9.1 (b)

Rearing conditions	Feeding rate (FR)	Absorption rate (AR)	Conversion rate (CR)	Metabolic rate (MR)	Absorption efficiency (AE) (%)	Gross conversion efficiency (GCE) (%)	Net conversion (NCE) (%)
Rohu & combination							
Individual	1981.47± 39.81	1653.31± 19.78	406.57± 11.73	1233.83± 31.54	82.84± 3.62	16.52± 2.16	24.76± 2.86
Mrigal + Rohu	1387.86± 17.61	1013.86± 40.93	132.39± 18.47	881.89± 26.47	73.04± 2.87	9.52± 1.37	14.97± 2.11
Mrigal + Tilapia	1488.63± 43.58	1234.55± 32.18	392.38± 19.63	846.49± 38.18	82.93± 3.66	24.14± 1.09	31.18± 1.99
Mrigal + Rohu + Tilapia	-	-	341.24± 18.47	-	-	-	-

Table 9.1 (c)

Rearing conditions	Feeding rate (FR)	Absorption rate (AR)	Conversion rate (CR)	Metabolic rate (MR)	Absorption efficiency (AE) (%)	Gross conversion efficiency (GCE) (%)	Net conversion (NCE) (%)
Rohu & combination							
Individual	2793.91± 29.36	2468.82± 18.33	488.03± 10.36	2079.03± 14.62	85.27± 2.11	17.46± 2.13	19.84± 1.49
Mrigal + Rohu	1807.22± 49.87	1461.39± 32.48	523.31± 40.43	938.64± 47.32	80.85± 2.27	27.96± 3.62	36.96± 3.87
Mrigal + Tilapia	1784.38± 18.93	1498.31± 45.36	508.47± 37.21	991.84± 37.28	83.97± 2.58	28.54± 3.29	33.96± 3.72
Mrigal + Rohu + Tilapia	-	-	511.83± 43.52	-	-	-	-

the growth rate of tilapia increased significantly under mixed rearing condition(Elangoven and Lethi, 1994). The same trend was noticeable when tilapia was stocked with *Cyprinus carpio*. the growth of fishes were higher in combined rearing rather than in monoculture (Bardach et. al., 1972). When catla was grown with *Ctenopharyngodon idella* an increase in the weight of catla was noticed compared to its monoculture (Basavaraju et. al., 1987). Stocking tilapia with predatory fishes decreased tilapia production only due to reduced number of recruits but increased the average size/mass of tilapia (Manzano,1990). Under combined culture improved growth performance was noticed in combination of Indian major carps and exotic carps than individual rearing (Jena et. al., 1998)

Under mixed rearing conditions tilapia cultured with rohu and mrigal showed an elevated mean growth rate but these increase was not highly significant ($p>0.05$) over rohu+ tilapia or mrigal+ tilapia combinations. A similar trend was also reported by Williams et. al.,(1987) in channal cat fish and tilapia combination and Elangoven and Lethi, (1994) in catla+ tilapia combinations.

The present study indicated that in combined rearing conditions one or the main reasons for the better growth even at low feeding level of fishes was due to the reduction in metabolic rate. Tilapia could be included in culture pond along with other carps provided its density and size should be considered and maintained in such a way that it does not exploit the culture pond. Monosex of tilapia with preferred density could also be practiced with carps which could nullify the proliferation of tilapia. Therefore inclusion of tilapia with rohu and mrigal under mixed rearing condition is economically viable.

REFERENCES

Bardach, H.E., Ryther, J.H., and William, O.M., (1972) Aquaculture The Farming and Husbandry of Fresh Water and Marine Organisms John Wiley and Sons, New York, pp 364 - 369.

Basavaraju, Y., Devaraj, K.V., and Keshave (1987). Culture of grass carp and catla in combination. Paper presented at the First Indian Fisheries Forum. Mangalore. Dec.4-8, 1987.

Beem, M.D., and Gebhart, G.E., (1988). Winter polyculture of channel cat fish and rainbow trout in cages. Prog. Fish.Cult.,50.pp 49-51.

Chiayvareesajja, S.C., Wongwit, A.,Cronin, K., Supamataya, C., Tantikitu and Tansakul, R.,(1989). Utilization of aquatic weed mixture pellet as feed for nile tilapia *(Oreochromis niloticus)* and pig pp 143 - 147. In S. De Silva (ed.) Fish Nutrition Research in Asia, Asian Fisheries Society, Manila, Philippines,

Elangoven, A., and Lethi., C.D., (1994) Growth rate comparisons of two Indian major carps - catla (Catla catla) and rohu *(Labeo rohita)* with tilapia *(Oreochromis mossambicus)* J.Aqua. Trop. 9 pp 241 - 246.

Elizabeth, A.R.,(1986). Physiological studies on the composite culture of carp and murrel. M.Phil dissertation, Bharathidasan University, Tiruchirapalli. India.

Holm, J.C.,(1985) Mono and duoculture of Juvenile Atlantic salmon (Salmo salar) and Arctic charr *(Salvelinus alphinus)* Can. J.Fish. Aquat. Sci. 46: pp 697 - 704.

Jauncey, K., (1982 b) Carp *(Cyprinus carpio)* nutrition. A review in J.F.Muir and R.J. Roberts (eds.) Recent advances in Aquaculture. Groom Helen. London, pp 215 - 263.

Jena, J.K., Aravindakshan, P.K., Suresh Chandra, Muduli, H.K., Ayyappan, S., (1998). Comparative Evaluation of Growth and survival of Indian major carps and exotic carps in raising fingerlings. J.Aqua. Trop. 13(2). PP. 143 - 150.

Maynard,A.L., and Loosli, J.K., (1962). Animal Nutrition. McGraw Hill, New York. pp. 563.

Manzano,V.B.,(1990). Poly culture systems using grouper *(Epinephaelus tauvina)* and Tilapia *(Oreochromis mossambicus)* in brackishwater ponds pp. 213 216. In R. Hirano and I. Hanyu (eds.). The Second Fisheries Forum. Asian Fisheries Society, Manila, Philippines.

Mohanty, U.K., (1995);. Comparative evaluation of growth and survival of Indian major carp fry in aerated vis-a-vis non-aerated ponds under different stocking densities. M.F.Sc. thesis. Orissa University of Agriculture and Technology. Bhubaneswar. India.pp.123,

Mork,O.I., (1982),. Growth of three salmon id species in mono and double culture *(Salmo salar. S.trutta* and *S.gairdneri)* Aquaculture, 27 *t* pp. 141 - 147.

Murty, D.S., Dey, R.K., and Reddy, P.V.G.K., (1978). Experiments on rearing exotic carp fingerlings in composite fish culture in India. Aquaculture, 13(4), pp. 331 - 337.

Nitithamyong,C., Chiayvareesajja, J.,Chiayvareesajja, S., Wongwit, C., and Tansakul, R., (1991). Production of Nile Tilapia (Oreochromis niloticus) in different culture and Harvesting systems. In Fish Nutrition Research in Asia (Ed.) S.S.De Silva. Proceedings of IVth Asian Fish Nutrition Workshop, pp. 169 - 174.

Nortvedt, R., and Holm, J.C., (1991). Atlantic salmon in duoculture with Arctic charr. Decreased aggression enhances growth and stocking density potential. Aquaculture., 98: pp.355 - 361.

Petrusewicz, K., and MacFadyen, A., (1970). Productivity of Terrestrial Animals. Oxford Blackwell Scientific Publication (IBP Handbook

Schroeder, G., and Hepher, B., (1979). Use of Agricultural and Urban wastes in fish culture, pp. 487 - 489. In TVR Pillay and W.M.Dill (eds.) Advances in aquaeulture. Fishing News Book. Surrey, England.

Sen, P.R., and Chakraborty, R.D., (1979). Rearing of major carp fry to fingerlings in fresnwater ponds. Proc. Sem. Inland Aqua. CIFRI, Barrackpore. pp.. 43.

Sethuramalingam, T.A., and Haniffa, M.A., (2002). Effect of formulated diets on digestive enzyme of *Labeo rohita* (Ham.). Indian Journal of Experimental Biology. 40 : pp. 83 - 88.

Snedecor, G.W., and Cochran,W.G., (1967). Statistical method. Oxford and IBH Publishing Co. New Delhi, pp. 135.

Tripathi, S.D., (1990). Freshwater aquaculture in India. In : Aquaculture in Asia. M.Joseph. ed., pp. 191 - 222. Asian Fisneries Society, Indian Branch. Mangalore. India.

Williams, K., Gebhart,G.F., and Manghan,O.E., (1987). Enhanced growth of cage cultured channel cat fish through polyculture with blue tilapia. Aquaculture, 62: 207 - 214.

Woynarovich, E., (1979). The feasibility of combining animal husbandry with fish farming with special reference to duck and pig production. pp. 203 - 208. In TVR Pillay and W.M.Dill (eds.) Advances in aquaculture Fishing News Book, Surrey, England.

CHAPTER 10

EFFECTS OF PLANTS AND ANIMAL DIET ON FOOD UTILIZATION AND ENERGY BUDGET OF THE FRESH WATER FINGERLINGS *CYPRINUS CARPIO VAR. COMMUNIS* (LINNAEUS) AND *CATLA CATLA* (HAMILTON)

Hongray Howreliaj, J.*, A.T. Fleming, Bharat Bhusan Patnaik and M. Selvanayagam***

**Present Address: Environmental Sciences And Biotechnology Unit, P.G. and Research Department of Zoology, Loyola College, Chennai-600034, Tamilnadu, India*

ABSTRACT

Good quality, right quantity and adequate calorific value of diet are primary factors that determine the growth of fishes and cost of production in aquaculture. The Energy budget and food utilization on the fingerlings of *Cyprinus carpio* and *Catla catla* were conducted for a period of 30 days. The experimental troughs were supplied with 100% Plant diet (I series), viz., Beans, Cucumber, Radish and Carrot and 100% Animal diet (II series), viz., Egg albumin, Goat liver, Chicken and Mutton each 2gm. Dry weight/trough everyday. The feed conversion rates (FCR) were the highest for Cucumber and Goat liver in both *Cyprinus carpio* and *Catla catla* respectively. The absorption efficiency, were the highest for Cucumber (71.89%) and Goat liver (80.40%) in *Cyprinus carpio* and (74.97%) and (77.41%) for Cucumber and Chicken in *Catla catla.* The conversion efficiency of *Catla catla* is more or less same as in the case of Plant and Animal diet. In Plant food the conversion efficiency ranged from 4.5% to 8.23% and in Animal diet it was 5.6% to 8%. In general the conversion efficiency of *Cyprinus carpio communis* and *Catla catla* are very low due to high metabolic demand.

***Key words:** Energy budget, Food efficiency, Fish feed.*

**Corresponding author. +91-44-3721784/3721785. E-mail address. gowanna44(a)sifv.com

INTRODUCTION

In aquaculture, the feed forms the most expensive input and accounts to 57-87% of the total recurring expenditure (Nandeesha, 1993). Fish meal is used as the major source of Animal protein in the commercial fish and crustacean feed. However, due to escalating cost and uncertain availability of reliable quality of fish meal, it is essential that alternative sources of protein be identified for its partial or complete replacement (Gallagher, 1994). With the increasing research thrust on this aspect during recent years there has been much interest on the possibility of using various non-conventional feed sources. The inclusion of Animal and Plant protein sources with high protein contents in fish feed has been increased substantially due to its lower price and wider availability. The quality of a particular protein source not only depends on its digestibility but also on the aminoacid profile (Law et al., 1990; Kaushik and Cowey, 1991). When Animal or Plant protein sources are used endogenous antinutritional factors are the main constituents limiting the use of Plant protein feed stuff in fish feed (Satoh et al., 1989). But by proper treatment many of the factors such as protein inhibitors from ingredients could be inactivated by moist heat treatment or mild chemical treatment (Grant, 1989) and used as fish feed.

When a fish consumes food a protein of it is digested and absorbed and the rest is ejected out as faeces. The absorbed energy is channelized towards various physiological processes such as maintenance, specific dynamic action (SDA) and growth. Growth performances are done in fin and shell fishes by using squid mantle meal, shrimp head meal, (Akiyama, 1992), legumes and other oil cakes and leaf meals (Hughes, 1991; Saha and Ray, 1993) as alternative source of fish meal in fish feed. Soybean meal has been used as a protein source for fish at different growing stages because of their protein content and excellent aminoacid make-up (Reigh and Ellis, 1992; El Sayed, 1994). Experiments with the incorporation of materials of animal origin like Silkworm pupae (Nandeesha et al., 1988; Hossain et al., 1992), earth worm meal (Keshavappa et al., 1989) and slaughter house waste (Khan and Jafri, 1992; El Sayed, 1994) as substitutes for fish meal have also been encouraging. Curtailing the cost of fish feed through use of these alternative and less conventional raw materials would be ideal for producing farm based aqua feeds, thereby leading to economic success of the culture operation. Dietary nutrient requirements in a number of fishes have been studied by Chakrabarty et al. (1973), Bryant and Matty (1981), Viola et al. (1982), Charles et al. (1984), Silva and Gunasekhara (1991), Mazid et al. (1997), Srivastava et al. (1997), Hasan et al. (1997), Sarkar (1997), Chakrabarti and Jana (1998) and Erfanullah (1998).

The present experiment involved an attempt to understand the influence of different types of Plant and Animal feeds on the Food conversion efficiency and biomass production rate of two species of carps, *Cyprinus carpio* and *Catla catla* fingerlings. The rate of Consumption, Absorption, Conversion and Metabolism are also calculated along with Absorption and Conversion efficiencies. This facilitates the estimation of growth of fishes and enables to find out the most economical one for culturing on a large scale. The fishes being in the stressed conditions of the laboratory the mortality rates give a clue to the sensitivity of the fishes to the various physical and chemical parameters of the immediate surroundings. This is in compliance with the suggestions of Jhingran, 1977 as expressed in his book, "Fish and Fisheries".

Test fishes

The fishes selected for the present study were the exotic carp, *Cyprinus carpio communis* and the major carp *Catla catla*. These fishes are ideal for culture and are excellent food fishes. They are also known for their fast growth and meat production. These are also comparatively hardy fishes. These are capable of tolerating low oxygen and high carbon-dioxide content of water (Huet, 1970).

MATERIALS AND METHODS

Fish procurement

Fingerlings of *Cyprinus carpio communis* and *Catla catla* were collected from Poondi reservoir, Poondi, Thiruvallur district. They were transported to the laboratory in polythene bags containing oxygenated water, and were acclimatized in laboratory conditions for 15 days by keeping them in rectangular glass aquarium.

Fish feed

For the present stuuy two kinds of feed were selected i.e., 100% Plant and 100% Animal food. The Plant food selected were Beans, Radish, Cucumber and Carrot. Egg albumin, Chicken, Goat liver and Mutton were chosen as the Animal food items.

Experimental Design

Ten fingerlings of the fishes were selected at random and weighed to estimate the initial dry weight of the test fishes (Maynard and Loosely, 1962), followed by sacrifice method. The first series of troughs were supplied with 100% Plant food viz., Beans, Radish, Cucumber and Carrot 2gm/trough everyday, in the second series fishes were supplied with 100% Animal food viz., Egg albumin, Chicken, Goat liver and Mutton 2gm/trough respectively.

The unconsumed food and faeces egested by the fish were removed immediately before each feeding, dried and weighed. The experiment was continued for 30 days.

Bioenergetics scheme

The food consumed (C), egestion (F) and growth or conversion (P) were estimated following the gravimetric method of Waldbaeur, 1968. The scheme of energy budget followed here is slightly modified IBP formula (Petruse witz and McFodyen, 1970), represented as:

$$C=P+R+F$$

Absorption was estimated as the difference between the food consumed and that of faeces:

$$A=C-F$$

Metabolism was calculated as the difference between the absorption energy and that of growth:

$$M=A-P$$

RESULTS

Overall Energy budget

Table-10.1 reported the energy budget *of Cyprinus carpio communis* fed on four different types of Plant diets viz., Beans, Radish, Cucumber and Carrot for a period of 30 days. Among the Plant feed, the test fish had maximum consumption (7.5±0.09gm. dry weight) when fed on Cucumber. Food consumption was minimum when fishes were fed on Radish (6±0.011gm. dry weight). Amount of food absorbed by the test fish fed on Beans, Radish, Cucumber and Carrot were (4.65±0.023, 4.5±0.024,

5.4±0.78 and 5.1±0.05 gram dry weight) respectively. The amount of food converted was the highest for fishes fed with Cucumber (0.697±0.002gm. dry weight) and corresponding values for Beans, Radish and Carrot fed by fishes were (0.205±0.12, 0.183±0.12 and 0.207±0.004gm. dry weight) respectively. Data obtained for the amount of food metabolized were (3.857±0.067, 4.297±0.057, 5.247±0.006 and 4.893±0.77gm. dry weight) respectively for the feed Beans, Radish, Cucumber and Carrot.

CYPRINUS CARPIO COMMUNICES

Table 10.1

Overall Energy budget of *Cyprinus carpio communis* fed on different types of Plant diets for a period of 30 days. Each value represents the average performance of two groups. All values are expressed in gram dry weight.

Sl. No.	Food Provided	Food Consumed	Food Absorbed	Food Converted	Food Metabolized
1.	Beans	6.45±0.016	4.65±0.023	0.205±0.12	3.857±0.067
2.	Radish	6±0.011	4.5±0.024	0.183±0.12	4.297±0.057
3.	Cucumber	7.5±0.09	5.4±0.78	0.697±0.002	5.247±0.006
4.	Carrot	6.9±0.034	5.4±0.78	0.207±0.004	4.893±0.77

± Values indicate the Standard deviation.

Table-10.2 represented the amount of Feeding rate, Absorption rate, Conversion rate and Metabolic rate of *Cyprinus carpio communis*. The feeding rate of the test fishes fed on Beans, Radish, Cucumber and Carrot were (2.621±0.22, 2.4±0.67, 3.06±0.045 and 2.85±0.034mg. dry weight/ gm. live fish/ day) respectively. For the same order absorption rate were (1.890±0.012, 1.8±1.45, 2.20±0.74 and 2.10±0.93mg. dry weight/ gm. live fish/ day) respectively. The highest food conversion rate was in Cucumber (0.284±0.001 mg. dry weight/ gm. live fish/ day). The lowest was in Radish (0.0732±0.045mg. dry weight/ gm. live fish/ day). The intermediate was (0.083±0.001 and 0.085±0.005mg. dry weight/ gm. live fish/ day) for Beans and Carrot respectively. The values for metabolic rate were (1.560±0.03, 1.716±0.56, 2.130±1.02 and 2.021±0.81mg. dry weight/ gm. live fish/ day) for Beans, Radish, Cucumber and Carrot respectively.

Table 10.2

Energy budget of *Cyprinus carpio communis* fed on different types of Plant food for a period of 30 days. All values are expressed as mg dry weight/gm live fish/day.

Sl. No.	Food Provided	Feeding Rate	Absorption Rate	Conversion Rate	Metabolic Rate
1.	Beans	2.621±0.022	1.890±0.012	0.083±0.001	1.560±0.03
2.	Radish	2.4±0.67	1.8±1.45	0.073±0.045	1.716±0.56
3.	Cucumber	3.06±0.045	2.20±0.74	0.284±0.001	2.130±1.02
4.	Carrot	2.85±0.034	2.10±0.93	0.085±0.005	2.021±0.89

± Values indicate the Standard deviation.

Table-10.3 shows the absorption efficiency and conversion efficiency of *Cyprinus carpio communis* when provided with Plant diets. The absorption efficiencies ranged between 71.89±3.45% and 75±0.34%. Conversion efficiencies ranged between 3±0.23% and 9.28±1.78%.

Table 10.3

Absorption and Conversion efficiency of *Cyprinus carpio communis* fed on different types of Plant diets. Values are expressed in Percentage.

Sl. No.	Food Provided	Absorpiton efficiency	Conversion efficiency
1.	Beans	72.10±0.062	3.2±0.005
2.	Radish	75±0.34	3.05±0.67
3.	Cucumber	71.89±3.45	2.20±0.74
4.	Carrot	73.68±0.44	3±0.23

± Values indicate the Standard deviation.

Table-10.4 showed that among the four different types of Animal feed namely Egg albumin, Chicken, Goat Liver and Mutton supplied to the *Cyprinus carpio communis*, the maximum food consumption was noticed for the fishes fed on Goat Liver (7.65±0.009gm. dry weight). The lowest were for those fed on Egg albumin (6.3±0.15gm. dry weight). The data obtained for food absorption were (4.5±1.02, 6±2.8, 6.15±1.09 and 4.8±2.79gm. dry weight) respectively for the fishes fed on Egg albumin, Chicken, Goat Liver and Mutton. The amount of food converted was the highest for fishes fed with Goat Liver (0.517±0.045gm. dry weight) and corresponding values for Egg albumin, Chicken and Mutton were (0.223±1.2, 0.482±0.009 and 0.447±0.24gm. dry weight) respectively. Data obtained for food metabolized were (4.277±0.001, 5.518±1.22, 5.633±1.5 and 4.353±0.885gm. dry weight) respectively for feeds Egg albumin, Chicken, Goat Liver and Mutton.

Table 10.4

Overall Energy budget of *Cyprinus carpio communis* fed on different types of animal diets for a period of 30 days. Each value represents the average performance of 2 groups. All values are expressed in gram dry weight.

Sl. No.	Food Provided	Food Consumed	Food Absorbed	Food Converted	Food Metabolized
1.	Egg albumin	6.3±0.15	4.5±1.02	0.223±1.2	4.277±0.001
2.	Chicken	7.5±0.1	6±2.8	0.482±0.009	5.518±1.22
3.	Goat Liver	7.65±0.009	6.15±1.09	0.517±0.045	5.633±1.5
4.	Mutton	6.6±0.506	4.8±2.79	0.447±0.24	4.353±0.88

± Values indicate Standard deviation.

Table-10.5 shows that among the animal feed, Egg albumin, Chicken, Goat Liver and Mutton, *Cyprinus carpio communis* showed the highest rate of feeding for Chicken (3±0.35mg. dry weight/ gm. live fish/ day). It was minimum for Mutton (2.17±0.03 lmg. dry weight/ gm. live fish/ day). Absorption rate of the fishes fed on Egg albumin, Chicken, Goat Liver and Mutton were (1.931±0.003, 2.4±0.12, 2.38±0.34 and 1.97±0.31mg. dry weight/ gm. live fish/ day) respectively. The conversion rates for the same order were (0.095±0.002, 0.192±0.18, 0.2±0.023 and 0.18±0.036mg. dry weight/ gm. live fish/ day) respectively.

Table 10.5

Energy budget of *Cyprinus carpio communis* fed on different types of Plant food for a period of 30 days. All values are expressed as mg dry weight/gm live fish/day.

Sl. No.	Food Provided	Feeding Rate	Absorption Rate	Conversion Rate	Metabolic Rate
1.	Egg albumin	2.70±0.74	1.931±0.003	0.095±0.002	1.80±0.009
2.	Chicken	3±0.35	2.4±0.12	0.192±0.18	2.19±0.34
3.	Goat Liver	2.96±0.89	2.38±0.34	0.2±0.023	2.17±0.34
4.	Mutton	2.17±0.031	1.97±0.31	0.18±0.036	1.79±0.43

± Values indicate Standard deviation.

Table-10.6 shows the maximum absorption efficiency of *Cyprinus carpio communis* in case of Goat Liver (80.40±0.97%) and minimum in Egg albumin (71.51±4.51%). The maximum conversion efficiency was found in Goat Liver (6.7±0.55%) and minimum in Egg albumin (3.51±0.89%).

Table 10.6

Absorption and Conversion effciency of *Cyprinus carpio communis* fed on different types of animal diets. Values are expressed in Percentage.

Sl. No.	Food Provided	Absorpiton efficiency	Conversion efficiency
1.	Egg albumin	71.51±4.51	3.51±0.89
2.	Chicken	80±3.23	6.4±0.02
3.	Goat Liver	80.40±0.97	6.7±0.55
4.	Mutton	72.69±1.25	6.64±1.23

± Values indicate Standard deviation.

Table-10.7 reported the Energy budget of *Catla catla* fed on four different types of Plant food viz., Beans, Radish, Cucumber and Carrot for a period of 30 days. Among the Plant feed the fish had maximum consumption (11.7±3.89gm. dry weight) when fed on Cucumber. Food consumption was minimum when fishes were fed on Radish (7.2±2.67gm. dry weight). The amount of food converted was the highest for fishes fed on Cucumber (0.965±0.001gm. dry weight) and the intermediate for those fed on Carrot (0.475±0.001gm. dry weight) and the lowest for those fed on Radish (0.451±0.67gm. dry weight).

CATLA CATLA

Table 10.7

Overall Energy budget of *Catla catla* fed on different types of Plant diets for a period of 30 days. Each value represents the average performance of 2 groups. All values are expressed in gram dry weight.

Sl. No.	Food Provided	Food Consumed	Food Absorbed	Food Converted	Food Metabolized
1.	Beans	8.85±0.34	6.45±1.004	0.473±0.056	5.977±1.56
2.	Radish	7.2±2.67	5.1±1.007	0.451±0.67	4.649±0.67
3.	Cucumber	11.7±3.89	8.7±3.67	0.965±0.001	7.735±0.005
4.	Carrot	10.5±1.09	7.8±0.89	0.475±0.001	7.325±0.56

± Values indicate Standard deviation.

Table-10.8 explains that among the Plant food the maximum feeding, absorption, conversion and metabolic rates (4.775±0.56, 3.58±0.74, 0.396±0.07 and 3.182±0.94 mg. dry weight/ gm. live fish/ day) respectively was when fed with Cucumber. The minimum feeding, absorption, conversion and metabolic rates (2.88±0.078, 2.04±0.21, 0.181±0.001 and 1.858±0.56mg. dry weight/ gm.live fish/day) respectively were when fed on Radish.

Table 10.8

Energy budget of *Catla catla* fed on different types of Plant diets for a period of 30 days. All values are expresed as mg dry weight/gm live fish/day.

Sl. No.	Food Provided	Feeding rate	Absorption Rate	Conversion Rate	Metabolic Rate
1.	Beans	3.597±0.89	2.621±0.006	0.191±0.006	2.429±0.002
2.	Radish	2.88±0.078	2.04±0.21	0.181±0.001	1.858±0.56
3.	Cucumber	4.755±0.56	3.58±0.74	0.396±0.07	3.182±0.94
4.	Carrot	4.338±1.56	3.22±0.87	0.195±0.04	3.026±2.67

± Values indicate Standard deviation.

Table-10.9 shows that the absorption efficiency in *Catla catla* was maximum in Cucumber (74.97±0.13%) and it was minimum in Radish (70.83±0.78%). Conversion efficiencies were (5.32±0.59%, 6.29±0.04%, 8.29±0.07% and 4.51±0.24%) when fed with Beans, Radish, Cucumber and Carrot respectively.

Table 10.9

Absorption and Conversion efficiency of *Catla catla* fed on different types of Plant diets. Values are expressed in percentage.

Sl. No.	Food Provided	Absorpiton efficiency	Conversion efficiency
1.	Beans	72.86±3.29	5.32±0.59
2.	Radish	70.83±0.78	6.29±0.04
3.	Cucumber	74.97±0.13	8.29±0.07
4.	Carrot	74.22±0.99	4.51±0.24

± Values indicate the Standard deviation.

Table-10.10 shows that the maximum food consumed, absorbed, converted and metabolized (11.55±0.073, 8.55±3.89, 0.785±0.55 and 7.765±2.54gm dry weight) in *Catla catla* fed on Goat Liver. The minimum food consumed, absorbed, converted and metabolized (8.7±2.9, 6.3±0.61, 0.491±0.11 and 5.809±4.59gm dry weight) respectively in *Catla catla* fed on Egg albumin.

Table 10.10

Overall Energy budget of *Catla catla* fed on different types of Animal diets for a period of 30 days. Each value represents the average performance of 2 groups. All values are expressed in gram dry weight.

Sl. No.	Food Provided	Food Consumed	Food Absorbed	Food Converted	Food Metabolized
1.	Egg albumin	8.7±2.9	6.3±0.61	0.491±0.11	5.809±4.59
2.	Chicken	9.3±6.67	7.2±2.39	0.75±0.02	6.45±2.11
3.	Goat Liver	11.55±0.073	8.55±3.89	0.785±0.55	7.765±2.54
4.	Mutton	10.2±1.99	7.6±0.98	0.715±0.01	6.885±0.66

± Values indicate Standard deviation.

Table-10.11 explains the rate of feeding, absorption, conversion and metabolism of *Catla catla* fed on Animal food. The maximum feeding, absorption, conversion and metabolic rates (4.476±1.72, 3.313±0.34, 0.303±0.12 and 3.009±0.11mg. dry weight/ gm. live fish/ day) were when fed on Goat Liver. The minimum feeding rate was witnessed in the *Catla catla* fed on Chicken (3.72±1.11 mg. dry weight/ gm. live fish/ day).

Table 10.11

Energy budget of *Catla catla* fed on different types of animal diets for a period of 30 days. All values are expressed as mg dry weight/gram live fish/day.

Sl. No.	Food Provided	Feeding Rate	Absorption Rate	Conversion Rate	Metabolic Rate
1.	Egg albumin	3.733±0.67	2.703±0.45	0.209±0.54	2.492±0.44
2.	Chicken	3.72±1.11	2.88±1.21	0.3±0.32	2.58±0.14
3.	Goat Liver	4.476±1.72	3.313±0.34	0.303±0.12	3.009±0.11
4.	Mutton	4.197±0.99	3.123±0.92	0.293±0.12	2.833±0.71

± Values indicate Standard deviation.

Table-10.12 shows that the absorption efficiency in *Catla catla* fed on Animal diets was maximum in Chicken (77.41±5.6%) and minimum in Egg albumin (72.40±6.2%). The conversion efficiency was maximum in Chicken (8.06±1.77%) and minimum in Egg albumin (5.62±3.01%) fed diets.

Table 10.12

Absorption and Conversion efficiency of *Catla catla* fed on different types of Animal diets. Values are expressed in percentage.

Sl. No.	Food Provided	Absorpiton efficiency	Conversion efficiency
1.	Egg albumin	72.40±6.2	5.62±3.01
2.	Chicken	77.41±5.6	8.06±1.77
3.	Goat Liver	74.01±5.2	6.77±2.1
4.	Mutton	74.41±2.9	7.00±2.4

± Values indicate Standard deviation.

Figure-10.1 and Figure-10.2 depicts the absorption efficiency of *Cyprinus carpio communis* & *Catla catla* fed on different types of Plant diets and Animal diets respectively. In case of Plant diets, the absorption efficiency *of Cyprinus carpio communis* is maximum in case of Radish but it is minimum in *Catla catla* when fed with Radish. *Cyprinus carpio communis* has a better absorption efficiency in case of Animal diets (Chicken and Goat Liver) compared to *Catla catla.*

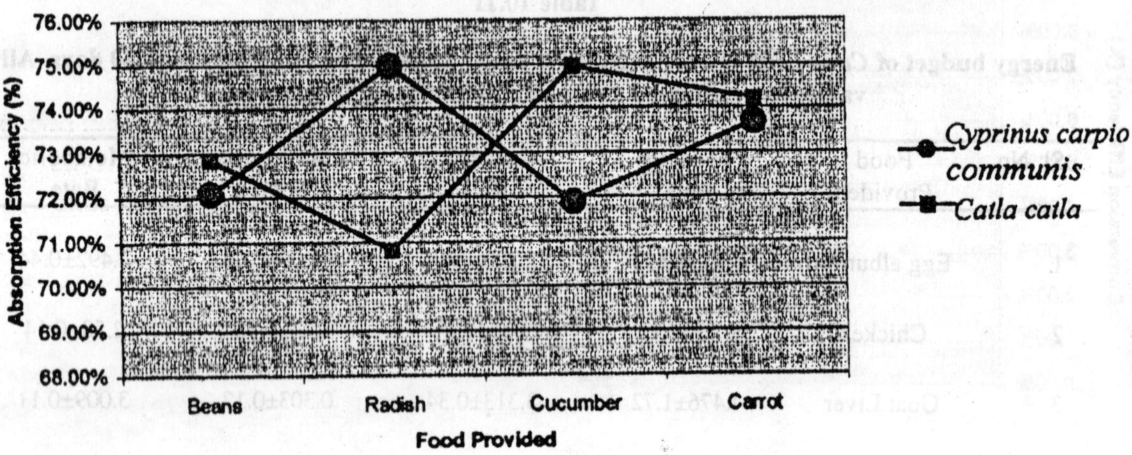

Fig. 10.1

Absorption Efficiency of *Cyprinus carpion communis* & *Catla catla* fed on different types of plant diets. Values are expressed in percentage.

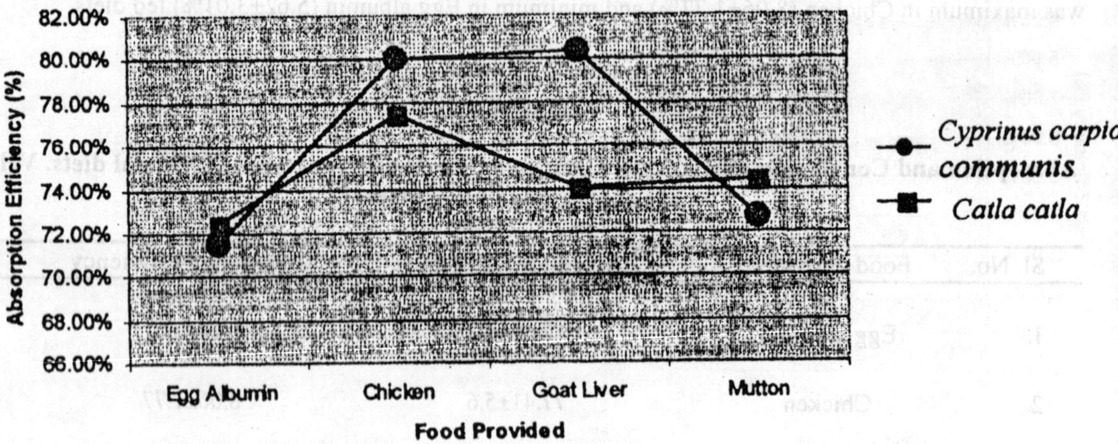

Fig. 10.2

Absorpiton Efficiency of *Cyprinus carpio communis* & *Catla catla* fed on different types of Animal diets. Values are expressed in percentage.

Figures 10.3 & 10.4 depict the conversion efficiency of the two fishes fed with Plant and Animal diets respectively. Conversion efficiency is the highest when fed with Cucumber in both the test fishes. Conversion efficiency in case of Animal diets doesn't show much variation in both the test fishes, though *Catla catla* has a better conversion ratio.

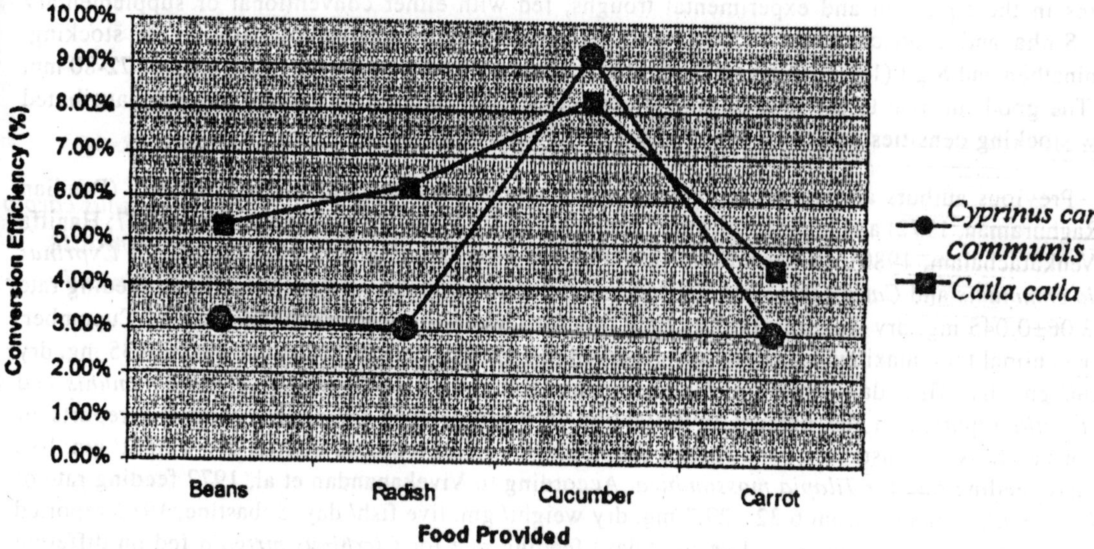

Fig. 10.3

Conversion Efficiency of *Cyprinus carpio communis* & *Catla catla* fed on different types of Plant diets. Values are expressed in percentage.

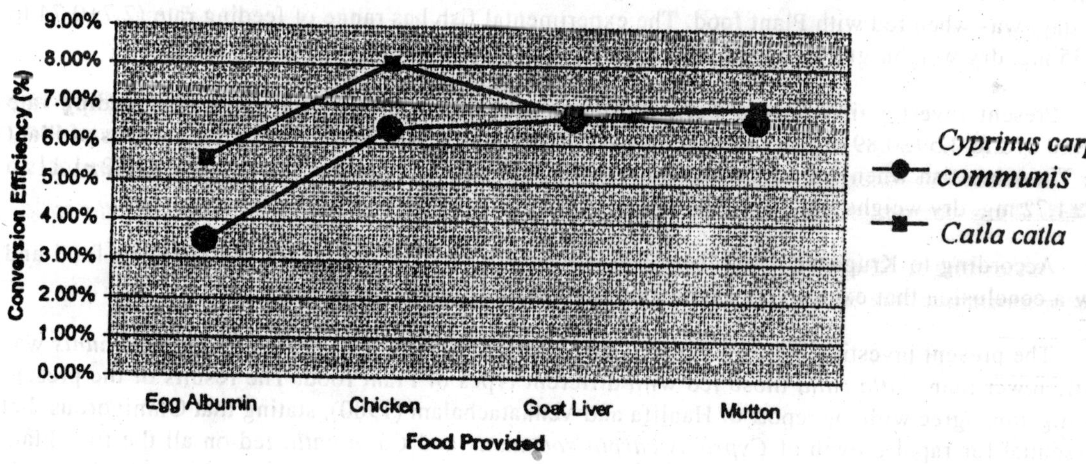

Fig. 10.4

Conversion Efficiency of *Cyprinus carpio communis* & *Catla catla* fed on different types of Animal diets. Values are expressed in percentage.

DISCUSSION

The results showed that the survival of *Cyprinus carpio communis* larvae was excellent (86-100%) during culture in the aquarium and in the experimental troughs, fed with either conventional or supplementary diets. The survival of *Catla catla* showed cent percent (100%) survival during both cultures in the aquarium and experimental troughs, fed with either conventional or supplementary diets. Sinha and Ramachandran (1985) recorded upto 87% of survival of carp fry at stocking. Swaminathan and Sigit (1982) obtained survival as high as 60% from spawn to fingerlings of 72-80 mm size. The good survival to the tune of almost cent percent obtained in this study is mostly attributed to low stocking densities and complete eradication of other materials.

Previous authors also reported variations in Energy budget of fishes due to quality (Pandian and Raghuraman, 1972) and quantity of food (Mathavan et al., 1976; Vivèkanandan et al., 1977; Haniffa and Venkatachalam, 1980). In the present investigation also variation in Energy budget of *Cyprinus carpio communis* and *Catla catla* were noticed as a function of quality of food. Maximum feeding rate was 3.06±0.045 mg. dry weight/ gm. live fish/ day for *Cyprinus carpio communis* fed on Cucumber. Among animal food maximum feeding rate observed in *Cyprinus carpio communis* was 3±0.35 mg. dry weight/ gm. live fish/ day in case of Chicken. The feeding rate of *Cyprinus carpio communis* and *Catla catla* reported in the present investigation is more or less compatible with the reports of previous authors, for instance (Pandian and Raghuraman, 1972) reported 28 mg. dry weight/ gm. live fish/ day feeding rate for *Tilapia mossambica.* According to Vivekanandan et al, 1977 feeding rate of *Anabas scandens* ranged from 6.32 - 29.3 mg. dry weight/ gm. live fish/ day. Sebastine, 1982-reported (43.4 to 57.4 mg. dry weight/ gm. live fish/ day) feeding rate for *Cirrhinus mrigala* fed on different types of molluscan flesh.

Fleming et al., 1983 *reported* (14.9 to 19.4 mg. dry weight/ gm. live fish/ day) feeding rate for *Labeo rohita* fed on different kinds of Plant, Animal and Mixed diets. Haniffa and Venkatachalam, 1980 suggested Omnivorous diet for a rapid growth rate in fishes. The present experimental fish *Cyprinus carpio communis* is an Omnivore as the feeding rate (2.4±0.67 to 3.06±0.045 mg. dry weight/ gm. live fish/ day) was when fed with Plant food. The experimental fish has range of feeding rate (2.7±0.74 to 310.35 mg. dry weight/ gm. live fish/ day) when fed on Animal diet.

Present investigation proved *Catla catla* fed with Plant and Animal diets, the feeding rate (2.88±0.078 to 3.59±0.89 mg. dry weight/ gm. live fish/ day) was when fed on different types of Plant diet. The same fish when fed with different types of Animal diets had the feeding rate (3.72±1.11 to 4.47±1.72 mg. dry weight/ gm. live fish/ day).

According to Krupauer, 1967 there was no growth rate for grass carp fed on Plant food and drew a conclusion that exclusively Plant food must lead to death of the fish.

The present investigation also stressed that conversion rate of *Cyprinus carpio communis* was a little lower than *Catla catla* those fed with different types of Plant food. The results of the present investigation agree with the report of Haniffa and Venkatachalam (1980), stating that Omnivorous diet is essential for rapid growth of *Cyprinus carpio communis* and *Catla catla* fed on all the four Plant diet and Animal diet.

According to Haniffa and Venkatachalam (1980) the absorption efficiency of *Ctenopharyngodon idella* was more in Plant diet (84%) when compared with Animal diet (75%). The present investigation proved that the absorption efficiency for *Cyprinus carpio communis* ranged from 71.89% to 75% in case of Plant diet and 71.51% to 80.4% in case of Animal diet. So therefore, in the present investigation *Cyprinus carpio communis* showed less absorption efficiency in Plant diet than that of the Animal diet.

Rosamma (1996) reported *Hypothalmichthys molitrix* showed maximum food consumption (7.44 gm. dry weight) and maximum food metabolized (6.04 gm. dry weight) for that fed on *Rana hexadactyla.* In similar way in *Catla catla,* maximum food consumption (7.55 gm. dry weight) and maximum food metabolized (6.21gm. dry weight), for those fed on *Rana hexadactyla* (thigh muscle). In *Labeo rohita,* maximum food consumed and metabolized were (8.4 gm. dry weight) and (7.5 gm. dry weight) respectively for that fed on *Rana hexadactyla.* The present investigation showed of the Animal diets, maximum food consumption and maximum food metabolized for *Cyprinus carpio communis* were (7.65±0.009 gm. dry weight) and (5.633±1.5 gm. dry weight) fed on Goat Liver. In *Catla catla,* maximum food consumption and food metabolized were (11.55±0.073 gm. dry weight) and (7.765±2.54 gm. dry weight) respectively when fed on the same Goat Liver. This showed that among the Animal diets, Goat Liver was an easy source of consumption and conversion in both *Cyprinus carpio communis* and *Catla catla* respectively.

Bharat Bhusan Patnaik et al., (2002) elaborated that absorption efficiency was highest in *Labeo rohita* when fed on *Daucus carrota* (Carrot) among the Plant diets, but the conversion efficiency was minimum. This study also indicates that the conversion efficiency in case of *Catla catla* and *Cyprinus carpio communis* was minimum (4.51% and 3% respectively) when fed with Carrot compared to other Plant diets. In general, the conversion efficiency of *Cyprinus carpio communis* and *Catla catla* are very low due to high metabolic demand.

CONCLUSION

Cyprinus carpio communis and *Catla catla* being well known Omnivores easily consume a variety of food items offered. Such Omnivores feeding habits permit the use of varied types of feed items. Present study confirms the above findings. In case of *Cyprinus carpio communis,* Cucumber provided the maximum Overall Energy budget among Plant diets and Goat Liver among the Animal diets. This concludes that Cucumber and Goat Liver are the best as supplementary feeding diets in *Cyprinus carpio communis.* In case of *Catla catla,* Cucumber as the Plant diet was more productive for its growth efficiency. Among the Animal diets though Chicken showed much more growth efficiency in *Catla catla* than Goat Liver. These studies help us to portray the Bio-energetic schemes in carps like *Cyprinus carpio communis* and *Catla catla,* which will help us to relate to its maximum productivity in aquaculture farms.

ACKNOWLEDGEMENTS

We deeply acknowledge the kind support of P.G and Research Department of Zoology, Loyola College for their concerted effort in providing all the facilities to carry out this work.

REFERENCES

Akiyama, D.M., 1992. Future considerations for the feed industry. *In*: Proceedings of Aquaculture Feed Processing and Nutrition Workshop. Akiyama, D.M., Taw, R.K.H. (Eds.). Processing and Nutrition Workshop. Thailand and Indonesia 19-25 Sep., 1991. American Soybean Association, Singapore, 5-9 pp.

Bharat Bhusan Patnaik., Fleming, A.T., Selvanayagam, M. 2002. Effects of Plant and Animal diets on Food utilization of the Fresh Water Carp *Labeo rohita* (Hamilton). In: Ecology and Ethology of Aquatic Biota Arvind Kumar (eds.) Daya Publishing House, N Delhi, 207-211 pp

Bryant, P.L., Matty, A.J., 1981. Adaptation of a carp *(Cyprinus carpio)* larvae to artificial diets, optimum feeding rate and adaptation age for a commercial diet. Aquaculture 23, 275-286.

Chakrabarty, R.D., Sen, P.R., Chatterjee, D.K., Kowtal, G.V., 1973. Observations on the relative usefulness of different feed for carp spawn and fry. J. Inland fish. Soc India. 21(1). 1-6.

Chakrabarty, R., Jana, B.B., 1998. Effects on growth and water quality of feeding exogenous plankton compared to use of manure in culture of mrigal, *Cirrhinus mrigala* and rohu *Labeo rohita* fry in tanks. J. Applied. Aquaculture 8(2), 87-95.

Charles, P.M., Sebastian, M.C., Victor Raj and Marian, M.P., 1984. Effect of feeding frequency on growth and food conversion *of Cyprinus carpio fry.* Aquaculture 40, 293-300.

El-Sayed, A.M., 1994. Evaluation of Soybean meal, Spirulina meal and Chicken offal meal as protein sources for silver sea bream *(Rhabdosargus sarba)* fingerlings, Aquaculture 127, 169-176.

Erfanullah, J.A.K., 1998. Growth rate, feed conversion and body composition of *Catla catla, Labeo rohita* and *Cirrhinus mrigala* fry fed on diets of various carbohydrate to- lipid ratios. J. World Aquacul. Soc. (Baton rouge LA). 29(1), 84-91.

Fleming, A.T., 1983. Effects of Plant and Animal food combinations on food utilization of the fresh water fish *Labeo rohita.* M.Sc., dissertation submitted to St. Xavier's College, Palayamkottai (unpublished).

Gallagher, M.L., 1994. The use of Soybean meal as a replacement for fish meal in diets for hybrid stripped bass *(Morone saxatilis x Morone chrysops).* Aquaculture 126, 119-127.

Grant, G., 1989. Antinutritional effects of Soybean: a review. Prog. Food Nutr. Sci. 13, 317-348.

Haniffa, M.A., Venkatachalam., 1980. Effects of food quality on energy budget and chemical composition of the grass carp *Ctenopharyngodon idella.* Proceedings of the International Symposium on conservation. In: Life science University, Kebamgsaan, Malaysia, 163-173 pp.

Hasan, M.R., Macintosh, D.J., Jauncey, K., 1997. Evaluation of some plant ingredients as dietary protein sources for common carp *(Cyprinus carpio* L.) fry. Aquaculture (Amsterdam) 151(1-4), 55-70.

Hossain, M.A., Islam, M.N., Alim, M.A., 1992. Evaluation of Silkworm pupae meal as dietary protein source for cat fish *(Heleropneustes fossilis).* In: Fish nutrition in practice. Kaushik, S.J and Luquen, P. (Eds.), 785-791 pp.

Huet, M., 1970. Textbook of Fish culture. Fishery News books Ltd. London, 435 pp.

Hughes, S.G., 1991. Use of lupin flour as a replacement for full-fat soy in diets for rainbow trout *(Oncorhynchus mykiss).* Aquaculture 27, 97-109.

Jhingran, V.G., 1977. Fish and Fisheries in India. Hindustan Publishing corporation, India, 689-697 pp.

Kaushik, S.J., Cowey, C.B., 1991. Dietary factors affecting nitrogen excretion by fish. In Cowey, C.B and Cho, C.Y. (Eds.), Nutritional Strategies and Aquaculture Waste Guelph, Ontario, 3-20 pp.

Keshavappa, G.Y., Devaraj, K.V., Seenappa, D., 1989. Evaluation of three animal based feeds for rearing catla fry. Proc. Nat. freshwater Aqua. Bhubaneswar, India, pp 143-144.

Khan, M.A., Jafri, A.K., 1992. Replacement of fish meal with offal meal in formulated diets for the carp fingerlings *Labeo rohita* (Hamilton). J. Aqua. 2, 56-62.

Krupauer, V., 1967. Vybcrovost V postrave U dvouletych bilych amru (Food selection of two year old grass carp), Bulletin VYZK-UST. RYB. Vodnary. 3, 7-17.

Law, A.T., Cain, K.S.S., Aug, K.J., Kamarudin, M.S., 1990. Digestibility of low cost ingredients in pelleted feeds by *Macrobarachium rosenbergii* (Deman). In: Riramo, R and Hanyu, L. (Eds.). The second Asian Fisheries Forum. Asian Fisheries Society, Manila, Philippines, pp. 323-336.

Mathavan, S., Vivekanandan, E., Pandian, T.J., 1976. Food utilization in the fish *Tilapia mossambica* fed on plant and animal foods. Helgolander Wiss Messeresunters. 28, 66-70. Maynard, A.I., Loosely, J.K., 1962. Animal nutrition. McGraw Hill, NewYork, 553pp.

Mazid, M.A., Zaher, M., Begum, N.N., Ali, M.Z., Nahar, P., 1997. Formation of cost effective feed from locally available ingredients for carp polyculture system for increased production. Aquaculture 151(1&4), 71-78.

Nandeesha, M.C., Srikanth, G.K., Varghese, T.J., Keshavanath, P., Shetty, H.P.C., 1988. Influences of Silkworm pupae based diets on growth, organoleptic quality and biochemical composition of Catla- Rohu hybrid. In Aquaculture research in Asia: Management techniques and nutrition. Proc. Asian Sem. Aquaculture. Huisman, E.A., Zanneveld, N., Boumans, A.H.M. (Eds.), pp. 211-220.

Nandeesha, M.C., 1993. Aqua feeds and feeding strategies in India. In: Farm made Aqua feeds, FAO Fisheries Technical paper 343. New, M.B., Tacon, A.G.J., Csavas, I. (Eds.), Rome, pp. 213-254.

Pandian, T.J., Raghuraman, R., 1972. Effects of feeding rate on conversion efficiency and chemical composition of fish *Tilapia mossambica*. Mar. Biol. 12(2), 129-136.

Petruse Witz. K., McFodyen., 1970. Productivity of terrestrial animals. IBP Hand Book. No. 13. Blackwell Scientific Publication, Oxford.

Reigh, R.C., Ellis, S.C., 1992. Effects of dietary Soybean and fish protein ratios on growth and body composition of red drum *(Scianeops oscellatus)*. Aquaculture 127, 169-176.

Rosamma, M., 1996. Effects of plant and animal diets on food utilization and energy budget of the fresh water carps, *Hypothalmicthys molitrix* (Val.), *Catla catla* (Ham.) and *Labeo rohita* (Ham.). M.Phil dissertation submitted to Loyola College, Chennai (Unpublished).

Saha, A.K., Ray, R.K., 1993. Nutrient digestibility of three oil cakes by fingerlings of an Indian major carp, *Labeo rohita* (Ham.). In Proceedings of the Third Fisheries Forum. Mohan Joseph (Eds.). Asian Fisheries Society, Indian Branch, Mangalore, India, pp. 1-4.

Sarkar, S.K., 1997. Comments on artificial feeds used in fish culture. Fishing Chimes 16(11), 15-17.

Satoh, S., Poe, W.E., Wilson, R.P., 1989. Effect of supplemental phytate and/or tricalcium phosphate on weight gain, feed efficiency and zinc content in vertebrate channel catfish. Aquaculture 80, 155-161.

Silva, S.S., Gunasekhara, R.M., 1991. An evaluation of the growth of Indian and Chinese major carps in relation to the dietary protein content. Aquaculture 12(2&3), 237-241.

Sinha, V.R.P., Ramachandran, V., 1985. Freshwater Fish Culture. Indian Council of Agricultural Research, New Delhi, 30 pp.

Srivastava, P.P., Shrangare, U.P., Jain, K.K., Sinha, P.S.R.K., 1997. Studies on the growth in early growing stage of common carp *(Cyprinus carpio)* on dietary supplementation of Vitamin H. Proc. Nat. Acad. Sci. India 67(2), 21-29.

Swaminathan, V., Sigit, G.S., 1982. Massive fish seed production in pens. Fishing chimes 2(7), 42-45.

Viola, S., Mokady, V., Rappoport, V., Arieli, Y., 1982. Partial and complete replacement of fish meal by Soybean meal in feeds for intensive culture of carps. Aquaculture 26, 223-226.

Vivekanandan, E., Pandian, T.J., Visalam, C.N., 1977. Effects of algal and animal food combinations on surfacing activity and food utilization in the Climbing perch, *Anabas scandens*. Pol. Arch. Hydrobiol. 27, 555-562.

CHAPTER 11

Food and Feeding Habit of The Cyprinid Fish *Semiplotus manipurensis* From Manipur

Laishram Kosygin* and Waikhoim Vishwanath

Department of Life Sciences, Manipur University, Canchipur- 795003 India.

**Loktak Development Authority, Manipur, Lamphel - 795004, India.*

ABSTRACT

Semiplotus manipurensis feeds on phytoplankton, zooplanktons, crustaceans, insects, detritus and sands. The gut content reveals that 70.5 - 74.0 % of food items encountered are algae and diatoms. Alimentary canal of the fish is described and discussed. Feeding intensity was high during November and low during March. The nature of gut length and content, structure of mouth, and nature of gill rakers suggests that *Semiplotus manipurensis* is an omnivorous bottom feeder, feeding predominantly on planktonic materials.

Key Words: *Semiplotus manipurensis, food and feeding habit, Manipur.*

INTRODUCTION

A detailed knowledge on the biology of fish is useful in fishery management, conservation and selecting the fish of high productivity for fish culture. Now-a-days emphases are being laid more and more on studies of vital statistics of fishes for development and conservation of fishery. This aspect is scarcely touched for many of the riverine fishes of India, many of which incidentally are also favourite fishes among pisciculturists.

Semiplotus manipurensis Vish. and Kosygin, 2002 is a delicious carp which has a high demand for its good taste. This fish is known only from Ukhrul district of Manipur. Due to difficult hill terrain and lack of proper means of communication from the fishing areas to the fish landing centres, fishermen smoke the fish and bring to Ukhrul market where it fetches high prices. However, in this region, fishing operations are usually carried out during the late winter and early summer months (November to May) as fishing is practically difficult during other months. This is because of the following reasons: (i) rough weather conditions and huge volume of water with whirlpools and great velocity during rainy seasons; (ii) hill streams are shallow and do not hold enough water except in the localised deep pools during late premonsoon; (iii) use of chemicals and explosives which causes over-exploitation during the premonsoon period which result in scarcity of fish. [This commercially important fish belongs to a rare genus *Semiplotus* that distributed only in South East Asia and very little is known about the fishes of this genus (Vishwanath & Kosygin 2000). Menon (1989) included the type species of the genus *S. semiplotus* in the list of endangered freshwater fishes of India. Similarly, Talwar & Jhingran (1991) also remarked that *S. modestus* was a rare species which distributes only in Mayanmar. However, there is no information on the biology of these fishes. Thus a preliminary study on the biology of these fishes is considered necessary for conservation and development of fishery. The present preliminary investigation has undertaken on the food and feeding habit of *S. manipurensis* in the Ukhrul district of Manipur.

MATERIALS AND METHODS

Specimens of *S. manipurensis* were collected from the Challou river during the period from May, 1997 to March, 1998. A total of 65 specimens ranging from 60.0 to 150.0 mm in TL of *S. manipurensis* were examined for the analyses of gut contents. The food of the fish was identified by the "points method" (Hynes, 1950). The gut contents were removed, sorted and identified up to generic level with a compound microscope.

The total length of the gut was measured for determining 'Relative gut length" (RLG) which was calculated as the ratio of intestinal length to total body length. Feeding habits were observed in the field.

For the study of alimentary Canal, the head of the fish were severed from the body, a few inches behind the operculum and cut across the mouth passing through one of the angles of the jaw. The specimen was fixed on tray and structure, and arrangement of the bucco-pharynx was studied. The types of gill rakers were also observed. Intestine was taken out from the body of the fish and studied.

OBSERVATIONS AND RESULTS

The alimentary canal of *S. manipurensis* is modified according to its feeding habit. It is morphologically distinguished as a non-tubular part consisting of mouth and bucco-pharynx and tubular region comprising of oesophagus, intestinal bulb, intestine and rectum. A brief description of the alimentary canal is given below:

Mouth

Mouth of the fish is of great wide and located inferiorly. The lower jaw has a exposed cornified cutting edge and free to move anterio-porteriorly.

Buccal cavity

The mouth leads into a spacious buccal cavity, which is lined by numerous membranes. The anterior part of the roof shows feeble longitudinal markings. Whereas the floor of the cavity is lined by a thick transverse mucous membrane in the middle. The root of the cavity is deeply concave while the floor is nearly flat. Tongue is rudimentary, almost absent in small specimens (<80 mm in TL.). Teeth are completely absent in the buccal cavity of this species. Behind the upper lip, there is a fold of the velum, which is papillated.

Pharynx

The buccal cavity leads into a short pharynx which is provided with minute phyaryngeal teeth on the floor. These teeth are borne by the inferior pharyngcal bones and work against a pharyngeal pad on the roof. The plateau of the floor of pharyngeal cavity is provided with numerous taste buds.

Gill rakers

The ventro-lateral wall of the pharynx is perforated by oblique gill slits, separated by means of four pairs of gills. Gill raker is short, thin, densely packed and arranged in two rows. The rakers of neighbouring arches meet with one another to form a sieve across the gill slits, providing a very efficient filtering apparatus by which the outgoing current of water is filtered.

Oesophagus

The pharynx leads into a short indistinct oesophagus. While entering the body cavity it become slightly expanded.

Intestine

This fish do not posses a true stomach. The intestine is swollen behind the oesophagus for the storage of food and is known as intestinal bulb or intestinal swelling. The intestinal bulb gradually narrow down into the thin walled intestine proper and finally opens outside through the anus. The intestine of the fish is very long and highly coiled.

GUT LENGTH IN RELATION TO BODY LENGTH OF FISH (RLG)

The relationship of gut length to the total length of the fish of 60.0 - 150.0 mm was ranges between 3.2865583 and 5.6910344. The value of "r" was found to be 0.9201377, showing a positive correlation of the gut length and total length of the fish.

FOOD ANALYSIS

The percentage of different groups of food items was calculated from the gut content of about 65 fishes in the size group 60.0 - 150.0 mm in TL. The food contents of *S. manipurensis* reveal the occurrence of algae, diatoms, zooplanktons, detritus, sand and insects, and its larvae. However, the stomach contents may be divided into the following broad categories.

Plant matter (Phytoplankton)

The result shows that plant matters constitute the main part of the gut content, which constitute about 70.5 - 74.0%. It comprises algae and diatoms, which comes under chlorophyceae (green algae), myxophyceae (blue green algae) and bacillariae (diatoms). Desmids have been treated under chlorophycea. The generic constituents are *Zygnema, Gonatozygon, Spirogyra, Cladophora, Mougeotia* of chlorophyceae; *Nostoc, Anabaena, Lyngbya* of Myxophyceae; and *Epithemia, Navicula, Nitzschia, Diatoma* of bacillariae.

Detritus and Sand

Detritus consist of unidentified plant as well as animal matter, occurring regularly in guts throughout the study period, forming about 16.3 - 20.2%. Sand was found to occur in small amount along with the detritus.

Animal matter

Zooplankton, Rotifera and crustaceans mainly represented the animal matter. They were recorded in small percentage (4.3 - 7.5%).

Miscellaneous

Miscellaneous matter was composed mainly of dipteran larvae, insect appendages, oligochaetes. These items were occasionally recorded in insignificant percentage (1.5 - 3.2%) in the gut content.

Feeding intensity

The condition of feed was determined from the degree of distension of the stomach and the same has been classified as 'gorged' 'full', '3/4 full' '1/2 full', '1/4 full', 'trace' and 'empty'. It is obvious that the feeding intensity of the fish is appreciably high, a good number of stomachs being either in 'gorged' or full conditions in the month of November. The few empty stomachs observed were from specimens collected during March (Table 11.1).

Table 11.1

Conditionof food (in percentages) of *S. manipurensis* in three different months during 1997-1998.

Month	Gorged	Full	3/4 full	1/2 full	1/4 full	Trace	Empty	N
July	-	2.4	31.3	49.9	15.2	1.2	-	20
November	15.8	57.7	22.4	4.9	-	-	-	14
March	-	-	-	10.2	49.2	37.8	2.8	31

DISCUSSION

The anatomical features of the alimentary canal in a fish are related to the food and feeding habit. The alimentary canal of *S. manipurensis* is structurally and functionally well adapted to their food and feeding habits. Presence of "Sector mouth" in this carp is of special interest in its feeding habit. According to Howes (1982) "sector mouth" is an inferior mouth which is wide (probably the widest mouth in relation to head length of any cyprinid fish) and lower jaw with an exposed corrified mandibular cutting edge. The lower jaw of this fish has a mobile structure, which is free to move anterio-posteriorly and appears to have the basic function of scrapping or ploughing epilithic material. They scrap off the algae in small instalments making a series of crescentic impressions on the rocks.

It is well known fact that, the stomachless fishes are provided with pharyngeal chewing device, which permit ingested food reach the intestine in a fragmented condition. In *S. manipurensis*, the pharyngeal teeth born by the inferior pharyngeal bones seems to work against a bony callous pad, thereby forming on effective grinding apparatus for crushing the food. Various workers have divided the buccopharyngeal region of fishes into an anterior part, the buccal cavity extending upto the first gill arch and a posterior portion, the pharynx. Sinha (1993) while studying the histo-morphology of the alimentary canal in Indian freshwater teleosts remarked that there is a little morpho-histological difference in the two parts and probably there is no hard and fast line to demarcate them. Hence, in the present study the entire region has been described as the buccopharyngeal region. The densely packed gill rakers of this fish seem to form an effective sieve to stain plankton from the irrigating water.

The length of the gut in relation to that of the total length varies greatly in different species. In general, the carnivorous species have a shorter intestine, the omnivores a longer one and the herbivores the longest due to obvious reasons emanating from digestion of specific type of food (Sinha, of.cit.). The fish has a very long, highly coiled and stomachless alimentary canal which confirms that it is a herbivorous fish. However presence of intestinal bulb compensates the absence of stomach in this fish as described by Lalger (1984). According to many workers (Das & Moitra 1963; Biswas, 1985; Biswas & Phukon, 1996), RLG value has direct relationship with the feeding types of fish. A lower value indicates intake of animal matter in the diet whereas higher value suggests dependence on plant matter in large quantities. In the present study RLG value of the fish was high and the value tends to increase as the fish grow. This finding shows herbivorous nature of the fish and increasing of RLG value as the fish grow suggest more dependence on plant matter in large quantities as the fish grow.

The feeding intensity of the species was low during March. It may be attributed to the developments of gonads, which occupy the major space of the abdominal cavity, which is described as 'spawning fast' by many workers (Frost, 1954; Lawler, 1965). The present observations on the nature of gut content, RLG values, structure of mouth, nature of gill rakers suggests that *S. manipurensis* is an omnivorous bottom feeder, feeding predominantly on planktonic materials.

REFERENCES

Biswas, S.P. 1985. Studies on the gut length in relation to feeding habit of five commercially important fishes from Assam. *J. Assam Sc. Soc.* 28(1): 10-13.

Biswas, S.P. and Phukon, J. 1996. On some aspects of an ornamental fish, *Erethistes pussilus* from the Brahmaputra River System. *J. Freshwater Biol.* 8(1): 27-32.

Das, S.M. and S.K. Moitra 1963. Studies on food and feeding habit of some freshwater fishes of India, Part-IV. A review on the food and feeding habits with general conclusion. *Ichthyologica.* 2 (.1-2): 107-115.

Frost, W.E., 1954. The food of pike *Esox lucius* L., in Windermer. *J. Anim. Ecol.* 23(2): 339-360.

Howes, G.J. 1982. Anatomy and evolution of the jaws in the semiplotine carps with a review of the genus *Cyprinion* Heckel, 1843 (Teleostei: Cyprinidae). *Bull. Brit. Mus. Nat. Hist. (Zool).* 42(4): 299 - 335.

Hynes, H.N.B. 1950. The food of freshwater strickle backs (*Gasterosteus acculatus* and Pygosteus pungitius) with a review of methods used in studies of the food of fishes. *J. Arim Ecol.* 39: 36 - 58.

Lalger, K.F. 1952. Freshwater fishery biology. Win.C. Brown Company (Publishers) Dubuque Lowa.

Lawler,G.H. 1965. The food of the pike, Esox lucius, in Hemming Lake, Manitoba. *J. Push Res. Board Can.*, 22(6): 1357 -1377.

Menon, A.G.K. 1989. Conservation of the ichthyofauna of India. In: Jhingran A.G. and V.V. Sugunam, (Eds). Conservation and management of Inland Capture fisheries resources of India. Inland Fisheries Society of India, Barrackpore: 25 - 33.

Sinha, G.M. 1993. Histo-morphology of the alimentary canal in Indian freahwater teleost. *Adv. Fish Res.* 1: 193 - 220, pl. 2.

Talwar, P.K. & A.G. Jhingran, 1991. Inland Fishes of India and Adjacent Countries, I. Oxford and IBH Publ. Co. Pvt. Ltd. New Delhi, 541 pp.

Vishwanath, W. & L. Kosygin, 2000. Fishes of the cyprinid genus *Semiplotus* Bleeker, 1859 with description of a new species from Manipur, India. *J. Bombay nat. Hist. Soc.* 97(1): 92 - 102.

CHAPTER 12

Effects of Pituitary Hormone Administration on Post - Embryonic Development in *Cirrhina Mrigala* (Ham).

Rita Rani, K.K. Gaur and Vishal Tiwari
Department of Zoology, R.B.S. College, Agra-282 002.

ABSTRACT

Pituitary treatment given to the larvae of *Cirrhina mrigala* (Ham.) for about one month revealed faster post-embryonic development as against the control experiments. 2 mg/lit. dose of pituitary proved most suitable. Mortality figure comes to 4, 5, 7 and 12 for control, 1 mg/lit, 2 mg/lit and 3 mg/lit. dose treatments respectively. Morphometrical parameters showed marked changes in all pituitary treated experiments, suggesting that hormone therapy may prove useful in the culture *of Cirrhina mrigala* (Ham.).

Key Words: *Pituitary hormone administration, Post-embryonic development, Cirrhina mrigala,.*

INTRODUCTION

Pituitary hormone plays an important role in the maturation of gonads and body growth of fish. The growth hormone (somatotropin) of the pituitary gland also influence protein, carbohydrate and fat metabolism to a great extent (Knobil *et al.*, 1957). The effect of pituitary hormone treatment on the post-embryonic development of *Cirrhina mrigala* is not yet fully known and needs further investigation.

With the view point, in the present communication, an effort is made to work out the optimum dose of hormone which could be able to affect some aspects of post-embryonic development in a positive direction, as pituitary hormone treatment has been shown to promote post-embryonic development, growth and survival in several fish species.

Some of the important contributions on the said field of studies are from Landgrebe, F.W. (1941); Baldwin, F.M. *et al.*, (1942); Swift, D.R. (1954); Knobil, E. *et al*, (1957); Atz, J.W. *et al.*, (1959); Das, S.M. *et al.*, (1962); Tata, J.R., (1968); Liley, N.R. *et al.*, (1969); Yamazaki, F., (1976); Clarke, W.E. *et al*, (1977); Adelmeu, I.R., (1982); Joss, J.M.P., (1985); Mchean, E. *et al.*, (1990); Lakra, W.S. *et al*, (1996); Wilson, R.P. *et al.*, (1998); Brzuska, E. *et al.*, (1999); and that of Roy, P. *et al* (2000).

MATERIALS AND METHODS

The hatchlings (3-4 days old) of *Cirrhina mrigala* (Ham.) were collected from fish farm in the close vicinity of Agra. The hatchlings for consecutive development stages were brought to laboratory and transferred to plastic trough under observation. In each experimental trough, 20 specimens were placed on the basis of proper stocking density (approximately 2 fish/lit.) The physico-chemical conditions of the experimental troughs were maintained nearly natural by the use of sand particles, hydrilla plants, zoo and phytoplanktons. Hitachi readymade food pellets in equal quantity were also given at regular intervals to all the experimental larvae. The pituitary hormone doses of 1 mg/lit., 2 mg/lit., and 3 mg/lit. concentrations were prepared by pituitary extract alcoholic method from the common carp, and in each trough one kind of them was added.

By random sampling five specimens from each kind of experiment and control were taken out for derivation of mean standard under different aspects of studies on post-embryonic development at a successive time period viz., 5, 7 and 15 days. The same course was applied after lapses of 21 and 30 days in order to study later developmental stages. Morphometrical parameters of the larvae were recorded with the help of Vernier calipers at the end of the experiments.

RESULTS AND DISCUSSION

On the perusal of parameters indicated in the Table 12.1, it can be concluded that a dose of 2 mg/lit. proved to be the best in terms of post-embryonic development (Fig 12.1, 12.2, 12.3, 12.4, 12.5 & Table 12.1).

Table 12.1

Morphological parameters of *Cirrhina mrigala* (Ham.) larvae treated with pituitary dose (mg/lit.).

S.No.	General Measurements	000 mg/lit.	1 mg/lit.	2 mg/lit.	3 mg./lit.
		Average in mm ± S.D.			
1	Standard length	21.897±.048	23.712±.050	30.867±.119	30.067±.039
2	Body length	13.397±.010	14.507±.397	19.847±0.45	19.067±0.39
3.	Head length	6.307±.036	6.857±.057	7.392±0.94	7.067±0.39
4.	Distance between snout and anal pore.	14.892±0.61	16.475±.136	20.0417±.143	20.117±0.67
5.	Distance between snout and anal fin base.	16.677±.090	16.807±.053	24.192±.044	23.562±.302
6.	Distance between snout and dorsal fin base.	13.552±.068	16.457±.147	18.622±.205	19.067±.039
7.	Height of the body of anal fin base.	4.887±.109	5.432±.150	6.437±.088	6.262±0.37
8.	Height of the body at dorsal in base	6.747±0.74	6.897±.105	8.832±0.35	8.362±0.83
9.	Mortality	Four	Five	Seven	Twelve

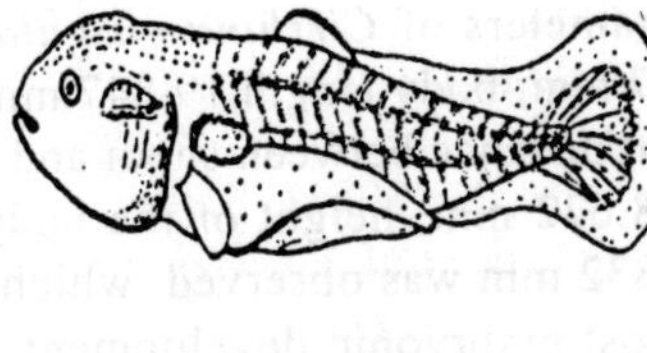

Fig. 12.1
Post larvae of *C. mrigala*, 5th day after hatching.

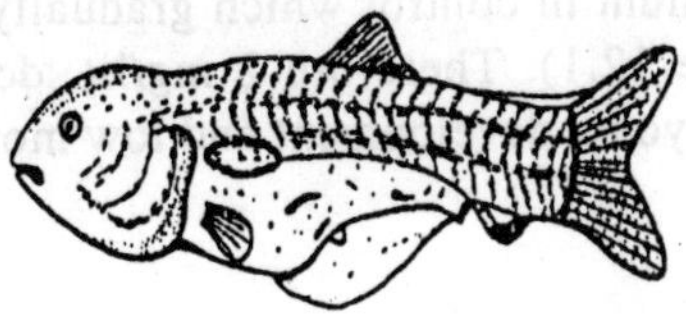

Fig. 12.2
Post larvae of *C. mrigala*, 5th days after hatching.

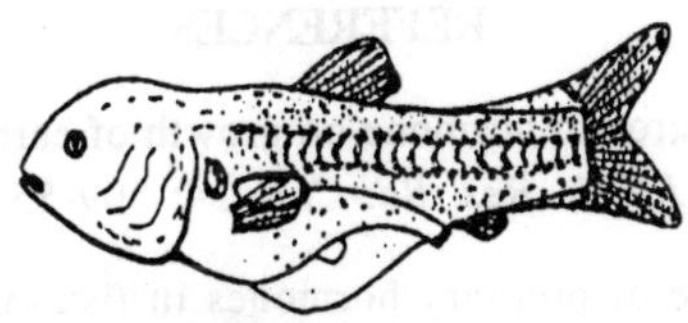

Fig. 12.3
Post larvae of *C. mrigala*, 15th days after hatching.

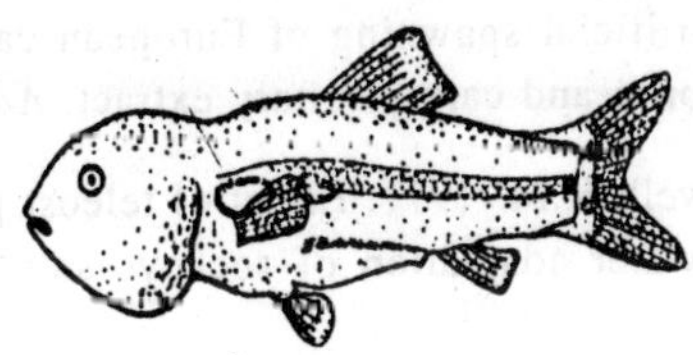

Fig. 12.4
Post large of *C. mrigala*, 21st days after hatching.

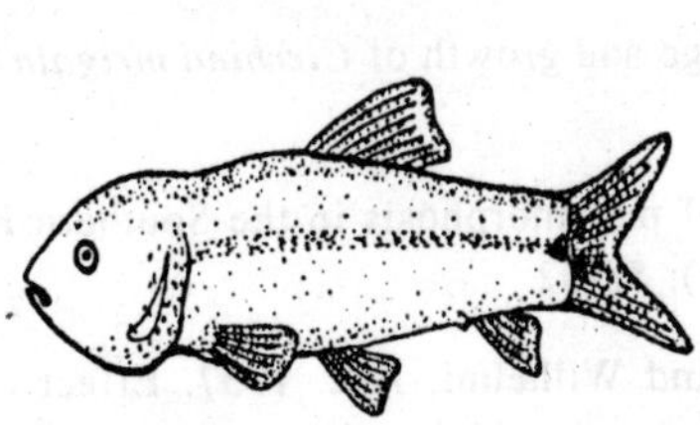

Fig. 12.5
Post large of *C. mrigala*, 30th days after hatching.

The general morphometrical parameters of *Cirrhina mrigala* (Ham.) larvae revealed that at a dose of 2 mg/lit., standard length 30.867 mm; body length 19.847 mm; head length 7.392 mm; distance between snout and anal pore 20.417 mm; distance between snout and anal fin base 24.192 mm; distance between a snout and dorsal fin base 18.622 mm; height of the body at anal fin base 6.437 mm; and height of the body at dorsal fin base 8.832 mm was observed, which is significant in terms of the best dose estimated in correspondence to post-embryonic development.

It has also been marked that during early developmental periods when morphogenesis is on its full swing, the pituitary treatment resulted into maximum increment in standard body length at 2 mg/lit. dose. Survival percentage was maximum in control which gradually decreased at 1 mg/lit., 2 mg/lit., and 3 mg/lit. dose treatments (Table 12.1). Therefore 2 mg/lit. dose was calculated to be the best effective dose in respect to post-embryonic development and low mortality *of Cirrhina mrigala* (Ham.) larvae.

ACKNOWLEDGEMENTS

The authors are thankful to the Principal, R.B.S. College Agra for providing the essential research facilities during this investigation.

REFERENCES

Adelmen, I.R. 1982. Effect of bovine growth hormone on growth of carp, *Cyprinus carpio* (L.) and influence of temperature and photoperiod. *J.Fish. Res. Bd. Can.* 34: 509-557.

Atz, J.W. and Pickford. 1959. The use of pituitary hormones in fish culture. *Endeav.* 18(7): 125-129.

Baldwin, F.M. and Hsin-Li, M. 1942. Effects of gonadotropic hormones in fish, *Xiphophorous helleri* (Heckel). *Proc. Soc. Exp. Biol. Med.* **49:** 601-604.

Brzuska, E. and Adamek, J. 1999. Artificial spawning of European catfish, *Siluris glanis* (L) stimulation of ovulation using LHRH a, ovaprim and carp pituitary extract. *Aqua, Res.* 30(1): 59-64.

Clarke, W.C.; Farmer, S.W. and Hartwell, K.M. 1977. Effect of teleost pituitary growth hormone on growth of *Tilapia mossambica* and sea water adaptation of sockeye salmon, *Oncorhynchus nerka. Gen. Comp. Endocrinol.* 33: 174-178.

Das, S.M. and Khan, H.A. 1962. The pituitary and pisciculturte in India with an account of the pituitary of some Indian fish and a review of techniques and literature on the subject. *Ichthyolog.* 1:43-58.

Jhingran, V.G. 1959. Studies on the age and growth of *Cirrhina mrigala* (Ham.) from the river Ganga. *Proc. Nat Sci. Indian.* 25: 107-137.

Joss, J.M.P. 1985. Pituitary control of metamorphosis in the Southern Hemisphere lamprey, *Geotria australia. Gen. Comp. Endocrinol.* 60(1): 58-62.

Knobil, E.W.; Greep, R.O. and R. and Wilhelmi, A.E. 1957. Effect of a primate pituitary growth hormone preparation on nitrogen metabolism in the hypophysectomized Rhesus monkey. *Gen. Comp. Endocrinol.* 60: 166-168.

Lakra, W.S.; Mishra, A.; Dayal. R. and Pandey, A.K. 1996. Breeding of Indian major carps with the synthetic hormone ovaprim in Uttar Pradesh. *J. Advan. Zool.* 17(2): 105-109.

Landgrebe, F.W. 1942. The role of pituitary and thyroid in the development teleosts. *J. Exp. Biol.* 18: 162-169.

Liley, N.R. and Donaldson, E.M. 1969. The effects of fish pituitary material on the behaviour of hypophysectomized female guppies, *Poecilia reticulata* (Peters). *Can. J. Zool.* 47: 569-573.

Mchean, E.; Donaldson, E.M.; Dye, H.M. and Souza, L.M. 1990. Growth acceleration of coho salmon, *Oncorhynchus kisntch* following oral administration of recombinant bovine somatotropin, *Aquacult.* 91(1): 197-203.

Roy, P.; Datla, M.; Dasgupta, S. and Bhattacharya, S. 2000. Gonadotropin releasing hormone stimulates thyroid activity in a fresh water murrel, *Channa gachua* (Ham.) and carps, *Catla Catla* (Ham.) and *Cirrhina mrigala* (Ham.) *Gen. Comp. Endocrinol.* 117(3): 456-463.

Swift, D.R. 1954. Influence of mammalian growth hormone on the rate of growth of the fish. *Nat.* 173:1096.

Tata, J.R. 1968. Hormonal regulation of growth and protein synthesis. *Nat.* **219:** 331-337.

Wilson, R.P.; Poe, W.E.; Nemetz, T.G. and Macmillan. 1998. Effect of recombinant bovine growth hormone administration on growth and body composition of channel catfish. *Aquacult.* 73(1): 229-236.

CHAPTER 13

Observations on Mixed Fish Farming of Indian and Exotic Carps in a Seasonal Pond of Jodhpur

Anita Jhajhria

Department of Zoology, Government P.G.College, Jalore- 334001, Rajasthan

ABSTRACT

Mixed fish farming in Shekhawat pond (0.736 ha) was carried out for a period of 8 months from August 1999 to March 2000 using *Catla catla, Labeo rohita, Cirrhinus mrigala* and *Cyprinus carpio.* No supplementary feed and fertilizers were supplied to the fishes reared. The growth of *Cyprinus carpio* (1173.50 gm.) was best followed by *Catla catla* (578.30 gm.), *Labeo rohita* (263.40 gm.) and *Cirrhinus mrigala* (252.50 gm.). The fish production (4127 kg./ha/8 months) was achieved which is higher than the average fish production (100-400 kg/ha/yr) in similar water bodies of this region.

Key words: *Mixed fish farming, Shekhawat pond, Indian and exotic carps, production.*

INTRODUCTION

The goal of achieving productivity from pond ecosystem is to sustain high yield of fish rich in protein. In aquatic ecosystems energy culminates at its different trophic levels with considerable loss of energy at each link of food chain. For augmenting high fish productivity of carps, the knowledge of bioecology of fresh water ponds is utmost important as ecology plays vital role in regulating productivity of these fishes.

The basic principle of mixed fish farming or polyculture is that when compatible fishes of different feeding habits are stocked together, they secure for themselves in the most efficient manner, all life requisites available in the pond for fish production without harming each other.

In North-Western Rajasthan, there are 176 seasonal water bodies with water spread area of 5700 ha. in which the fish production is as low as 100-400 kg/ha/yr (Pragati Prativedan, 1995). In Jodhpur township, as well as in its vicinity, there are many ponds of varied dimensions where knowledge on ecology and related fish productivity of fresh water pond ecosystems for utilization of appropriate fish fauna is still lacking. With this consideration, the studies where conducted for assessing the fish production of Shekhawat pond.

MATERIALS AND METHODS

The fries of three Indian major carps namely, *Catla catla, Labeo rohita* and *Cirrhinus mrigala* alongwith an exotic carp *Cyprinus carpio* were obtained from the induced spawning exercises using the synthetic hormone. These fries were then transported in oxygen packed system of polythene bags in August 1999 for stocking in the pond. The number of fries to be stocked in the pond was computed by adopting the formula (Jhingran, 1975) :

$$\text{Number of fries to be stocked per unit area} = \frac{\text{Total expected increase in weight}}{\text{Expected increase of weight of individual fish}} + \text{mortality (not more than 10\%)}$$

The Shekhawat pond (26°49'02" North Latitude and 73°47'03" East Longitude) is about 10 Km. from the Jodhpur railway station. The area of the pond is 0.7366 ha having a mean depth of 1.89 mt. It lies in the vicinity of the army and the army mess is located upstairs to it.

The harvesting was done using cast and drag net in March, 2000. The pond water was collected monthly during the stocking period between 8:30 a.m. - 11.30 a.m. and was analysed using the methods of APHA (1985) and Welch (1948).

RESULTS AND DISCUSSION

The results of stocking and production are summarised in Table 13.1. The abiotic parameters analysed are shown in the Table 13.2. The production of 4127 kg/ha/8 months was achieved in the pond. The average weight of (1173.50 gm.) *Cyprinus carpio* was highest followed by *Catla catla* (578.30 gm.), *Labeo rohita* (263.40 gm.) and *Cirrhinus mrigala* (252.50 gm.).

Table 13.1

Species-wise stocking and production of Indian and exotic carps in Shekhawat pond (0.076 ha.)

Fish species	Fin Formula	Habitats	Feeding Habits	Stocking density of fries	Total weight (Kg.)	Average weight (gm.)	Average length (cm.)
Catla catla (Catla)	Diiii-iv, Aiii5, PI20, Via 8	Surface feeder	Planktoni	900	580	578.3	25.23
Labeo rohita (Rohu	Diii-iv 12-14; Aii-iii5; Pi 16-18; Vi8	Colum feeder	Herbivorous	2000	450	263.4	19.69
Cirrhinus mrigda (Mrigal)	Diii-iv 12-13; Aiii5; Pi 17; Vi8	Bottom feeder	Omnivorous	3200	410	252.5	20.65
Cyprinus carpio (Common carp)	Diii-iv 18-20; Aiii, Pi 15, Vi8	Bottom feeder	Omnivorous	4000	1600	1173.5	40.23

Table 13.2
Physico-chemical properties of pondwater during the mixed culture period

Abiotic Parameters	August 1999	September 1999	October 1999	November 1999	December 1999	January 2000	February 2000	March 2000
Water temperature (^{0}C)	24.1	24.5	23.6	20.1	16.5	15.5	15.9	20.4
pH	7.6	7.7	7.8	7.8	7.9	8.0	7.8	8.0
Transparency (cm)	14.0	20.0	22.0	24.0	19.0	17.0	26.0	30.0
DO (ml/l)	5.643	4.559	4.453	4.749	4.997	5.088	4.640	4.159
Free CO_2 (ppm)	9.0	8.0	7.0	6.0	4.0	3.0	5.0	4.0
CO_3^{-2} (ppm)	-	-	-	-	-	-	-	-
HCO_3^{-2} (ppm)	84.0	98.0	95.0	102.0	110.0	115.0	120.0	124.0
Total Alkalinity (ppm)	84.0	98.0	95.0	102.0	110.0	115.0	120.0	124.0
Total Hardness (ppm)	70.0	89.0	86.0	110.0	120.0	125.0	135.0	140.0
Chloride (ppt)	0.302	0.526	0.496	0.574	0.608	0.649	0.797	0.803
Salinity (ppt)	0.054	0.950	0.895	0.104	0.109	0.117	0.144	0.145
Phosphate (µg/l)	0.212	0.354	0.410	0.287	0.279	0.250	0.398	0.559
Nitrate (µl/l)	3.336	3.685	4.528	4.104	3.910	3.432	4.728	6.223

Valuable constributions in the field of fish productivity of ecologically different ponds with reference to carps has been done by various workers, Lakshmanan *et.al.* (1971) and Singh *et.al.* (1972) conducted year long experiments at the pond culture, division of CIFRI combining seven Indian and exotic species of fish and obtained a production varying from 2234-4210 kg/ha/yr Sukumaran (1976) obtained a production of 6191-7332 kg/ha/8months at Karnal. Sharma *et. al.* (1979) reported a production of 4323 kg/ha/yr in fish-cum-duck rearing without adding any fertilizer to the pond and giving any supplementary feed.

Mathew *et. al.* (1979) obtained productions with six species combination as 2199-2242 kg/ha/8 months at Poona.

Kumar (2001) recorded maximum growth of catla (9.3 kg.) followed by common carp (3.1 kg.), rohu (1.9 kg.) and for mrigal (320 gm.) in one year from deep ponds of Patna. Baghel and Saxena (2002) reported a production of 4965.97 kg/ha/9 months in Tarai region of Uttaranchal.

In the present study, the fish production (4127 kg/ha/8 months) relied only on the natural food and the kitchen wastes of the army mess. At least 50% of the total food of carps in the fish ponds comes from the pond itself (Schaeperclaus, 1952). Gaudet (1967) reported that in Israel out of 2000 kg/ha, probably 400 kg. comes from the natural food.

Thus, it may be concluded that on the basis of ecobased planning, judicious management and by incorporating the application of supplementary feed and fertilizers a many fold enhancement in fish production of the ponds of this region is possible.

ACKNOWLEDGEMENTS

The author expresses sincere gratitude to Dr. Devendra Mohan. Associate Professor, Department of Zoology, Jai Narain Vyas University, Jodhpur for evincing keen interest and guidance in the research work.

REFERENCES

APHA. 1985. Standard methods for the examination of water and waste water. AWWA, WPCE, New York, 16th edition.

Baghel, D.S. and Saxena, V. 2002. Composite fish culture in Tarai region of Uttaranchal. Geobios. 29: 101-104.

Gaudet, J.L. 1967. The status of warm-water pond fish culture in Europe. FAO, Fish, Rep. pp.174.

Jhingran, V.G. 1975. Fish and fisheries of India. 1st edition, Hindustan Publishing Corporation, New Delhi.

Kumar, D. 2001. Fish productivity of ecologically different ponds with reference to carps and air breathing fishes J. Inland. Fish. Soc. India. 33: 8-16.

Lakshmanan, M.A.V., Sukumaran, K.K., Murty, D.S., Chakraborty, D.P., and Philipose, M.T. 1971. Preliminary observations on intensive fish farming in fresh water ponds by the composite culture of Indian and exotic species. J. Inland Fish Soc. India. 2: 1-21.

Mathew, P.M., Singh, B.K.and Chakraborty, D.P. 1979. Stocking of fry in composite fish culture. In : Symposium on Inland Aquaculture, CIFRI, Barrackpore. pp. 1-44.

Pragati Prativedan (Raj.) 1995. Inter State Fishery Planning Cell (1994-95). State Fisheries Department, Rajasthan, Jaipur.

Schaperclaus, W. 1952. *Die* karpfenteichswirtschaf: in der Deutsche Demokratischen Republik. Sitz. Ber. VII (7), Berlin, Deutsche Ahad. Landwirtschaftwiss.

Sharma, B.K., Kumar, D., Das, S.R. and Chakraborty, D.P. 1979. Observations on swine-dung recycling in composite fish culture. In : Symposium on Inland Aquaculture (Abstract) CIFRI, Barrackpore, February 12-14. pp. 99.

Singh, S.B., Sukumaran, K.K., Chakrabarti, P.C. and Bagchi, M.M. 1972. Observations on composite culture of exotic carps. J. Inland Fish. Soc. India. 4: 38-50.

Sukumaran, K.K. 1976. Progress report of the karnal subcentre of Haryana for the period November 1972 to August 1975. In : Third workshop on the All India Coordinated Research project on composite fish culture and fish seed production. February 26-27, CIFRI, Pun.

Welch, P.S. 1948. Limnological methods. The Blakiston Co., Philadelphia.

CHAPTER 14

EFFECT OF LETHAL CONCENTRATION OF METASYSTOX (OXYDEMETON-METHYL) ON BLOOD PARAMETERS OF AN AIR BREATHING FISH *HETEROPNEUSTES FOSSILIS* (BLOCH)

Meenaxi Das and Shambhu Prasad

Biotoxin Research Laboratory, University Department of Zoology
L. N. Mithila University, Darbhanga-846004

ABSTRACT

The accumulation of pesticide residues in different component of aquatic ecosystem and their detrimental impact is of significant environmental concern. In the present investigation *Heteropneustes-fossilis* exposed to lethal concentration of metasystox exhibited striking alteration in blood parameters. The observed effect of exposure to 48h LC_{50} reflected, hyporcromasia, elevated erythrocytic count with increase in size of cremation of cell and their membrane, decrease in total leucocyte count and haemoglobin percent.

It has been concluded that the present fish in response to metasystox stress faces with a serious physiological crisis due to alteration in several blood parameters that ultimately tells upon the survival of the fish.

Key words: *Metasystox, blood profile, Heteropneustes fossilis, pesticides.*

INTRODUCTION

Pesticides have became essential to modern agriculture but they are also a potential source of chemical degradation of soils. The 1992-93 data shows that India used 47.62 million tonnes of insecticide, 19.45, millions tonnes of fungicides, 5.79 million tonnes of herbicides, 0.90 million tonnes of rodenticides, and 0.08milion tonnes of fumigants (employment news 7-13 August 1999) Bulk of pesticides used in India are organochloride compounds. Half life of some of the organo-chloride pesticides like HCH, chlordane, and heptachlore, in cultivated soils is as high as 80 to 110 days. This means that these pesticides are highly resistant to biological degradation and are potential toxicants. Case studies have indicated that pesticides applied to soil can enter the food chain. Pesticides have undesirable effects on the environment. Some of the effects are excessive mortality and reduced reproductive potential of organism, reduction in the productivity of natural resources and development of resistance to pesticides in target and non target species.

The increased use of pesticides have helped in controlling insects and pests but have also created undesired environmental deterioration specially hazardous influence in aquatic media causing fatal and long term effects on fishes as well as aquatic fauna (Basak et. al. 1980) as a consequences, fishes are being used as indicator of aquatic pollution. Like other tissues, blood is an important physiological reflector of animal body and serves as primary target of many pesticidal action.

There are many reports related to toxicity to chemicals organophosphorus and chlorinated hydrocarbon pesticides on the blood of some Indian fishes (Shrivastva et. al 1985). But recently some new improved type of systemic pesticides have been developed in the light of growing resistance in the pest of previous insecticides (Sharma *et. al* 1991).

Metasystox is a systemic pesticide and widely used by the farmers of Bihar to control different crops including paddy,mango, the two main crops. It is already a proved fact that the systemic toxicant have biaccumulative properties and are metobolically degraded through enzymatic action at low pace. This is why an animal is recommended for slaughtering or for milching after a safe period of about 60 days from the time of application or administration of the toxicant.

Indiscriminate use of metasystox may cause greater damage to fresh water fishes as well significant alteration in aquatic ecosystem. The present communication highlights the effect of matasystox on some blood parameters of *H. fossilis*.

MATERIAL AND METHODS

The fish *Heteropneustes fossilis* (28± 2g) obtained from local fish market were thoroughly washed in 2% $KMNO_4$ solution and acclimatised for 15 days. Fishes were fed with egg white cut into small pieces once a day. The feeding was stopped one day before the experiment.

48-h LC_{50} value was calculated follwing recommend procedure (APHA-AWWA & WPCF-1971). Five groups of twenty fishes were kept in separate aquaria. Group I was treated as control. Four different concentration of metasystox (2, 3, 4, and 5ml/L.) was used for treating the test animals. LC_{50} was found to be 3ml/L since 50% population of the fishes died in group II exposed to the said concentration. Surviving fish were used for further studies.

Blood was sampled from the caudal peduncle of fishes using a heparinized tuberculin syringe. Blood film was stained with Leishman's stain. Parameters like total count, differential count, haemoglobin per cent were studied following routine techniques. Cells were measured by 'Sico' make micrometer.

RESUTLS AND DISCUSSION

The present effects of metasystox 3ml/L. on *H. fossilis* included cremation, hypochromasia in erythrocytes and significant increase in size of erythrocyte including nucleus. Total erythrocyte count and haemoglobin percentage revaled a decreasing trend. Similar observations of hypochromais, a cremation and shrinkage of cell membrane were observed in Anabus exposed to Rogor (Singh, 1990). Chakarbarty and Banarjee, (1988) also reported simulating findings in *Channa punctatus* exposed to lethal exposure of organophosphate pesticides and gradual increase in toxicity of Mahua oil cake respectively. Chauhan et. *al* (1983) reported hypochromasia and vaculation of erythrocytes in *Puntius-ticto* exposed to herbicide but in present case inspite of hypochromasia no vacuolar degeneration could be marked.

Increase in erythrocyte surface area finds support with the findings of Chakrabarty and Banerjee (1988) in *Channa-punctatus* exposed to thimet 10G and Elsan 2% dust.

Increase in total erythrocyte count (TEC) and Haemoglobin percentage (HB%) as found in present case have also been reported earlier by Mahajan & Juneja (1979) in *Channa punctatus*; Dhilon & Gupta (1983) in *Clarias batrachus* and Ghosh & Chatterjee (1986) in *Sarotherodth-mossambicus* due to Aldrin exposure.

In contrast to the finding of Kumari *et. al* (1989) who reported significant increase in TLC (Total Leucocyte count) in *Clarias-batrachus* exposed to organochlorine pesticide (DDT and Lindane). The present pesticide (Metasystox) causes significant decrease in TLC. Mishra & Shrivastava (1985) have also reported increase in TLC of H. *fossilis.* The fish were treated with dichlorous and lindane. *Clarias-batrachus* exposed to the same pesticide also reflected increase in TLC (Natrajan 1981).

In lethal exposure of metasystox (5ml/L. for48-h) *Channa straitus* reflected decreased TEC, Haemoglobin perecentage and TLC where as when the same fish were exposed to 48h LC_{50} of Thimet 10G and Banzanon 5G the same fish reflected significant increase in TEC but a fall in haemoglobin level. (Natrajan 1981).

Thus metasystox induces significant changes in TLC, TEC and Hb% reflecting its toxic effects in *Heteropneustes fossilis.*

Table 14.1

Blood parameter of *H. Fossilis* following 48-L LC_{50} exposure to Metasystox

Parameters	Control	Treated
Erythrocyte length (μm)L	9.12±0.01	9.17±0.03
Erythrocyte breadth (μm)B	8.5±0.01	8.72±0.04
Nucleus length (μm)L'	3.42±0.01	4.08±0.02
Nucleus breadth (μm)B'	3.28±0.01	3.58±0.01
L/B (Cell)	1.07:1	1.05:1
L'/B' (Nucleus)	1.03:1	1.1:1
Hb%	63% to 76%	60% to 62%
Small lymphocyte	12%	10%
Large lymphocyte	18%	20%
Monocyte%	00	00
Neutrophil%	06	08
Eoshophil%	10	1
Basophil%	04	00
TLC	1632 to 1964 mm^3	1600 to 1820 mm^3
TEC	$169x10^3xmm^3$ to $189x10^3mm^3$	$143x10^3mm$ to $161x10^3mm^3$

TLL = Total Lecocyte count, TEC = Total Erythrocyte count.

Hb%= Haemoglobin percentage.

ACKNOWLEDGEMENTS

The authors are grateful to Dr. N. K. Dubey Prof. & Head for provided laboratory facilities during the investigation and Dr. Shishir K. Verma Principal Investigator, I. C. A. R Project, University Department of Zoology, L. N. Mithila University, Darbhanga for having critically gone through the manuscript and suggesting improvements.

REFERENCES

APHA- AWWA & WPCF (1971). Standard Method, American Public Health Association, Washington D. C.

Basak, P. K and Konar, S. K. (1980) Proc. Nat. Acad. Sci. India 50(B) 218.

Chauhan, M. S.; Verma, D. and Pandey, A. K. (1983). Herbicide Induced haem a to begical changes and their recovery in fresh. water fish, *Puntius ticto* (Ham.) Comp. Physiol. Ecol. 8 (A): 249-251.

Chakraborty, P. and Banerjee, V. (1988). Effect of orgnophosphorus pesticides on the peripheral haemogram of the fish, Channel punctatus, environ & Ecol., 6(2): 390-394.

Dhilon, S. S. & Gupta, A. K. (1983). A clinical approach to study the pollutants intoxication in fresh water *clarias batrachus*, water air and soil pollution, 20: 63-68.

Employment news courtesy, Vigyan Prasar, 7-13 Aug. 1999.

Ghosh T. K. & Chatterjee, S. K. (1986). Effect of Aldrin on some haematological variables of a fresh water fish Sarotheryodon mossambicus, Environ, Ecol., 4(3): 427-429.

Hoemechaudhuri,S.; Pandit.T.; Poddar, S. & Banerjee, S.(1986). Proc. Ind. Acad. Sci.,9;617.

Kumari, M.; Kumari, G. & Yadav, S.C. (1989). Perils of environmental pollution by pesticides; an Assessment through leucocytic and haemostatic response of a fresh water fish *Clarias batrachus* (L) Him. J. Env. Zool., (3): 36-43.

Mahajan, C. L. & Juneja, C. S. (1979). Effects of Aldrin on peripheral blood of a fish *Channa punctatus*, Indian J. Environment Health, 21: 169-172.

Mishra, J. & Srivastva, A. K. (1985). Effect of dichlorous on the blood and tissue chemistry of the Indian catfish. Proc. Nat. Symp. Industry. Pollu. Pest., Bhopal tragedy, Gorakhpur.

Natarajan, G. M. (1981). Changes in the bimodal gas exchange and some blood parameters in the air breathing fish Channa Striatus (Blacker) following lethal (LC_{50} /46 hrs) exposure to metasystox (emeten), Curr. Sci., 50 (1): 40-41.

Sharma G. D. (1991). Toxicity of BHC on the gills of *H. fossilis* proceeding of the seventy eight session of Indian science Congress, Indore.

Singh, S. K. P. (1990). Changes in Haematological *parameters* and haemopoeticttissue of *Anabas festidineus* (Bloch) under influence of stress. Ph. D Thesis, Patna University, Patna, India.

Srivastava, A. K. & Mishra, J. (1985). Lindene induced haematological changes in catfish, Nat. Acad. Sci Letter., 8: 391-394.

CHAPTER 15

Anomalies in Haematological Parameters of an Indian Major Carp, *Labeo Rohita* (Ham.) Under Copper and Chromium Poisoning

Arun Kumar Singh, K. S. Rana and Vishal Tiwari*

Department of Zoology, Agra College, Agra-282002, U.P., India.

**Department of Zoology, R.B.S. College, Agra-282002, U.P., India.*

ABSTRACT

In the present investigations, the toxicity of copper sulphate and potassium dichromate on haematological parameters of an Indian major carp, *Labeo rohita* (Ham.) has been studied. RBC, haemoglobin percentage, PVC and MCHC values were significantly decreased while WBC, MCV & MCH were increased considerably upon treating the fish with both metallic compounds independently. Chromium proved more toxic than copper.

Keywords: *Industrial effluents, Labeo rohita, Haematology, Copper & Chromium,*

INTRODUCTION

The aquatic environment remains the traditional recipient of many industrial wastes containing heavy metals. Rapid industrialization has left behind various effluents which are added to the aquatic system. Metallic compounds in these effluents accumulate in various parts of animal or plant body (Barman *et al.*, 1994) causing huge mortification of inhabiting biomass including palatable fish. Toxicity of metals to fish has been reviewed by various workers (Pickering 1974, Agarwal *et al.*, 1980, Subramaniam *et al.*, 1993).

The problem of toxicity of metals and metallic compounds in water is very complex. The effluents from artifcial silk factories, smelters, paper mills and chemical industries are the main source of copper and chromium contamination. These toxic chemicals, if introduced in higher concentration, not only affect fish population but also the food web representatives of many commercially important food fishes (Panwar *et al.*, 1977). Blood is a pathophysiological reflector of whole body and therefore, blood parameters are an asset in diagnosis the structural and functional status of body organs exposed to toxicants (Bansal *et al.*, 1979.)

The present study reports some anomalies in haematological parameters of an Indian major carp, *Labeo rohita* under copper and chromium poisoning.

MATERIALS AND METHODS

A 20 days chronic exposure was conducted with copper and chromium with *Labeo rohita.* The test fish *Labeo rohita* was procured from local fishery resources of Agra and acclimated to laboratory conditions in big glass aquaria (60 X 30 X 30 cm) for a week before conducting an experiment. They were regularly fed and the water of aquaria was changed periodically to avoid any mortality due to accumulation of their excreta. Thirty healthy fish weighed 40-50 gm and 12-15 cm long were then selected for study. The fish were divided in three batches of 10 fish each.

The 1st and 2nd batches were injected by 0.01% solutions (randomly selected) of $CuSO_4$ and $K_2Cr_2O_7$ respectively in caudal region. The fish of 3rd batch were treated as control and injected by normal saline (0.038% KC1) (Hoar and Randall 1971). The injections were given on alternate days in all the three sets of fish. After 10 injections fish sacrificed and blood was collected from caudal vein in EDTA (an anticoagulant) coated vials. The RBC and WBC were counted by the Neubar haemocytometer. Haemoglobin percentage was estimated by Sahli's haemocytometer (Germany). Haematocritic value (Packed cell volume = PCV) was determined by standard Wintrobe method. The absolute blood parameters (MVC, MCH and MCHC) were calculated using standard methods of Dacie and Lewis (1977). The experiments were conducted at room temp (22±2°C). Student 't' test (Fisher 1983) was applied for statistical evaluation of the data.

RESULTS AND DISCUSSION

Comparative data of blood values of fish after copper and chromium treatment along with control are present in Table 15.1. It is clear from the table that there is a significant decrease in RBC, HB%, PCV and MCHC, while there is a considerable increase in WBC, MCV and MCH under stress of both toxicants as compared to control. The data revealed more significant changes with chromium treatment than copper.

Table 15.1 Haematological irregularities in Labeo rohita exposed to cu & cr.

Blood Values	Control	Copper	Chromium
RBC X 10^6/cmm.	3.38±0.02	2.12±0.01^c	1.68±0.03^c
WBC X 10^3/cmm.	19.96±0.52	41.12±1.19^c	42.30±0.50^c
PCV%	50.20±0.31	43.76±0.62^b	38.50±0.52^c
Hb/dl	13.68±0.17	11.60±0.12^b	8.40±0.13^c
MCV/um^3	157.39±3.71	226.23±4.96^c	258.62±5.22^c
MCH pg.	35.48±0.39	66.83±0.89^c	52.89±0.72^c
MCHC g/dl.	23.26±0.28	20.79±0.46^b	20.54±0.36^a

All values are mean ± S.E. (15 estimations); a, P<0.05; b, P < 0.01; c, P < 0.001.

Metallic ions present in river water enter the fish body through permeable surfaces and they get accumulated in various organs like liver and kidney (AL-Mohanna 1994). These metallic ions are the main probable cause of the physiological anomalies in fish. The fall in RBC count, HB% and PCV, upon treatment with both copper and chromium shows symptoms leading to hypochromic mycrolytic anaemia which is ascribed to iron deficiency and a subsequent reduction in haemoglobin synthesis, Christensen *et al* (1972) have also reported fall in RBC count, PCV and Hb% in brown bull head after metallic stress.

The increase in MCH, MCV and decrease in MCHC values, as observed in this study, is the sign of macrocytic anaemia (Lynch *et al* 1969). The chromium and copper produced leukocytosis (increased leukocytes) probably due to the tissue damage and subsequent removal of cell debris (MCLeay and Brown 1974).

The toxicity test help in evaluating the nature and degree of harmful effect produced on the aquatic organisms especially fish by toxic substances and provide valuable information to be used for protection of aquatic ecosystem as a whole (Annon, 1975). Therefore, industrial effluents containing these metals should be treated prior to release into environment.

REFERENCES

Agarwal, S.J. and Srivastava, A.K. 1980. Haematological responses in a fresh water fish to experimental manganese poisioning. *Toxicology.* 17: 97-100.

Al-Mohanna, M.M. 1994. Residues of some heavy metals in fishes collected from (Red Sea Coast) Jizan, Saudi Arabia. *J. Environ. Biol.* 15(2): 149-257.

Annon. 1975. In: *Report offish toxicity testing procedure.* EIFAC/T 24. European Inland Fisheries Advisory Commission, F.A.O., Rome.

Bansal, S.K.; Verma, S.R.; Gupta, A.K. and Dalela, R.C. 1979. Physiological dysfunction of the haemopoitic system in a fresh water teleost, *Labeo rohita* following chronic chlordane exposure. Part 1. Alteration in certain haematological parameters. *Bull. Environ. Contain. Toxicol.* 22: 666-673.

Barman, S.C. and Lal, M.M. 1994. Accumulation of heavy metal (Zn, Cu, Cd & Pb) in soil and cuttivated vegetables and weeds grown in industrially polluted fields. *J. Environ. Biol.* 15(2): 107-115.

Christensen, G.M. ; Mckin, J.M.; Brungs, W.A. and Hunt, E.P. 1972. Changes in the blood of the brown bull head, *Ictalurus nebulosus* following short and long term exposure to Copper (II). *Toxicol. Appl. Pharamacol.* 23: 417-427.

Dacie, J.V. and Lewis, S.M. 1977. *Practical Haematology,* ELBS, Churchill, Livingston.

Fisher, R. A. 1983. Statistical methods for research workers. 13th edn. Oliver and Boyd, London.

Hoar, W.S. and Randall, D.J. 1971. Fish Physiology Vol. VI-Environmental relations and behaviours. Academic press, New York, pp. 519-521.

Lynch, M.J.; Raphael, S.S.; Mellor, L.D.; Sapre, P.D. and Inwood, M.J.H. 1969. Medical laboratory technology and clinical pathology. W.B. Saunders Co., Philadelphia.

McKim, J. M.; Christensen, G.M. and Hunt, E.P. 1970. Changes in the blood of brook trout, *Salvelinus fontinalis* after short and long term exposure to copper. *J. Fish. Res. Bd. Can.* 27: 1883-1889.

McLeay and Brown, D.J. 1974. Growth stimulations and biochemical changes in juvenile coho salmon, *Onchorhynchus kistuch* exposed to bleached kraft pulpmill effluent for 200 days. *J. Fish. Res. Bd. Can.* **31**: 1043-1049.

Panwar, R.S.; Gupta, R.A. and Mohamed, M. Peer. 1977. Bioassay to determine toxicity of selected pesticide to bivalve, *Parreysia favidens* (Benson). Paper presented in the seminar on "Industrial Wastes", Joint collaboration of NEERI and Metropolitan Calcutta, Dec. 8-9, 223-234.

Pickering, O.H. 1974. Chronic toxicity of Nickel to fathead minnow. *J. Water. Poll Control Fed.* **46**: 760.

Sastry, K.V. and Sharma, K. 1980. Mercury induced haematological and biochemical anomalies in *Ophiocephalus punctatus. Toxicol. Lett.* 5: 245-249.

Shivraj, K.M. and Patil, H.S. 1998. Toxicity of cadmium and copper to a freshwater fish, *Punctius arulius. Environ & Eco.* 6: 1.

Subramaniam, M.A. and Varadraj, G. 1993. Impact of Tannery effluents on biochemical constituents in the haemolymph of the fresh water prawn, *Macrobrachium idella* (Heller). *J. Environ. Biol.* 14(3): 255-259.

CHAPTER 16

COMBINED EFFECTS OF THYROID AND PITUITARY HORMONE ADMINISTRATION ON POST-EMBRYONIC DEVELOPMENT IN *CIRRHINA MRIGALA* (HAMILTON)

Rita Rani, K.K. Gaur and Vishal Tiwari

Department of Zoology, R.B.S. College, Agra-282002.

ABSTRACT

Combined (thyroid and pituitary hormone) treatment given to the larvae of *Cirrhina mrigala* for about one month revealed faster post-embryonic development as against the control experiments. 100+2 mg/lit. dose of combined (thyroid and pituitary hormone) proved most suitable. Mortality figure comes to 4, 6, 5 and 10 for control, 75 +1 mg/lit., 100 + 2 mg/lit. and 125 + 3 mg/lit. dose treatments respectively. Morphometrical parameters showed marked changes in all thyroid and pituitary hormone treated experiments, suggesting that hormone therapy may prove useful in the culture of *Cirrhina mrigala.*

Key words: *Thyroid hormone (T), Pituitary hormone (P), Post-embryonic development, Cirrhina mrigala.*

INTRODUCTION

The thyroid and pituitary hormones have been known as the most important biochemical substances, which are responsible for metamorphic differentiation during larval development along with many physiological process of prime importance during growth periods too (Belsare *et al.*, 1966).

The maintenance of adequate balance of thyroid and pituitary hormone levels are important prerequisite for proper management of reproductive biology, development, growth and it may prove important regulating factor. The effect of thyroxine hormone treatment along with the pituitary in a combined form on the post-embryonic development of *Cirrhina mrigala* is not yet fully known and needs further investigation. With this view point, in the present communication, an effort has been made to work out the optimum combined dose of thyroid and pituitary hormones, which could be able to affect some aspects of post-embryonic development in a positive direction.

Some of the important studies on the said field are from Hoar *et al.*, (1955); Lostroh *et al.*, (1958); Scow (1959); Sneed *et al.*, (1961); Nicall *et al.*, (1965); Belsare *et al.*, (1966); Yamazaki(1976); Holder *et al.*,(1980); Miwa et al., (1985); Tay *et al.*, (1997); and that of Nayak *et al.*, (2000).

MATERIALS AND METHODS

The hatchlings (3-4 days old) of *Cirrhina mrigala* were collected from fish farm in the close vicinity of Agra. The hatchlings for consecutive development stages were brought to laboratory and transferred to plastic trough under observation. In each experimental trough, 20 specimens were placed on the basis of proper stocking density (approximately 2 fish/lit.). The physico-chemical conditions of the experimental troughs were maintained nearly natural by using sand particles, hydrilla plants, zoo and phytoplankton's. Hitachi readymade food pellets in equal quantity were also given at regular intervals to all the experimental larvae. The combined (thyroid and pituitary hormone) doses of 75 +1 mg/lit., 100 + 2 mg/lit. and 125 + 3 mg/lit. concentrations were prepared and in each trough one kind of them was added.

By random sampling five specimens from each kind of experiment and control were taken out for derivation of mean standard under different aspects of studies on post-embryonic development at a successive time period viz., 5,7 and 15 days. The same course was applied after lapses of 21 and 30 days in order to study later developmental stages. Morphometrical parameters of the larvae were recorded with the help of vernier calipers at the end of the experiments.

RESULTS AND DISCUSSION

On the perusal of parameters indicated in the Table 16.1, it can be concluded that a dose of 100 + 2 mg/lit. proved to be the best in terms of post-embryonic development.

Table 16.1

Morphometry of *Cirrhina mrigala* (Ham.) larvae treated with combined doses of thyroid (T) and Pituitary (P) hormone (mg/lit.)

S. No.	Morphometrical Parameters	Control	Treated Experiments		
		000 mg/lit.	75 + 1 mg/lit. (T+P)*	100+2 mg/lit (T+P)*	125+3 mg/lit. (T+P)*
			(Average in mm. ± S.D.)		
1.	Standard length.	21.879±.048	33.247±1.564	37.535±1.533	37.067±1.337
2.	Body lenngth	13.397±.010	22.242±0.892	25.295±1.652	25.66±.843
3.	Head length	6.307±0.36	8.322±0.867	9.562±0.907	8.55±1.232
4.	Distance between snout and and pore	14.892±0.061	21.987±1.793	26.170±1.580	23.685±0.963
5.	Distance between snout and anal fin base.	16.677±.090	26.382±3.176	29.025±.883	29.46±.812
6.	Distance between snout and dorsal fin base.	13.552±0.068	21.332±.996	22.852±1.317	22.705±.811
7.	Height of the body at anal fin base.	4.887±.109	7.222±.925	7.617±1.091	7.522±.659
8.	Height of the body at dorsal fin base.	6.747±.074	9.405±.965	9.690±.818	8.24±1.305
9.	Mortality.	Four	Six	Five	Ten

*Thyroid & Pituitary dose

The general morphometrical parameters of *Cirrhina mrigala* larvae revealed that at a dose of 100+2 mg/lit., Standard length 37.535 mm; body length 25.295 mm; head length 9.562 mm; distance between snout and anal pore 26.170 mm; distance between snout and anal fin base 29.025 mm; distance between snout and dorsal fin base 22.852 mm; height of the body at anal fin base 7.617 mm; and height of the body at dorsal fin base 9.690 mm was observed, which is significant in terms of the best dose estimated in correspondence to post embryonic development.

It has also been marked that during early developmental periods when morphogenesis is on its full swing, the combined thyroid and pituitary hormone dose (100+2 mg/lit.) resulted into maximum increment in standard body length. Survival percentage was maximum in control,which gradually decreased at 75+1 mg/lit., 100+2 mg/lit. and 125+3 mg/lit. dose treatments (Table 16.1). Therefore it can be infered that 100+2 mg/lit., dose seems to be the best effective dose in respect to post-embryonic development and low mortality of *Cirrhina mrigala* larvae.

ACKNOWLEDGEMENT

The authors are thankful to the principal, R.B.S. College Agra for providing the essential research facilities during this investigation.

REFERENCES

Belsare, D. K.; Zargar, V.N. and Vermar, S.M. 1966. The effect of thiourea and thyroxine on the developing larvae of *Channa punctatus* (Bloch.). *Zool. Polom.* **16:** 144-154.

Hoar, W.S.; Keenleyside, M.H. and Goodall., R.C. 1955. The effect of thyroxine and gonadal steroids on the activity of salmon and gold fish. *Can. J. Zool.* 33: 428-439.

Holder, A. T.; Walls, M.; Biggs, P. and Preece, M.A. 1980. Effect of growth hormone, prolactine and thyroxine on the body weight-somatomedin-like activity in vivo sulphation of cartilage in hypopituitary dwarf mice. *J. Endocrinol* 84: 35-47.

Lostroh, A.J. and Choh, H.L. 1958. Effects of growth hormone and thyroxine on body weight of hypophysectomized C_3H mice. *Gen. Comp. Endocrinol.* 62: 484-491.

Miwa, S. and Inui, Y. 1985. Effect of L-thyroxine and bovine growth hormone on smoltification of a mago-salmon, *Onchorhynchus rhodurus. Gen Comp. Endocrinol.* 58 (3): 436-442.

Nayak, P.K.; Mishra, T.K. and Mishra, J. 2000. Combined application of cortisol and thyroxine in the culture of larvae of freshwater catfish, *Heteropneustes fossilis. J. Aquacult. Trop.* 15 (4): 323-328.

Nicall, C.S.; Bern, H.A.; Dunlop and Strohman, R.C. 1965. Prolactin, growth hormone, thyroxine and growth in Tadpoles of *Rana Catesbeiana. Amer. Zool.* 5: 738-739.

Scow, R.0.1959. Effect of growth hormone and thyroxine on growth and chemical composition of muscle, bone and other tissue in thyroidectomized rats. *Am. J. Physiol.* **196:** 859-865.

Sneed, K.E. and Dupree, H.K. 1961. The effect of thyroid stimulating hormone combined with gonadotropic hormones on the ovulation of goldfish and green sunfish. *Prog. Fish. Culturist.* 23: 179-182.

Tay, H.C.; Young, A.N.; Goh, J.; Him, H.S.; Chao, T.M.; Chou, R. and Lam, T.J. 1997. Effects of treatment of eggs with triiodothyroxine and cortisol on larval morphology and survival in the greasy grouper. ***Aquacult. Internat.*** **54:** **189-195.**

Yamazaki, F. 1976. Application of hormone in fish culture. *J. Fish. Res. Bd. Can.* 33: 948-958.

CHAPTER 17

HAEMATOLOGICAL CHANGES IN A HILL STREAM FISH *GARRA GOTYLA GOTYLA* (GRAY) INFECTED WITH TRYPANOSOMES

J.V.S. Rauthan, Geeta Rauthan and S.P. Grover

Fisheries Research Laboratory, Department of Zoology
D.A.V. (P.G.) College, Dehradun — 248001

ABSTRACT

Infected hillstream fish *Garrra gotyla gotyla* revealed the presence of Trypanosomes in the blood which causes conspicuous fall in total erythrocyte count (TEC), packed cell volume (PCV), Haemoglobin content (Hb), MCHC content and rise in total leucocyte cound (TLC), Erythrocyte sedimentation rate (ESR), Mean corpuscular volume (MCV). The differential leucocyte count (DLC) revealed a rise in the small lymphocyte, basophils and monocylcs, while the neutrophils, throinbocytes, large lymphocytes and eosinophils fall in the infected fish.

Key Words : *Fish Garra gotyla gotyla, Trypanosomes, TEC, TLC, Hb, PCV, ESR, MCHC, MCV **and DLC.***

INTRODUCTION

Presence of tiypanosomes in fishes has been reported from all over the world. Smirnova (1971), Tandon and Joshi (1973, 1977), Joshi (1979), Joshi and Dabral (1981), Shnan, et al. (1984), Gupta and Agarwal (1985), Gupta and Gupta (1987), Joshi (1989), Sharma and Joshi (1991), and Rauthan (1993). The paper present results on some haematological values obtained from the hill stream fish *Garra gotyla gotyla* **infected with trypanosomes under natural condition** of Doon Valley.

MATERIALS AND METHODS

Live fishes were collected from the river, Song near Railwala and Lacchiwala. The fishes were brought to laboratory in plastic containers and transferred to glass aquaria for 10 days acclimatization. During acclimatization the fishes were fed on usual diet. For all kind of haematological investigation blood was collected from branchial vein using a microsyring with 24 gauge needle. The fishes were sacrificed and studies on the peseted haematological parameters. Out of these 30 specimens. 12 were

Table 17.1
Haematological changes in hillstream fish *Garra gotyla gotyla* (Gray), All values are mean ± SE for six specimens each.

PARAMETERS	HEALTHY FISH	INFECTED FISH
TEC (x 10^6/cmm)	2.89±0.22	2.06±0.20
TLC(x10^3/cmm)	8840.81±120.2	10742.00+205.0
Hb (gm %)	10.4±1.03	7.5±1.06
PCV (%)	32.0±2.0	27.0+1.80
ESR (mm/hour)	1.12±0.49	2.30±0.20
MCV (um^3)	170.0±10.0	202.0±13.0
MCHC (%)	34.2+1.4	27.3+1.6

Table 17.2 :
Differential Blood Cell Counts (%)

PARAMETERS	HEALTHY FISH	INFECTED FISH
Large lymphocyte	7.2±0.52	5.3±1.2
Small lymphocyte	25.51±2.3	30.0±3.0
Monocyte	1.31±0.56	2.40±0.30
Neutrophil	20.2±3.5	16.2±2.6
Eosinophil	0.38±0.2	2.2±0.4
Basophil	1.60±0. 4	1.90±0.6
Thrombocyte	42.2	30.3±4.6

found trypanosomes. The studies were made on the selected parameters viz. total erythuocyte count (TEC), Total leucocyte count (TLC), Haemoglobin contents (Hb), Packed Cell Volume (PCV), Erythrocyte Sedimentation Rate (ESR.), Mean Corpuscular Haemoglobin Concentration (MCHC), Mean Corpuscular Volume (MCV) and Differential Leucocyte Count (DLC) following the standard methods as used Rauthan(1993).

RESULTS AND DISCUSSION

TEC was found lowered in diseased fish as compared to the healthy ones. The fall in diseased fish was 28.0% against the normal healthy fish (2.89 ±0.22x10^6/cmm) The Haemoglobin content (Hb) were also found depleted by 27.8% in diseased fish against the healthy fish (10.4±1.03 gm%).

The TLC value was found to have increased under disease by about 16.1% against the healthy fish value of 8840.81±120.2 x 10^3/cmm. The PCV value also showed a fall of about 15.6%. The MCHC was found lower in infected fish as compared to the healthy fish 20.1%. The MCV value had also rise in the infected fish by about 15.8% and the ESR. value rose by about 51.3% as compared to the healthy fish values. The differential cell counts also revealed conspicuous change. The number of small lymphocyte, bisophils and eosinophils rose due to infection, but the number of thrombocytes and neutrophils full in the diseased fishes, as compared to the healthy fishes. The number of large lymphocytes, like those of the neutrophils also fell by a little margin in infected fish as compared to the health ones (Table 17.2). The onono-cytes were found higher in infected fishes as compared to the healthy fishes.

Parasitic infections have been found to destroy erythrocytes ami cause anemia (Ven Brand 1973). Decrease Hb percentage, high value of ESR. and fall in PCV value is infected fishes in the present study are clear indications of anaemia. The present study tends support to Englel and Davis' (1961), Smirnova (1971), Tandon and Joshi (1973), Joshi (1979), Joshi and Dabral (1981) and Joshi (1989).

As many as ten species of fresh water teleosts having haemoflagellate trypanosomes in their blood fully revealed that trypanosomiasis cause definite alterations in various blood component TEC, Hb and PCV values fall due to these haemaflagellates while the TLC usually rise (Tandon and Joshi 1973) but Joshi and Dabral (1981) reported the TLC value fell sharply and significantly between 60-100% under heavy infection. The RBC, WBC, Hb and PCV values were found to have depleted in cat fish. Kumar, et. al. (1984) reported TLC and ESR values are highest in *Schizothorax rachardsonii* severally suffering from black spot disease. Smirnova (1970) in the perch *Lota lota* infected with trypanosomes noted decrease TEC, and Hb but increased number of leucocytes.

Similarly a recent report by Gupta and Gupta (1990) following experimental inoculation of trypanosomes in the fresh water shark as *Wallago attu* also found severe anaema 39 days post infection, where TEC had fallen from 2.4 to 1.6 x 10^6/cmm and Hb from 11.4 to 8.0 gm%. The differential blood cell counts also revealed noticeable changes. The small lymphocytes, basophils, monocytes and Eosinophils rose in the infected fish (Table 17.1), while the number of neutrophils and thromocyte and large lymphocyte fall (Table 17.2).

ACKNOWLEDGEMENTS

The authors wish to express their indebtedness to Dr. J.K. Srivastava, Head, Department of Zoology, D.A.V. (P.G.) College, Dehradun for valuable suggestions and critical reading of the manuscript.

REFERENCES

Engel, DW and Davis, EM, 1964 : Relationship between activity and blood composition in certain teleosts. *Copeia,* 31 : 586-587.

Gupta, N. and Gupta, D.K., 1986 : Trypanosome infectivity and changes in the glucose level of two fresh water fishes. *J. Parasitology* 10(2): 213-213.

Joshi, B.D., 1979 : Hyperlactemia in the fish blood infected with trypanosomes. *Sci.* Cult., 45(8): 320-321.

Joshi, B.D. and Dabral, R., 1981 : Some haematological changes in a fresh water cat fish *Heteropneustes fossilis* infected with trypanosome. *Proc. Ind. Acad Sci.,* 90:295-301.

Joshi, B.D., 1989 : Physico-pathological studies on the blood of few hill stream Teleosts. *Proc. Nan. Symp.* : 127-137.

Kumar, A., and Singh, H.R., 1984 : Haematological changes in cold water fish *Schizothorax plogiostomus* (Heckel), infected with black spot disease. *Ind. J. Parasit,* 8(1): 129-131.

Rauthan, J.V.S., 1993 : *Studies of the Haematological variations in some fishes of Doon Valley.* Thesis submitted to H.N.B. Garhwal University. Srinagar (Garhwal), 1-221.

Radhakrishnan, S., Nair, N.B., and Balasubramanian N.K., 1984 : *Gymnorhycus gigas* plerocercoid infection of the liver of *Diodon hystrix* (Pisces : Diodontidae) II. Haematological changes infected fish. *Fish* and *unwelt,* 13 : 27-39.

Smrihova, L.I., 1970 : Trypanosome in the blood *of Lota lota* L. (Tryponosoma lotai sps. Nov.) *Parasitologiya* 4 : 296-297.

Tandon, R.S. and Joshi, B.D., 1973 : Studies on the physiopathology of blood of fresh water fishes infected with two new forms of trypanosomes. Z. *wies* Zool. *Leipzg.,* 185 (314) : 207-221.

Von. Brand, T. 1973 : *Biochemistry of Parasites.* 2nd Ed. Academic Press, New York.

CHAPTER 18

TOXICITY OF AN ORGANOCHLORINE INSECTICIDE, BHC TO COMMON CARP, *CYPRINUS CARPIO* LINN (FAMILY : CYPRINIDAE)

R. Punitha, N. Krishnan and G.N. Subbiah*

Department of Zoology, Madura College, Madurai-6250011, India.

**Department of Zoology, Thiagarajar College, Madurai - 625 009, India.*

ABSTRACT

Toxicity of an organochlorine insecticide, BHC (50% WP) to fresh water cultivable carp *Cyprinus carpio* Linn was determined by static bioassay test. The LC_{50} value for 96 hr exposure duration is 10 ppm.

Keywords : *Cyprinus carpio*, Organochlorine pesticide, BHC, Toxicity,

INTRODUCTION

Aquatic pollution in fresh water ecosystems is caused by a wide range of chemical pollutants. Insecticides employed in pest control in India like agricultural country have contaminated the fresh water habitats remarkably. The problem of insecticidal pollution and environmental hazards is quite serious in India with reference to organochlorine insecticides. Acute toxicity studies in fish models have provided basic data such as nature of the toxicant, susceptibility of individual species, lethal concentration to 50% of the exposed fish species (LC_{50}) and determination of sublethal concentration of toxicant for further investigations. A large number of publications were available on acute toxicity of several insecticides (in terms of mortality and LC_{50}) to cultivable fishes (Sivaprasad Rao and Ramana Rao, 1982, Khillare, 1990; Ramesh et. el., 1992; Ramalingam et. al., 1992 ; Choudhary et. al., 1993). Similar kind of investigations were made on fishes with reference to other aquatic pollutants such as detergents (Jayasuriya et.al., 1991) agriculture fertilizers (Saxena and Chauhal, 1993) herbicides (Pulla Rao and Mishra, 1997) and tannery effluent (Jeyachandran and Chockalingam, 1986). The present study evaluates the acute toxicity of an organochlorine insecticide, BHC to fresh water carp, *Cyprinus carpio* adapting static bioacsay method.

MATERIALS AND METHODS

Healthy fingerlings of *C. carpio* (4 + 0.5g / live weight) were obtained from local fish farm and acclimatized to the laboratory conditions at 29 +1 C° for 7 days before experimentation. They were fed daily with chopped goat liver but food was withheld 24 hrs before bioossey test. The water was changed daily with an intermittent aeration. For the present study BHC, an organechlorine insecticide was selected as experimental toxicant. The commercial grade BHC (50% wp; trade name Hexidole) is obtained from Rallis India Ltd. Its stock solution in distrilled water (1g in 500ml) is prepared and then diluted to required test concentrations (8-15 ppm). The acclimatized fishes, ten in each group were exposed to test concentration along with suitable control. Three replicated were maintained for each treatment. Water and concentrations of BHC in aquaria were renewed every 24hr. Mortality rate was observed for 96 hr in each test concentration. LC_{50} for 96 hr was computed by plotting the percentage mortality against the concentration of BHC (Abbott, 1925). The relationship between the concentration of BHC and probit values is also analysed further by simple correlation and linear regression.

RESULTS AND DISCUSSION

Both percentage mortalities and probit values were increased with increasing test concentration of BHC as well as exposure duration. (Table 18.1). It is confirmed by high positive correlation coefficients for the relationship between concentrations of BHC and probit values and also by linear regression equations (Table 18.2). The LC_{50} for 96 hr. exposure duration was 10 ppm. Acute toxicity of BHC to fishes varies with species of model fish. For instance the 96 hr LC_{50} was 5.25 ppm to *Sarotherodon mossambicus* (Ramalingam et.al., 1992), 4.2 ppm to *Rasbora daniconius*. (Choudhary et. al., 1993) and 10 ppm to carp in the present study. In contrast, variations in LC_{50} were recorded for the same fish species with reference to diferent kinds of toxicants. According to Ramalingam et. al., (1992) 96 hr. LC_{50} value to *C. carpio* differs like 0.039 ppm for DDT (organochlorine pesticide). 3.8 ppm formalathion (organophosphorus pesticide) and 0.38 ppm for mercuric chloride (heavy metal). The LC_{50} values of chlorpyriphos and endosulfan to *S. mosambicus* were 0.14 and 0.0042 ppm respectively (Kamble et al., 1999). In addition, variations in LC_{50} to the same fish with reference to bioassay method were also reported. For instance,atrazine (herbicide) toxicity to *P.ticts* varies with static and residual oxygen methods (Pulla Rao & Mishra 1997). LC_{50} of a toxicant to a fish species depends upon several factors such as chemistry and mode of action of toxicant, genotype of the fish, exposure duration, effectiveness of detoxification meahcnism in fish and variations, in detoxifying enzyme systems. (Pritchard, 1993). In general organochlorine insecticides are highly toxic when compared to organophosphorous insecticides to fishes (Khillare, 1990 ; Ramalingam et.al., 1993). Several workers have also determined the LC_{50} of other aquatic pollutants such as herbicide (atrazine : Pulla Rao and Mishra, 1997), fertilizer (NPK and ammonium phosphate: Saxena and Chauhan, 1993), detergents (Nirma and Surf: Jayasuriya et.al., 1991) and tannery effluent (Jeyachandran and Chockalingam, 1986) to fishes.

ACKNOWLEDGEMENTS

This work was carried out by RP at Thiagarajar College under the guidance of Prof. GNS. We are grateful to Board of Madura College, HOD, Principal and all faculty members for encouragement. NK gratefully acknowledge Prof. Dr. Arvind kumar, Editor for his constant support and encouragement.

Table Table 18.1
Percentage mortality (PM in %) and probit values (PV) of *C. carpio* exposed to BHC.

Concen. of BHC (ppm)	Exposure duration (in hrs.) 24 PM	PV	48 PM	PV	72 PM	PV	96 PM	PV
8	10	3.718	10	3.718	20	4.158	30	4.475
9	10	3.718	20	4.158	30	4.475	30	4.475
10	20	4.158	30	4.475	40	4.746	50	5.000
11	30	4.475	40	4.746	60	5.123	60	5.123
12	40	4.746	50	5.000	70	5.521	80	5.841
13	50	5.000	60	5.123	70	5.521	80	5.841
14	60	5.123	70	5.521	80	5.841	90	6.281
15	70	5.521	80	5.841	80	6.281	100	8.090

Table 18.2
Relationship between concentration of BHC and probit values.

Exposure duration (hr)	Corrl. Coeffi. (r)	Slope (b)	Intercept (a)	Linear regression (Y = a + b x)
24	0.99	0.269	1.437	Y = 1.437 + 0.269 X
48	0.99	0.286	1.501	Y = 1.501 + 0.286 X
72	0.97	0.341	1.316	Y = 1.316 + 0.341 X
96	0.92	0.444	0.484	Y = 0.484 + 0.444 X

REFERENCES

Choudhary, M.V., Sherekar, P.Y. and Kulkarni, K.M. 1993. HCH impact on feeding and growth of fresh water fish *Rasbora daniconius. Uttar Pradesh.J.Zool.* 13:139-142.

Jayasuriya, S., Subramanian, M.A. and Varadaraj, G. 1991. Effects of detergents on the oxygen consumption of the cat fish *Mystus vittatus. J. Ecobiol.*, 3 : 217-220.

Jeyachandran, K.P.S. and Chokalingam, S. 1986. Effects of tannery effluent on respiratory parameters and biochemical constituents of fish *Channa punctatus. Comp. Physiol. Ecol.* 12 : 197-200.

Kamble, G.B., Muley, D.V. and Deshpande, V.Y. 1999. Enzymatic changes due to pesticidal stress in a fresh water fish *Sarotherodon mossambicus Indian.J.Env.Ecoplan.* 2 : 27 - 31.

Khillare, Y.K. 1990. Effects of pesticide on fresh water fish, *Puntius stigma.* In *Trends in Ecotoxicology*, Academy of Environmental Biology, Karad. 185-188.

Pritchard, J.B. 1993. Aquatic toxicology : past, present and prospects. *Environ. Health Persp.* 100 : 249 - 257

Pulla Rao, V.P. and Mishra, K.D. 1997. Acute toxicity of herbicides atrazine to fish *Punticus ticts* : a comparison of bioassay test methods. J.Environ.Poll. 4: 161 - 165.

Ramesh, M., Ramanujam, R. and Sivakumari, K. 1992. Effect of water hardner and the toxicity of malathion on haematological parameters of the fresh water fish. *Cyprinus carpio. J. Ecotoxicol. Environ. Monit.* 20 : 31 - 34.

Ramalingam, K., Abdul Rahiman, A. and Meera Bai, R. 1992. Biocides and carbohyderate metabolism in *Sarotherodon mossambicus. J.Ecotoxicol. Environ. Monit2* : 13 - 18.

Saxena, K.K. and Chauhan, R.R.S. 1993. Acute toxicity of agricultural fertilizer complex on food consumption and growth in *Heteropheustes fossilis. Poll. Res.* 12: 135- 138.

Siva Prasad Rao, K. and Ramana Rao, K.V. 1982. Toxicity of methyl parathian to the fresh water teleost, *Tilapia mossambica* (Peters). *Comp. Physiol. Ecol. 7*:27- 274.

CHAPTER 19

Studies on Soil Fertility, Productivity and Ecosystem as Influenced by Introduction of Fish and Prawn in Rice-Fish Integration System

Rajeeb K. Mohanty

Water Technology Centre for Eastern Region (ICAR), Bhubaneswar-751023

ABSTRACT

This experiment was aimed at to study soil fertility, productivity and ecosystem as influenced by introduction of fish and prawn in rice field. In soil, percentage increase in available-N, P and organic carbon in rice-fish culture field was 58.3%, 94% and 54.5% against control respectively while, the unit weight of top soil was 2.3% lighter than the control and porosity was higher by 1.2%. The benthic biomass, indicator of soil productivity, was higher in rice-fish culture field (0.44 ± 0.04 g m^{-2}) than control (0.21 ± 0.05 g m^{-2}). Faster growth rate of *Catla catla* and bottom dwellers were attributed to effective utilization of ecological niches and rich detrital food web. Highest average grain yield was recorded in rice-fish culture field (3595 kg ha-1) which was 20.3% higher against control (2988 kg ha^{-1}). To ensure better productivity, higher planting density must be avoided to provide greater ventilation and illumination, that helps in enhancing both plant and plankton growth which ultimately helps fish growth and yield.

Key words: *Rice-fish integration, ecosystem, water quality, soil fertility, productivity.*

INTRODUCTION

Being a traditionally old practice, rice-fish culture is very much popular in north-eastern parts of India and south-east Asian countries. In India, about 42 million ha of land is under rice cultivation, out of which 20 million is suitable for fish integration (Rao and Ram Singh, 1998). Unfortunately, the carrying capacity of these suitable lands in India have not been utilized to the fullest extent. However, if these lands are brought under integrated rice-fish system, would help to compensate the economic losses in rice production brought about by natural calamities. This will also optimize the water and land use without bringing about environmental degradation. As paddy fields are complex ecosystems where primary producers and consumers at different levels compete with rice for material and energy, decrease overall productivity. However, fish culture in paddy fields can turn and recycle the available material and energy in to fish production, accelerate the productivity of paddy fields and its yield for the benefit of human beings (Brahmanand and Mohanty, 1999). Although much works have been carried out on different aspects of rice-fish integration system (Likangmin, 1988; Huazhu, 1994 and Rao and Ram Singh, 1998), very little information are available on the mechanism of rice-fish culture and the relationship among rice, fish and other factors such as soil, water, fertilization etc. Keeping these in view, an attempt has been made to evaluate rice-fish integration system with special reference to soil fertility, productivity and ecosystem as influenced by introduction of fish and prawn in rice field.

MATERIAL AND METHODS

The present study was carried out at research farm of WTCER, Bhubaneswar, India (Lat. 20°30/ N and Long. 87°48'10" E) during 2000-2001 for two successive years. Three plots of 30 X 10 m size were selected for the proposed study with refuge (approximately 10.2% of field area) of 1.75 m deep and peripheral trench of 0.5 m wide and 0.3 m deep. Dyke height of rice fields were maintained at 20 cm. The excess water from the refuge was drained out through a surface drainage system provided with fine-meshed net to prevent escape of fish. Three plots of similar size and dyke height were kept as control with out refuge. No fish and prawn was introduced in control fields. *Swarna* variety of rice was tried in each plot of the present study including control at a spacing of 20 x 10 cm (between rows and plants) during second week of July each year. The seed and fertilizer application rate was 50 kg ha^{-1} and 80:40:40 (N:P:K) ha^{-1} respectively. 50% of N and full dose of P and K were given as basal at the time of transplanting. The rest of Nitrogen was given in two equal splits during tillering (20 days after transplanting) and panicle initiation stage (45 days after transplanting). Crop growth and yield parameters were recorded at regular intervals. However, no pesticide was used during the experiment.

After proper refuge preparation, liming @ 2000 kg ha^{-1}, manuring with raw cattle dung @ 5000 kg ha^{-1} and fertilization (Urea + SSP, 1:1) @ 3 ppm was carried out prior to stocking. Seven days after refuge preparation, fry of *Catla catla, Labeo rohita, Cirrhinus mrigala, Cyprinus carpio* and post larvae (PL_{30}) of freshwater prawn *Macrobachium rosenbergii* were stocked in the refuge with a species composition of 30:30:15:15:10 respectively. Stocking density of fish fry and post larvae of prawn was 25000 ha^{-1} and rearing was continued for 120 days. Supplemental feed (rice bran + groundnut oil cake, 1:1) @ 10%, 8%, 6% and 4% of mean body weight (MBW) was given twice a day, during 1st, 2nd, 3rd and 4th month to harvesting respectively. Periodic manuring @ 500 kg ha^{-1} and liming @ 200 kg ha^{-1} were carried out at every 15 days interval to maintain plankton population in the ecosystem. Plankton estimation, weekly observation on soil and water quality were recorded using standard methods (APHA, 1989 and Biswas, 1993). Field test instruments were also in use to analyze *in situ* water pH (Checker-1, HANNA, USA), Soil pH (DM-13, Japan), and dissolved oxygen (YSI-55, USA). Crop performance, fish/prawn growth parameters, condition factor of fish and prawn, feed conversion ratio were estimated using standard methods (Mohanty, 1999). To estimate the food preference and feed intake pattern of cultured species, gut content analysis, indices of electivity of different food components (Ivlev, 1961), feed abundance/frequency (Mohanty *et al.*, 2001) were carried out. Every year, during the experiments, 12 numbers of each species were sacrificed for this purpose.

RESULTS AND DISCUSSION

In this experiment, various hydro-biological parameters did not show any distinct trend between the treatment and control except in the cases of dissolved oxygen, total alkalinity, total suspended solid and total plankton count (2.4 x 10^2 - 9.1 x 10^3). Most of the parameters were within / nearly optimum ranges through out the culture period (Table 19.1). The dissolved oxygen content in rice-fish culture field showed a decreasing trend with the advancement of rearing period, attributed to gradual increase in biomass, resulting in higher oxygen consumption (Mohanty, 1995). Concentration of ammonia also showed an increasing trend as the days of culture increased, probably due to split application of nitrogen to paddy, higher metabolic deposition and organic load (Mohanty, 1999). Many times water in the rice fields were slightly acidic probably due to decomposition of weed and rice stalks. The available nutrients observed (Table 19.1) did not result in higher phytoplankton production and moreover it was also affected by shading and competition for nutrient with rice and weed. Density of zooplankton was however, not high (Fig 19.1) in comparison to composite fish culture ponds. In general, upper limit of planting density is usually practiced for rice. However, this practice must be avoided in rice-fish integration system as it would result in stunted plant growth, poor ventilation and illumination that adversely affect plankton population, a major source of natural food for fish. In contrast, at relatively lower density, plants grow fast, strong and become more resistant to lodging. In soil, percentage increase in available-N,P and organic carbon in rice-fish culture field was 58.3%, 94% and 54.5% against control respectively (Table 19.2). This was probably due to leftover supplemental feed, faecal matter of cultured species and stirring movement of fish / prawn that aerate the soil (Li Kangmin, 1988) and help in transforming insoluble nutrients in soil in to a soluble state. In rice-fish culture fields, the unit weight of top soil was 2.3% lighter than the control and porosity (vol. of pore space / vol. of soil x 100) was higher by 1.2%. Although, no pesticide was used, there was no substantial decrease in yield probably due to the fact that fish and prawn act as a biological controller of insect / pest and pathogen (Spiller, 1985). The benthic population, indicators of soil productivity, in both rice-fish culture field and control was consisted mainly of *Oligochaetes* and *Chironomids*, while their biomass was higher in rice-fish culture field (0.44 ± 0.04 g m-^{2}) than control (0.21 ± 0.05 g m-^{2}). This was probably due to periodic manuring, faecal matter of cultured species and increased organic load.

Table 19.1

Minimum and maximum average range of water quality parameters

Parameter	Rice Fish culture Plot		Control Plot	
	Min	Max	Min	Max
pH	6.9	8.3	6.7	7.6
Dissolved oxygen (ppm)	3.3	7.6	2.1	4.6
Temp (°C)	27.5	30.1	27.6	30.2
Total alkalinity (ppm)	56	109	41	87
Nitrite-N (ppm)	0.006	0.071	0.006	0.058
Nitrate-N (ppm)	0.06	0.52	0.06	0.48
Amonnia (ppm)	0.01	0.41	0.02	0.39
TSS (ppm)	116	319	105	211
Phosphate-P (ppm)	0.07	0.33	0.07	0.3

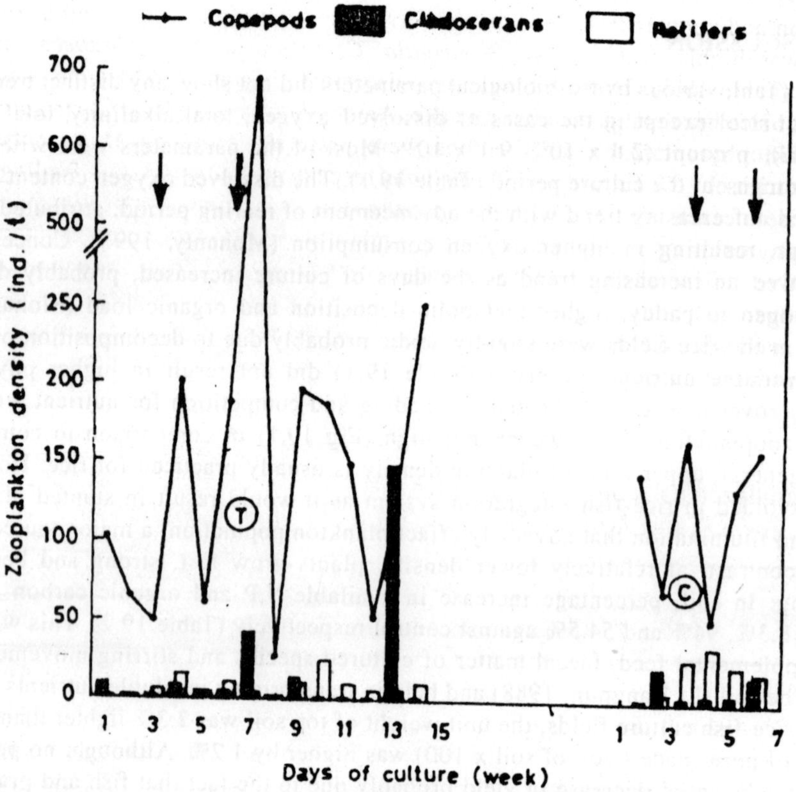

Fig. 19.1

Average zooplankton density in control (C) and treatment (T) fields in rice-fish integration system. Arrows indicate the time of nitrogen application to paddy.

Table 19.2
Changes in soil quality due to introduction of fish and prawn in rice field

Parameter	Rice Fish culture Plot		Control Plot	
	1st Week	15th Week	1st Week	15th Week
Soil pH	6.8	7.0	6.7	6.8
Available - N in soil (mg 100g^{-1})	8.8	10.7	8.4	9.6
Available - P in soil (mg 100g^{-1})	0.32	0.67	0.36	0.54
Organic carbon in soil (%)	0.19	0.53	0.17	0.39

Phytoplankton and zooplankton was most preferred food item for *C. catla* and *L. rohita*, while mud and detritus was highly preferred by *C. mrigala*, *C. carpio* and *M. rosenbergii* in rice fish integration system (Table 19.3). However, quantity-wise most consumed food items were artificial supplemental feed of rice bran and groundnut oil cake. Among bottom dwellers (C. *mrigala*, *C. carpio* and *M. rosenbergii*), phytoplankton and benthos were preferred more by *M. rosenbergii* while zooplankton and detritus by *C. Carpio* and *C. mrigala* respectively. Omnivorous feeding behaviour was observed in case of each species except *Catla catla*. Positive indices of electivity (0.02 - 0.31) were observed for phytoplankton in monsoon-winter, while it was negative for zooplankton during the same period. Negative indices of electivity for zooplankton (-0.14 to - 0.46) in case of all species was recorded during monsoon-winter (August - November) and improved thereafter. This was probably due to rich detrital food web in the initial phase of rearing where raw cattle dung was applied @ 5000 kg ha^{-1} for refuge preparation prior to stocking.

Table 19.3

Average % of analyzed fish and prawn in which mentioned food component was found (frequency) and % of individual gut content volume (abundance)

Food component	Frequency (%)					Abundance (%)				
	1	2	3	4	5	1	2	3	4	5
Phytoplankton	94.4	83.3	55.6	66.6	72.2	<11.2	<5.1	<2.3	<2.7	<4.3
Zooplankton	88.8	86.3	44.4	72.2	44.4	<5.9	<4.3	<1.4	<1.9	<1.6
Mud+Detritus	11.1	22.2	94.4	88.9	77.8	<5.6	<15.4	>29.1	>32.1	<21.0
Benthos/insect	-	5.5	44.5	55.6	61.1	-	<1.0	<12.2	<12.2	<16.4
Artificial feed	72.2	77.8	83.3	88.8	77.8	>56.7	>49.3	>45.8	>46.1	>61.7

1-Catla catla, 2- Labeo rohita, 3- Cirrhinus mrigala,, 4-Cyprinus carpio, 5-Macrobrachium rosenbergii

Faster growth performance under rice-fish culture system was recorded for *Catla catla* followed by *C. carpio*, *C. mrigala*, *L. rohita* and *M. rosenbergii* during 120 days of culture at stocking density of 25,000 ha^{-1} (Table 19.4). *C. carpio* and *C. mrigala* performed better growth rate against that *of L. rohita* probably due to the fact that being surface and column dweller, *L. rohita* is more sensitive to oxygen depletion, while being bottom dwellers *C. carpio* and *C. mrigala* are more tolerant to fluctuation of oxygen concentration (Vijayan and Verghese, 1986). Moreover, faster growth rate of *Catla catla* and bottom dwellers were attributed to effective utilization of ecological niches and rich detrital food web that was maintained through periodic manuring, liming and initial fertilization, which agrees to the findings of Mohanty (1995). Observations on feed conversion ratio (FCR) also supports the conclusion of effective utilization of ecological niches, as minimum and maximum FCR ranged between 0.94 - 1.13. Comparative daily growth performance (Table 19.4) was moderate to good as periodic organic manuring and liming was a regular practice instead of inorganic fertilization, that improved primary productivity (508.3 ± 71.8 mgC m^{-2} h^{-1}) which enhanced the growth rate. Similar observation was also made by Noriega-Curtis (1979). Encouraging growth rate of fish / prawn was also attributed by avoidance of periodic inorganic fertilization, that depress growth due to high ammonia concentration in water (Colt and Tehobonoglous, 1978). Condition factor (Ponderal index) of fish and prawn was less than 1.0 at the initial three weeks of rearing and improved thereafter with gradual

Table 19.4
Growth performance in rearing of fry to advanced fingerlings of fish and prawn in rice-fsh integration system at stocking density of 25000 ha⁻¹

Species stocked	Initial MBW (g)	Final MBW (g)	ADG (g)	Min.-max (K_n)	SR (%)	Productivity (Kg ha⁻¹ 120 DOC⁻¹)
C. catla	0.82±0.03	98.2±6.1	0.81	0.93-1.22	57.6	
L. rohita	0.88±0.03	76.2±4.9	0.63	0.84±1.17	61.5	
C. mrigala	0.92±0.06	87.7±6.3	0.72	0.93-1.2	61.5	1156.5
C. carpio	0.88±0.09	90.2±6.8	0.74	0.93±1.2	46.1	
M. rosengergii	0.04±0.01	22.8±0.08	0.19	0.99-1.33	50.0	

MBW - mean body weight, ADG-average daily growth, SR- survival rate, DOC-days of culture

improvement in water quality. Roy *et al.,(1990)* reported that, in traditional deepwater rice-fish system in India, yield of rice in a season range between 1.0-1.5 t ha-¹ and fish production between 50-200 kg ha-¹, while application of cowdung enhance the productivity of rice and fish to 3.1 and 0.67 t ha-¹ respectively. However, in the present study, an average productivity of 1156.5 kg ha-¹ of fish and prawn has been achieved with application of cowdung @ 5000 kg ha-¹, which was much higher than the earlier recorded productivity in a season. This high productivity of fish / prawn under rice-fish integration system was however, low in comparison to composite fish culture, probably due to shorter rearing duration and comparably less favourable environment as that in ponds (Rao and Ram Singh,, 1998).

Highest average grain yield (Table 19.5) was recorded in rice-fish culture field (3595 kg ha-¹) which was 20.3% higher against control (2988 kg ha-¹), contributed by higher number of panicles/m² (233.2) and number of filled grains/panicle (118.0). This was probably due to introduction of fish and prawn in rice field where, frequent locomotary movement of fish and prawn improves dissolved oxygen level and soil organic matter/ nutrient status by adding faecal matter; controls plankton population/ macro and micro aquatic insects/ bacteria/ organic detritus that compete with rice for material and energy. Hora and Pillay (1962) also reported that, introduction of fish has increased paddy yield by 15% in the Indo-Pacific countries due to better aeration of water, greater tillering effect and additional supply of fertilizer in form of leftover feed and fish excreta.

Table 19.5
Average rice yield and yield attributes as influenced by fish and prawn in rice field

Type of culture	Nos. of panicles/m²	Nos of grain/panicle	Test wt. (g)	Grain yield (kg ha⁻¹)	Straw yield (kg ha⁻¹)	Increase in grain yield
Rice + fish	233.2	118.0	19.9	3595.0	4303.0	20.3%
Rice (control)	204.2	110.2	19.5	2988.0	3572.0	

In brief, as discussed above, it is concluded that rice-fish culture is a feasible and efficient way for optimum utilization of land and water resources that improve soil fertility and productivity. Rice and fish when grow together, are not only compatible but also mutually beneficial. To ensure better productivity, higher planting density must be avoided to provide greater ventilation and illumination, that helps in enhancing both plant and plankton growth, which ultimately helps fish growth and yield.

ACKNOWLEDGEMENTS

I warmly thank Dr. H. N. Verma, Director, WTCER (ICAR), for providing me the necessary facilities for this experiment and Dr. S. K. Mohanty, Fisheries Consultant, CDA, for reviewing the manuscript in detail.

REFERENCES

APHA, 1989. Standard methods for examination of water and waste water. *American Public Health Association.* Washington, D.C., U.S.A, 1647 pp.

Biswas, K.P.,1993. Fisheries manual. *Bhagabat press and publishers,* Orissa, Indi. , 436 pp. Brahmanand, P.S. and Rajeeb K. Mohanty, 1999. Rice-fish Integration. *Yojana,* 43 (9): 11-12

Colt, J. and Tchobonoglous, 1978. Chronic exposure of channel cat fish *(Ictalurus punctatus)* to ammonia : effects on growth and survival. *Aquaculture,* 15 : 353-372.

Hora, S.L. and Pillay,T.V.R.,1962. Handbook on fish culture in the Indo-Pacific region. *Fish.Biol. Tech. Pap.* 14, FAO,Rome.

Huazhu Yang, 1994. Integrated rice-fish culture. In: Developments in aquaculture and fisheries science. *Elsevier Publication* Vol. 28 (1994): 368-385.

Ivlev,V.S., 1961.Experimental ecology of the feeding of fish. *Yale University Press,* New Haven. Likangmin, 1988. Rice-fish culture in China: a review. *Aquaculture,* 71 (1988)173-186

Mohanty, U.K. 1995.Comparative evaluation of growth and survival of Indian Major Carp fry in aerated vis-a-vis non-aerated ponds under different stocking densities. *M.F.Sc. thesis,* Orissa University of Agriculture and Technology, Bhubaneswar, India, 123pp.

Mohanty Rajeeb K.,1999. Growth performance of *Penaeus monodon* at different stocking density. *J. Inland fish Soc. India,* 31 (1): 53-59.

Mohanty, R.K.; H.N.Verma and P.S.Brahmanand, 2001. Rice-fish integration for enhancing land and water productivity. *Annual report,* 2000-2001, WTCER(ICAR), India:70-74.

Noriega-Curtis, P. 1979. Primary productivity and related fish yield in intensely manured fish pond. *Aquaculture,* 17: 335-344.

Rao, A.P. and Ram Singh, 1998. Rice-fish fanning system. In: S.H.Ahmad (ed.), *Advances in Fisheries and Fish Production. Hindustan Publishing Corporation,* NewDelhi, India, 309pp.

Roy, B.; Das, D.N. and P.K. Mukhopadhyay, 1990. Rice-fish-vegetable integrated farming : towards a sustainable ecosystem. *Naga,* ICLARM, October, 1990:17-18.

Spiller, G.,1985. Rice cum fish culture - environmental aspects of rice and fish production in Asia. Consultancy report, FAO, Bangkok, Thailand, 60pp.

Vijayan, M. M. and Verghese, T. J. 1986. Effect of artificial aeration on growth and survival of Indian Major Carps. *Proc. Indian Acad. Sci.* (Anim.sci.), 95 (4): 371-378

CHAPTER 20

On A Collection of Fishes From Khaji-Kotnoor Reservoir

Padmavathi and K. Vijaykumar

Environmental Biology Research Unit,
Department of Zoology,
Gulbarga University, Gulbarga-585 106, Karnataka
E-mail: katepaga_vijaykumar@yahoo.com
Padmachidri@rediffmail.com

ABSTRACT

The fish fauna of Khaji-Kotnoor reservoir consists of 27 species belonging to 9 families. The study of fish fauna is first from the Khaji-Kotnoor reservoir. Among the collections, one species *Notopterus notopterus* fall under the *Notopteridae*. Fifteen species of Cypriniformes, belonging to family Cyprinidae. Two species of Cypriniformes belonging to family Siluridae. One species belongs to order Cypriniformes and family Bagridae. Three species belongs to family Ophiocephalidae One species belongs to order Perciformes and family Gobiidae. One species belongs to order Cypriniformes and family Saccobrachidae. Two species belongs to family Centropomidae and order Perciformes. One species belongs to family Mastocembelidae and order Mastocembeliformes. The present paper deals with the total fish fauna availability in the Khaji-Kotnoor reservoir.

Keywords: *Fish fauna, Khaji-Kotnoor reservoir*

INTRODUCTION

Fishes are one of the most important sources of proteins and are being used by human beings since before prehistoric ages. The commercial fishery of a reservoir primarily depends on its size, shape and planktonic resources. The total reservoir area and total fish landed is proportionate. India is rich in fish fauna. India ranks next to China and Russia in World Inland Fish Production. The inland fisheries in India contribute nearly 1/3 of the total fish production (Srivastava and Vathsala, 1984). But in total domestic supply its share is 57% in contrast to marine sector with 43% though fish production from inland sector is greater significance in India. A series of attempts have been made to explore fish fauna (Moitra, 1956; Srivastava, 1968; Kamal and Ahsan, 1978; Singh, 1978; Mishra, 1983; Srivastava, 1984; Singh and Singh, 1989; Anjana and Singh, 1991; Pandey *et* al., 1996). Jayaram (1981) have studied on freshwater fishes of India, Pakistan, Bangladesh, Burma and Srilanka. Krishna *et al.*, (1999) studied on the limnology and status of the fishery of Nelligudda reservoir, Bangalore. The Khaji-Kotnoor reservoir used by Parkithwar Fisherman Co-operative Society, Gulbarga since 1996. To formulate sound scientific management measures for augmenting fish production in reservoirs, basic research on reservoir productivity is a must. This would provide clues to the production potential of the ecosystem and would help in setting targets of fish production from such water bodies.

MATERIALS AND METHODS

A preliminary survey of fish fauna of Khaji-Kotnoor reservoir was made at different intervals during the period of 1999-2001. Samples were collected from the different parts of the reservoir by using cast net, gill net, circular net and from the local fisherman collections. Samples were preserved in 10% formaline and brought to the laboratory for their identification. The identification was done according to Jayaram (1981), Dutta and Srivastava (1988), Menon (1988), Talwar and Jhingran (1991).

RESULTS AND DISCUSSION

A study of the commercial fisheries collected from Khaji-Kotnoor reservoir. The fish fauna of the reservoir consists of 27 species belonging to 9 families and which are as follows:

Fish-Fauna

The fish fauna of Khaji Kotnoor reservoir consists of 27 species belonging to 9 families and which are as follows:

Order: Clupeformes

Family : Notopteridae

- *Notopterus notopterus* (Pallas)

Order: Cypriniformes

Family : Cyprinidae

- *Chela (chela) cachius* (Hamilton - Buchanan)
- *Salmostoma clupeoides* (Bloch)
- *Puntius sophore* (Hamilton - Buchanan)
- *P. ticto* (Hamilton - Buchanan)
- *P. sarana sarana* (Hamilton - Buchanan)
- *P. chola* (Hamilton - Buchanan)
- *P. dorsalis* (Jerdon)
- *Parluciosoma daniconius* (Hamilton buchanan)
- *Catla catla* (Hamilton - Buchanan)
- *Cirrhina mrigala* (Hamilton - Buchanan).
- *Cirrhina reba* (Hamilton - Buchanan)
- *Labeo rohita* (Hemilton - Buchanan)
- *L. calbasu* (Hemilton - Buchanan)
- *L. bata* (Hamilton - Buchanan)
- *Cyprinus carpio* var *communis Linnaeus*

Order : Cypriniformes

Family : Siluridae

- *Wallago attu* (Schneider)
- *Ompok bimaculatus* (Bloch)

Order : Cypriniformes

Family : Bagridae

- *Mystus cavasius* (Hamilton - Buchanan)

Order : Ophiocephaliformes

Family : Ophiocephalidae (Channidae)

- *Channa marulius* (Hamilton - Buchanan)
- *C. striatus* (Bloch)
- *C. punctatus* (Bloch)

Order : Perciformes

Family : Gobiidae

- *Glossogobius giuris* (Hamilton- Buchanan)

Order : Cypriniformes

Family : Saccobranchidae (Heteropneustidae)

- *Heteropneustes fossilis* (Bloch)

Order : Perciformes

Family : Centropomidae (Ambassidae)

- *Chanda nama* (Hamilton - Buchanan)
- *Chanda ranga* (Hamilton - Buchanan)

Order : Mastacembeliformes

Family : Mastacembelidae

- *Mastacembelus armatus* (Lacepede)

Among the collections, one species *Notopterus notopterus* fall under the Notopteridae. Fifteen species of cypriniformes i.e., *Chela chachius, Salmostoma clupeoides, Puntius sophore, P. ticto, P.*

sarana sarana, P. chola, P. dosalis, P. doniconius, Catla catla, Cirrhina mrigala, Cirrhina reba, Labeo rohita, L. calbasu, L. bata, Cyrimus carpio var, Communis linnaeus, belongs to family cyprinidae. Two species of Cypriniformes belonging to family Siluridae namely *Wallage attu, Ompok bimaculatus. Mystus cavasius* belong to order Cypriniformes and family Bagridae. Three species of *Channa marulius, C. striatus, C. punctatus* belongs to family *Ophiocephalidae.* Glossogobius giuris belongs to order peciformes and family Gobiidae. One species of *Heteropneutes fossilis* belongs to family Saccobrachidae and order Cypriniformes. *Chanda nama* and *Chanda ranga* belongs to family Centropomidae and order Perciformes. One species of *Mastocembelus armatus* belongs to family Mastocembelidae and order Mastocembeliformes.

The survey of fish fauna has been done by a number of workers. Swarup (1953) gave an account of the fishes of Sagar lake. Soni (1959) reported 35 species in the lower lake of Bhopal. Malviya (1961) reported the fish fauna of Jabalpur. Dubey and Verma (1965) have made a detailed survey of the fish fauna of lake reservoirs, ponds and rivers of Madhya Pradesh.

It is known that indiscriminate disposal of domestic, agricultural and industrial wastes bring about a change in the hydrogen-ion-concentration of the receiving water body, which may adversely affect the survival, distribution, growth and physiology of fish life (Nimbergi *et al,* 1985). The study was undertaken for the first time in this reservoir and the study reveals that the Khaji-Kotnoor reservoir water is suitable and useful for fish culture as it has rich aquatic fauna

REFERENCES

Anjana, K. and D.K. Singh. 1991. Survey of icthyofauna of Nawada lake (Muzaffarpur, Bihar) Mendel. 8: 211-213.

Dubey, G.P. and M. Verma. 1965. Hydrobiological studies of Budhawari tank, Seoni Madhya Pradesh with reference to fish production. Icthyologica. V(1-2): 59-66.

Dutta, M.J.S. and M.P. Srivastava. 1988. Natural history of fish and systematic of freshwater fishes of India. Narendra Publishing House, New Delhi, pp. 10-1 5.

Jayaram, K.C. 1981. Freshwater fishes of the India. A Handbook of the Zoological Survey of India, Kolkata, pp. 1-475.

Kamal, M. and S.N. Ahsan. 1978. *Gorica biharenses* new species (Pisces: Clupeidae) river Ganges of Bihar. India. J. Inland fish Soc. India. 10: 28-31.

Malaviya, R.B. 1961. A list of fishes from Jabalpur (M.P.) Nat. Acd. Sci India, Allahabad 31: 349.

Menon, A.G.K. 1988. Checklist of freshwater fishes of India. Zoological Survey of India, Kolkata, pp. 65-66.

Mishra, S.K. 1983. Studies on Icthyo-fauna of Balmikinagar Tarai, West Champaran. Proc. 70th Ind. Sci. Cong. Abst. No. 142.

Moitra, S.K. 1956. Fishes of Lucknow (U.P.) Bharti. 3: 39-54.

Nimbergi, B.M., B.H. Annapurna, Y.M. Jayaraj and S.S. Rodgi. 1985. Pollution indicator bacteria in fishes inhabiting a polluted water body. Ind. J. Microb. 25: 181-185.

Pandey, B.N., S. Anupa, P. Rumana, and M. Kumar. 1996. Seasonal variations in biochemical composition in some freshwater fishes, part-III, *Channa punctatus* (Bioch). Advances in Fish and Wildlife Ecology and Biology. (Ed. B. Kaul). pp. 31-36.

Singh, D.K. and P. Singh. 1989. A study on the fish productivity of Motipur Ox-bow Lake (Muzaffarpur, Bihar). Mendel. 6: 49-51.

Singh, S.P. 1978. Fishes of Moradabad (U.P). Ind. J. Zootim. 15:75-76.

Soni, D.D. 1959. Fish fauna of Bhopal lower lake. Proceedings of First All India Congress of Zoology (Published 1962), pp. 639-641.

Srivastava, G.J. 1968. Fishes of the Eastern U.P. Vishwavidyalaya Prakashan, Varanasi. PP. 72.

Srivastava, M.P. 1984. Studies on the fish and fisheries of North. Eastern Bihar, 2. D.Sc. Thesis, Bhagalpur Univ. Bhagalpur, India.

Srivastava, U.K. and S. Vathsala. 1984. Strategy for development of Inland Fishery Resources in India. Indian Institute of Management, Ahmedabad. (Ed. Srivastava, U.K. and Vathsala, S.). pp. 630.

Swarup, H. 1953. Fish fauna of Sagar Lake. H.S. Gour University Journal, 1(2): 122-130.

Talwar, P.K. and A.G. Jingran. 1991. Inland fishes of India and adjacent countries. Oxford and IBH Publishing Co. Pvt. Ltd., New Delhi, pp. 1-322.

CHAPTER 21

Effect of Monocrotophos on Biochemical Markers DNA and RNA in *Cyprinus Carpio* During Exposure and Recovery Period

Maruthanayagam, C and G. Sharmila

A.V.C. College (Autonomous), Department of Zoology Mannampandal, Mayiladuthurai - 609 305.

ABSTRACT

The present study was designed for 15 days Monocrotophos exposure period followed by 15 days recovery period to understand the toxic effects of toxicants on the nucleic acid aspect of *Cyprinus carpio*. The pesticide leads to several changes in the biochemical markers like DNA and RNA may be due to the increased activity of the enzyme DNAase and the inhibition of RNA polymerase function. But during recovery period, the DNA and RNA level both progressively increased indicating a probable recovery from the disruption of internal organs.

INTRODUCTION

The indiscriminate use of various types of pesticides in the modern world has led to much greater emphasis on the probability of serious environmental contamination to the survival of mankind (Cremlyn, 1978). Such indiscriminate use of pesticides in agriculture and public health operations has increased the scope of disruption of ecological balance as many of the 'non-target' organisms which include important members of the food chain that are perishing (Saxena, 1989). Ahana Lakshmi (1992) has reported that pesticidal contamination of water bodies causes changes in the hydrobiological characters of water, ultimately affecting the aquatic biota in general and fishes in particular. In fact fishes are very sensitive to a wide variety of toxicants in water (Holden, 1972). The non-target species such as fishes from marine and freshwaters have been monitored for pesticide contamination (Couch, 1978).

Another toxic effect of the pesticide is seen in the alterations it brings in the nucleic acid content of the animal. The role of nucleic acids particularly RNA/DNA and protein/DNA ratios, which are used as an index of protein synthesis and cell size, are considered to be important and form an integral part of the growth and toxicity studies in animals. But such studies in fishes are scanty. The treatment with the pesticides causes variability in the nucleic acid content in different tissues and the degree of variability or extent of alterations caused by the pesticides is found to be dose dependent (Malla Reddy and Bashamohideen, 1988).

MATERIALS AND METHODS

The fishes were collected from the control and the experimental tanks at the end of 15 days of exposure to Monocrotophos and at the end of 15 days of recovery period and all the fishes were subjected to the nucleic acid analysis individually.

The nucleic acids were extracted from the tissues by the method of Schneider (1957). Subsequently, the DNA and RNA concentrations of muscle, liver and intestine were estimated by the method of Burdon (1957) and Ceriotti (1965) respectively.

RESULTS

When the fishes were exposed to presumable harmless concentration of 51.57 ppm of Monocrotophos, the tissues of muscle, liver and intestine showed significant amount of decreased in the nucleic acid content (RNA and DNA) during exposure period. However during recovery period, both the RNA and DNA level progressively increased (Table-21.1 and Table-21.2).

Table 21.1

Sublethal effect of pesticide on RNA content (mg/g) in different tissues of *Cyprinus carpio* during exposure and recovery periods. Each value in Mean ± S.E. (n=5)

Period of Study	Muscle (mg/g)	Liver (mg/g)	Intestine (mg/g)
Control (0 day)	5.4±0.33	6.3±0.41	4.8±0.26
Exposure (15 days)	3.4±0.24* (-37.03)	4.2±0.31** (-33.3)	3.1±0.18** (-35.41)
Recovery Control	5.5±0.53	6.8±0.63	4.9±0.09*
Recovery (15 days)	5.0±0.32** (-9.09)	5.9±0.33** (-13.2)	4.8±0.09* (-2.04)

(-) Values in parenthesis indicates percent decrease over control.
** Values are statistically significant at p < 0.5 (Student's 't' test)
* Valeus are statistically significant at P < 5

Table 21.2

Sublethal effect of pesticide on DNA content (mg/g) in different tissues of *Cyprinus carpio* during exposure and recovery periods. Each value is Mean ± S.E. (n=)

Period of Study	Muscle (mg/g)	Liver (mg/g)	Intestine (mg/g)
Control (0 day)	1.8±0.07	2.9±0.07	1.4±0.15
Exposure (15 days)	1.4±00.04** (-22.2)	1.9±0.13** (-34.4)	1.0±0.12** (-28.5)
Recovery Control	1.9±0.09	2.4±0.03	1.5±0.11
Recovery (15 days)	5.0±0.08** (-21.0)	1.98±0.13** (-17.5)	1.3±0.13** (-13.3)

(-) Values in parenthesis Indicates percent decrease over control.
** Values are statistically significant at $p < 0.5$ (Student's 't' test)

DISCUSSIONS

The pesticides lead to several changes in the biochemical markers such as RNA and DNA non-target agents of aquatic pollution resulting in secondary responses (Mazeaud *et al,* 1977). In the present study, the exposure of fishes to Monocrotophos resulted in the depletion of RNA and DNA. This decrease in nucleic acid suggests the decrease in protein synthesis and further damage to the liver, which is the major metabolic organ of drug detoxification (Ramalingam *et al.,* 2000). The decrease of RNA may be suggested that the daily addition of pesticides results in the swelling and chromatolysis of Nissle bodies which are rich in RNA (Watson, 1969). According to Mukhopadhyay and Dehadrai (1980), the decrease of RNA may also be due to interference in the incorporation of precursor in the nucleic acid synthesis or inhibiting the function of RNA polymerase. Dawood (1986) and Benjamin (1990) have suggested that the decrement of RNA may also be due to the non-coding for the process of protein synthesis, thereby decrease in the mRNA content, which in turn would have reduced the concentration of RNA. Organophosphorus compounds exhibit strong mutagenic and clastogenic potentiality (Patankar and Vaidya, 1980), which may be responsible for the alteration of DNA level. However the decrease of DNA is not very prominent when compared to RNA. The decrease may be attributed to the increased activity of DNAase as suggested by Tayyaba *et al.,* (1981). The decrement of DNA may also be due to the degeneration of hepatic cells due to the effect of Organophosphate compounds, which is in agreement with the work of Bachelard (1974).

The observed persistent significant decrease in both protein and nucleic acids would suggest that pollutant impair the process of protein synthesis in the tissues of fishes exposed to pesticides. Since RNA is the biochemical mid wife in the formation of proteins, the diminished RNA content also affects the cellular protein content. Eichorn (1995) have also suggested that the depression in DNA synthesis is not energy dependent and may be due to the disruption of the replication process. In the recovery period, the RNA and DNA level progressively increased indicating a probable recovery from the disruption of internal organ and decrease of hypoxic condition to the fish, as evidenced by the recovery of whole animal activity after their shift from pesticide water to normal water.

Hence in the present investigation, the reduction of nucleic acids can be taken as meaningful biochemical indices of pesticides toxicity to assess the extent of pollution of the aquatic environment.

ACKNOWLEDGEMENTS

We express our sincere thanks to Mr. S. Swetharanyam, Head of the Department, Department of Zoology, Dr. V. Radhakrishnan, Principal, A.V.C. College for giving constant support and encouragement and providing us the necessary laboratory facilities in doing this research work.

REFERENCES

Ahana Lakshmi, P. 1992. Pesticides in India. Risk assessment and policy. Netherlands.

Bachelard, H.S. 1974. Outline studies in biology Brain Biochem. Chapman and Hall, London : 8.

Benjamin, N. 1990. Biochemical changes due to learning in the brain of rat exposed to an Organophosphate pesticide. M. Phil. Thesis, University of Madras.

Burdon, R.M. 1957. In : Methods in Enzymology. S.P. Colowick and N.O.Kaplam (Eds). Academic press, New York. 1 : 267.

Ceriotti, G. 1965. A microchemical determination of deoxyribonucleic acid. J. Biol. Chem., 198: 297-303.

Couch A. J. 1978. Histopathological effects of pesticides and related chemicals on the liver of fishes. Nature., 273: 559-586.

Cremlyn, R. 1978. Pesticides: Preparation and Mode of action. John Wiley and Sons, New York, 240.

Dawood, N. 1986. Studies on the functional organization of the brain of rabbit, *Oryctolagus cuniculus*. Ph.D. Thesis, University of Madras.

Eichorn, G.L. 1995. Organohalogen compound. A.D. McIn tyre and C.F. Mills (Eds). Plenum press, New York Active sites of biological micromolecules and their interaction with Heavy metals. In: Ecological toxicology research : Effects of heavy metals 123-143.

Holden, A.V. 1972. The effects of pesticides on life in fresh water. Proc. R. Soc. Cond. 180: 383-394.

Malla Reddy, P and M.D. Bashamohideen. 1988 Toxic impact of Fenvelerate on the Protein metabolism in the branchial tissues of fish *Cyprinus carpio*. Curr. Sci., 57: 211-212.

Mazeaud, M.M., Mazeaud, F and E.M. Donaldson. 1977 Primary and secondary effects of stress in fishes : some new data with a general review. Trans. Am. Fish. Soc. 106: 201-212.

Mukhopadhyay, P.K and P.V. Dehadrai. 1980. Biochemical changes in the air-breathing cat fish, *Clarias batrachus* (Linn.) exposed to Malathion. Environ. Pollut. A. 22: 149-158.

Patankar, N and V.G. Vaidya. 1980. Evaluation of genetic toxicity of the insecticide Phosphamidon using invitro mammalian test systems. Ind. J. Exp. Biol. 18: 1145-1147.

Ramalingam, V., Vimaladevi., Narmadaraji, R and P. Prabakaran. 2000. Effect of the lead on the haematological and biochemical changes in the fresh water fish, *Cirrhina mrigala*. Poll Res. 19(1): 81-84.

Saxena, R.C. 1989. Detection of insecticides residues on fruits and vegetables sold in local market, Project report: 82, 8.

Schneider, R.P. 1957. Mechanism of inhibition of rat brain ($Na^{++}K^{+}$) Adenosine triphosphate by 2, 2-bis (P-chlorophyll)-1,1,1- Trichloroethane (DDT). Bio chem.. Pharmacol., 24: 939-946.

Tayyaba, K., Hasan, M., F and N.H. Khan. 1981. Organophosphate Metasustox - R, Induced regional alterations in brain nucleic acid metabolism., Indian J. Exp. Biol. 19: 688-690.

Watson, W.E. 1969. The response of motor neurons to intramuscular injection of botulinum toxin. J. Physiol. 202: 611-630.

CHAPTER 22

HAEMATO-BIOCHEMICAL VARIATIONS INDUCED BY MONOCROTOPHOS IN *CYPRINUS CARPIO* DURING THE EXPOSURE AND RECOVERY PERIOD

Maruthanayagam; C and G. Sharmila

A.V.C. College (Autonomous), Department of Zoology Mannampandal, Mayiladuthurai - 609 305.

ABSTRACT

The present study was designed for 15 days Monocrotophos exposure period followed by 15 days recovery period to understand the toxic effects of toxicant on the haemato- biochemical aspects of *Cyprinus carpio*. During the exposure period, there was a significant amount of increase in blood glucose and serum cholesterol may be due to increased glucogenesis and due to increased lipid content in blood respectively. However, the serum protein level decreased in fishes due to the inhibition of RNA synthesis or increased proteolytic activity. During recovery period, the restoration of glucose, protein and cholesterol level was found to be slow and gradual.

INTRODUCTION

Generally aquatic organisms are susceptible to pollution effects by pesticides as well as by industrial effluents. However, an organism tries to adapt itself to these changes by changing their metabolic activities. But at higher concentrations, these pollutants can cause damage to the physiological systems by affecting the organisms either at organ and cellular level or even at molecular level, which in turn cause changes in the biochemical composition and can be used to study the different protective mechanisms of the body to resist toxic substances and detoxification. The pesticides cause a number of subsidiary problems like affecting the ecosystem, affecting the growth, reproduction and behavior by causing pathological and physiological changes (Meenakshi, 1993; Holden, 1973) and alterations in biochemical constituents of fishes (Anon, 1962).

Besides this, the pesticides also bring some changes in the blood parameters. The metabolic status in an organism is very much reflected in its milieu interior (Lu, 1985). Blood being the medium of intercellular and intracellular transport which comes in direct contact with various organs and tissues of the body, thus the physiological state of an animal at a particular time is reflected in its blood. Pesticides rapidly bind to the blood proteins and induced haematological changes such as changes in blood glucose, serum protein and serum cholesterol levels are of some value in assessing the impact of exposure under natural conditions and may also serve as tools for biological monitoring.

MATERIALS AND METHODS

The fishes were collected from the control and the experimental tanks at the end of 15 days of exposure to Monocrotophos and at the end of 15 days of recovery period and all the fishes were subjected to the haemato-biochemical analysis individually.

The blood glucose, serum protein and serum cholesterol level of the normal as well as the Monocrotophos treated fishes were estimated by following the standard procedures given by Roe (1955), Lowry *et al.*, (1951) and Barnes and Blackstock (1973) respectively.

RESULTS

In the present investigation, the presumable harmless concentration of Monocrotophos to *Cyprinus carpio* is calculated as 51.57 ppm.

When the fishes were exposed to presumable harmless concentration of 51.57 ppm, the blood glucose level and serum cholesterol level increased significantly while serum protein level gradually decreased during exposure period. But during recovery period, they attained almost the normal level as depicted in the Table-22.1. The percentage change over control for all parameters has been depicted in the Fig. 22.1.

Table 22.1

Changes in selected Haemato - biochemical parameters in *Cyrinus carpio* during exposure and recovery periods. Each vaule is Mean ± S.E. (n=5)

Haemato-biochemcial parameters	Nature of tissues	Exposure period (15 days)	Recovery period (15 days)
Bloood Glucose (mg/100 ml)	Control	34.00±1.3	36.34±2.4
	Treated	40.00±1.18** (+17.6)	32.14±1.** (-11.5)
Serum Protein (mg/100 ml)	Control	514.13±1.43	523.14±1.34
	Treated	463.14±1.03* (-9.9)	480.13±1.43* (-8.2)
Serum Cholesterol (mg/100 mg)	Control	280.13±1.03	285.14±0.34
	Treated	314.14±1.14* (+12.14)	290.13±0.45** (+1.75)

(-) Values are in parenthesis indicates percentage decrease over control.

(+) Indicates percentage increase over control.

** Values are statistically significant at p < 0.5 (student's 't' test)

* Values are statistically significant at p<5

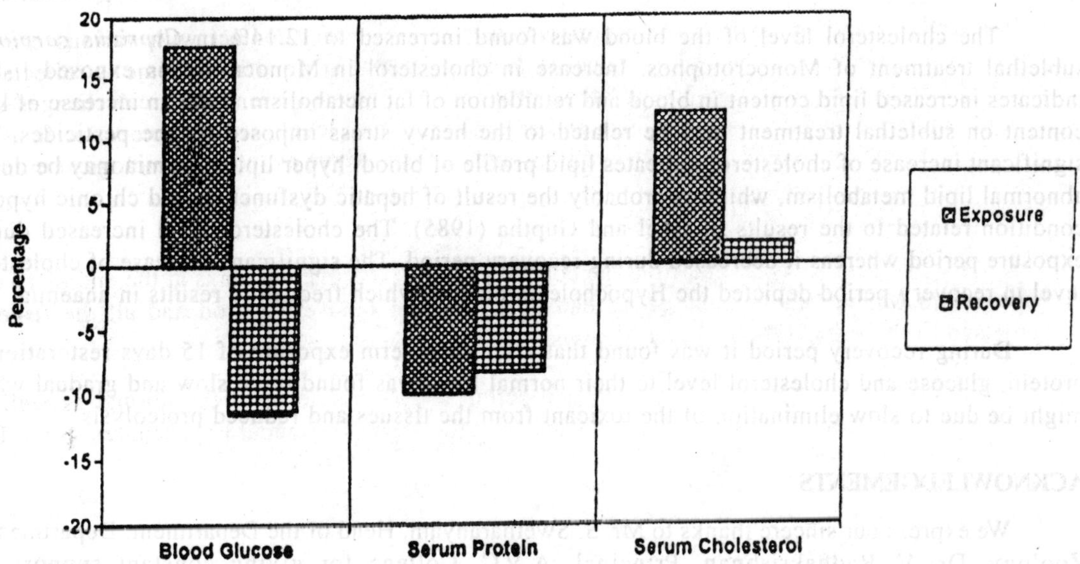

Fig. 22.1

Percentage change over control in the Selected Haemato-biochemcial parameters of *Cyprinus Carpio* during exposure and recovery periods

DISCUSSIONS

After exposure to sublethal concentration of Monocrotophos, the *Cyprinus carpio* showed alterations in the blood parameters. It was observed that in the blood of the fish treated with sublethal concentration of pesticide, the glucose level was increased by 17.6% than the control fishes. The elevation of blood glucose in the *Cyprinus carpio* exposed to the toxicant may be attributed to physiological stress caused by the Monocrotophos. The elevation in blood glucose level gains support from the works of Metelev *et al.*, (1981) and Amudha (1986). These authors have suggested that such an elevation is due to increased glucogenesis as well as due to inhibition of glycogenolysis and glyconeogenolysis during stress. In the toxicant medium the fish absorbs little oxygen from the environment. Their respiratory metabolism being depressed hence the stored intracellular glycogen is utilized. Under such conditions, the hyperglycemic hormone is released for the degradation of glycogen. Thus glucose leaks into the blood causing hyperglycemia (Bhattacharya, *et al.*, 1975).

In general, any toxic substances like pesticide or heavy metal are known to depress blood protein in fishes (Jana and Bandyopadhyaya, 1987). In the present study also, the decrease in protein content in blood was observed. The blood of *Cyprinus carpio* treated with sublethal concentration of Monocrotophos showed a decline in protein content (-9.9%) when compared with that of the protein content of -8.2% of recovered fishes, thereby indicating the occurrence of proteolytic activity in the tissues. Several similar findings have reported the depletion in the level of plasma protein (Goel and Guptha, 1985). The pesticide may inhibit the proteolytic activity resulting in low protein content or it may some how affect the uptake of amino acids in the polypeptide chain. It is possible that the pollutant stress influences the conversion of tissue protein into soluble fractions reaching the blood for utilization. The depletion in protein level as observed in *Channa striatus* (Baskaran, 1980) and *Barbus stigma* (Manoharan and Subbaiah, 1982) may be due to the diversification of energy to meet the impending energy demand caused by toxic stress.

The cholesterol level of the blood was found increased to 12.14% in *Cyprinus carpio* on sublethal treatment of Monocrotophos. Increase in cholesterol in Monocrotophos exposed fishes, indicates increased lipid content in blood and retardation of fat metabolism. Such an increase of lipid content on sublethal treatment may be related to the heavy stress imposed by the pesticides. The significant increase of cholesterol indicates lipid profile of blood; hyper lipid anaemia may be due to abnormal lipid metabolism, which is probably the result of hepatic dysfunction and chronic hypoxic condition related to the results of Goel and Guptha (1985). The cholesterol level increased during exposure period whereas it decreased during recovery period. The significant decrease of cholesterol level in recovery period depicted the Hypocholesterolamia, which frequently results in anaemia.

During recovery period it was found that after short term exposure of 15 days restoration of protein, glucose and cholesterol level to their normal level was found to be slow and gradual which might be due to slow elimination of the toxicant from the tissues and reduced proteolysis.

ACKNOWLEDGEMENTS

We express our sincere thanks to Mr. S. Swetharanyam, Head of the Department, Department of Zoology, Dr. V. Radhakrishnan. Principal, A.V.C. College for giving constant support and encouragement and providing us the necessary laboratory facilities in doing this research work.

REFERENCES

Amudha, R. Christy. 1986. Effects of Malathion on bloodglucose level and its histopathology on the endocrine adrenal gland of *Anabas testudineus.* M. Phil. Thesis, Bharathidasan University, Trichy.

Anon, M. 1962. The wealth of India. Raw materials of vol IV supplement fish and fisheries. Council of Scientific and Industrial Research, New Delhi. 132.

Barnes, J and J. Blackstock. 1973. Estimation of lipids in marine animals and tissues. Detailed investigation of the Sulphophos-phovanillin method for total lipids. J. Exp. Mar. Ecol. 12: 103-118.

Baskaran, R. 1980. Biological studies on a chosen thermo conformer, *Channel striatus.* Ph. D. Thesis, Madurai Kamaraj University

Bhattacharya, S., Mukherjee, S and S. Bhattacharya. 1975. Toxic effects of Endrin on Hepatopancreas of the teleost fish *Clarius batrachus.* Indian J. Exp. Biol., 13: 185-186.

Goel, K.A and K. Guptha. 1985. Haematobiological characteristics of *Heteropneustes fossils* under the stress of Zinc. Ind. J. Fish, 32: 256-259.

Holden, A.V. 1973. Effects of pesticides on fish. In: Environmental pollution by pesticide (Ed.C.A. Edwards). Plenum Press, NY. 213-253.

Jana, S and N. Bandopadhyaya. 1987. Effect of heavy metals on some Biochemical parameters in the freshwater fish, *Channa punctatus.* Environ. Col. 5(3): 488-493.

Lowry, D.H. Rosenbrough, N.J., Farr, A.L and R.J. Randall.1951. Protein measurement with folin phenol reagent. J. Biol. Chern. 193-275.

Lu, F.C.1985. Basic toxicology. Hemisphere Publishing Corporation, Washington, US A: 184-195.

Manoharan, T and G.N. Subbaiah, 1982. Toxic and Sublethal effects of Endosulfan on *Barbus stigma.* Proc. Ind. And. Acad. Anim. Sci. 91: 523-532.

Meenakshi, V. 1993. Histological and haematological studies on the effect of pesticide Phosphamidon in an air breathing fresh water fish, *Anabas testudineus,* M. Phil. Thesis, Annamalai University.

Metelev, V.V., Kanaev, A.I and N.G. Dzaskhova. 1981. Water Toxicology. Amerind Publ. Company. New Delhi. : 174-175.

Roe, J.H. 1955. The determination of sugar in blood and spinal fluid with Anthrone reagent. J.Biol. Chem., 212 : 335-343.

Sprague, J.B. 1971. The ABC's of pollutant bio-assay using fish: Biochemical methods for assessment of water quality. Amer. Soc. Test, Mat. 528:6-30.

CHAPTER 23

ENVIRONMENTAL APPRAISAL OF DEPLETING FISH DIVERSITY OF KASHMIR VALLEY

M. A. Khan

Division of Environmental Sciences, S.K. University of Ag. Sci. Tech. Care: GPO. Box. 726 Srinagar-190 001, Kashmir, India

ABSTRACT

Since Heckel's work in 1838, the ichthyofauna of hilly region of Kashmir has been fairly well researched. The valley, with its pristine water systems, harboured rich and diverse fish-fauna, and accounts for more than 80% of total fish production of the State. In view of taxonomic uncertainties, the exact/authentic census of fish species is far from complete; though Kashmir claims as many as 42 species in the check-list of S. Nath, 1986.

There has been drastic decline in the fish diversity of Kashmir region. The introduction of exotic fast growing species (e.g. common carp) has posed threat to indigenous Schizothoracine fish. The grass carp *(Ctenopharyngodon idella)* and silver carp *(Hypophthalmichthys molitrix)* introduced in some Kashmir fish-farms by J&K Fisheries Department merits proper environmental impact assessment. Anthropogenic stresses, over the years, have also remarkably affected the fish biodiversity. Habitat destruction, indiscriminate exploitation for commercial gains, deleterious effect of mechanical weed harvesters, use of pesticides in the catchment, introduction of exotic fish species, unabated eutrophication of waterbodies and environmental pollution are the major causative factors of depletion of fish diversity in the region.

The paper examines the various anthropogenic pressures endangering the Kashmir ichthyofauna and calls for measures aimed at wise-use concept towards the conservation of fishery resources of the valley.

INTRODUCTION

Kashmir Himalayan Valley is blessed with unparalleled water resources sustaining myriad of fish fauna. Maintaining this ichthyodiversity gains paramount importance in view of mounting anthropogenic perturbations in their natural environment. The general factors influencing fish distribution and diversity include, *inter alia,* consideration of evolution and biogeography, the tolerance range of the species to environmental factors and predatory action and inter-specific competition. Throughout the world, evidences are emerging of dwindling diversity of ichthyofauna. Warren and Burr (1994) report that about 33% of the North American native freshwater fishes are endangered. Dehadrai & Poniah (1997) mention that 79 fishes of India are threatened. Kashmir water ecosystems are no exception to this global phenomenon. The present communication attempts to focus on major factors responsible for affecting the fish diversity of Kashmir.

KASHMIR ICHTHYOFAUNA AND DEPLETION TRENDS

Status of Fish Diversity in Kashmir

Kashmir valley holds some unique clean, deep and large mountain lakes as well as the flatland lakes and wetlands-, majority of them home to diverse species of fish which accounts for more than 80% of the total fish production of the State. A number of stresses, mostly man-made, detrimental to fish species in their once ideal environments have been observed to operate in aquatic ecosystems. The principal factors identified for the decline of fishes (Dehadrai & Poniah, 1997) include physical habitat loss, degradation or alteration, and chemical pollution or alteration. Palearctic fishes belonging to Schizothoracinae, Botinae and Nemachelinae are well represented in the ichthyofauna of Kashmir valley. Exotic trout and endemic Cyprinids are found to be predominant groups in the mountain lakes (Zutshi & Vass, 1970). A checklist of fish of Jammu and Kashmir State prepared by Nath (1986) includes as many as 42 species of fishes for Kashmir. Yousuf (2000) recorded 37 number of fish species. However, Khan (2000a) while describing aquatic biodiversity of Kashmir documented 40 fish species. Over the past few decade there have been various factors, acting in unison affecting the fish fauna in the valley.

Impact of Exotic Fish on the Local Ichthyofauna

Introduction of the exotic species for commercial gains has resulted in the loss of diversity. The introduction, in 1956, of the exotic common carp, *Cyprinus carpio L.* caused a sharp decline in the population and almost exterminated the schizothoracine fishes in Kashmir valley (Dehadria & Poniah, 1997; Zutshi & Gopal, 2000). Three varieties of *Cyprinus carpio L. viz.*, scale carp, mirror carp and leather carp introduced in the valley led to the predominance of the exotic species over the indigenous fish population (Foteder & Qadri, 1974). The major reasons identified for such drastic shifts are:

- Food competition due to more or less identical food spectra. Studies (Yousuf, 2000) indicate that the similarity in food items varies from 31.3% between *Schizothorax esocinus* and *Cyprinus carpio specularis* to 51.7% between *S. niger* and *S. curvifrons.*
- Higher fecundity of exotic carp.
- Low fecundity of the Schizotnoracine fish causes low production number of young ones compared to the common carp.
- Spawning migration of Schizothoracine (except *S. niger)* from low reaches to high altitude streams.
- Better fertilization and growth rate

The introduction of grass carp *(Ctenopharyngodon idella)* and silver carp *(Hypophthalmichthys molitrix)* in some Kashmir waters (e.g. Trigam sar) by the J&K Fisheries Department needs proper ecological monitoring and their impact assessment on indigenous fish. Studies by Sandhu *et al.* (1994) also indicate that *Cyprinus carpio* and *Hypophthalmichthyc molitrix* have alarmingly reduced the Mahseer *(Tor putitora)* fishery in Govindsagar reserviour. According to Yousuf (2000) exotic species have also been found to feed on the fry and fingerling stages of the local fish, especially Schizothoracine. For instance *Salmo trutta* L. and *S. gairdnerii* L. are true carnivores and consume significant quantities of fish.

Stress Impact of Catchment

Reclamation of water-bodies and their catchment for construction of buildings and agricultural purposes has destroyed the natural habitat of fishes. The use of pesticides, in cultivable land in and around the catchment, has added harmful chemicals to water bodies. Many of these chemical pesticides are non-biodegradable and get accumulated at various trophic levels of food chain. Consumption of plankton by the fish adds some of the toxic substances into their system, which can be detrimental for the very survival of the fish population itself.

Due to the increase in human pressure, many of the natural water-bodies have either reduced in size or completely disappeared. Environmental deterioration of the water takes place by the addition of nutrients from anthropogenic perturbations in the catchment. Fishes are affected the most by the media in which they live. Because of deforestation and other detrimental human activities in the catchment areas of the flatland lakes like Dal, Anchar, Manasbal and Wular, most of the breeding grounds have undergone siltation. This has further aggravated the habitat conditions the consequence of which result in the decrease of fish population. The chemical and physical changes may induce alteration in the composition and population of plankton or the destruction of spawning grounds. The effluents may also affect through toxic derivatives in the sewage.

Impact of Eutrophication

Heavy accumulation of nutrients causes eutrophication. There are certain species of algae, noted for their rapid multiplication in waters, which impart specific colour (e.g. bluish- green, green, reddish brown) to water-bodies causing 'algal blooms' and this often is responsible for large scale fish-mortality. The native fish Schizothoracine in particular, find it difficult to cope up with the problem of eutrophication and quite often suffer heavy losses. The decay of organic matter with concomitant, consumption of the oxygen content of the water causing fish-asphyxiation. Little thought is being given to this problem in Kashmir despite frequent recurrence of algal blooms in some lake waters of Kashmir. The development of red-bloom with inherent potential of inducing toxicity in aquatic environments endangering fish is a sign of accelerated eutrophication (Khan, 2000b). The effluents coming from hotels and hospitals in the vicinity of water-bodies may also be harmful for the fish population.

Impact of Mechanical Weed Harvesters

Different lakes of Kashmir are characterized by the prolific growth of many macrophytes. *Ceratophyllum demersum, Myriphyllum spicatum, Hydrilla verticillala, Nelumbo nucifera* and many more form the major part of the. aquatic vegetation in the valley. Thick growth of these have been causing considerable hinderance to swimming, diving and other aquatic-sport. To overcome the situation, selective deweeding has been employed in Dal lake over the past decade or so. As a consequence of this practice some undesirable changes have taken place disturbing the balance of the lake ecosystems (Zutshi & Tikku, 1900). Furthermore, studies carried out (Ticku & Zutshi, 1991) show that for every 100 kg of harvested weeds about 500 g of fish get tangled and thus lost from the ecosystem. Estimates show that about 7-8 tonnes of fish i.e. 33% of annual production is removed each year because of deweeding alone.

One of the indirect effects of large scale non-selective deweeding is temporary reduction of plant cover and removal of the bottom sand and gravel from the beds which otherwise may be useful for feeding and breeding of some indigenous fish species like *Schizothorax*. Hence the fish loss may be due to the loss of habitat, of which aquatic plants form an important component. Weed harvesting exposes both invertebrates and small fishes to increased risk of predation and as such have deleterious effect.

CONCLUDING REMARKS

Apart from introduction of exotics,over-exploitation and stresses due to water pollution or anthropogenic pressures along the catchment have caused adverse impact on the fish diversity. Today the scarcity of the protein-rich food for the burgeoning population is haunting the mankind and urgent need is felt to conserve the ichthyofauna of Kashmir valley. Despite a number of causes identified for the perturbing decline in fish diversity it has not been possible to undo the damage caused by each factor. Thus, holistic approach integrating the concept of sustainable development and conservation measures would go a long way in arresting the depletion of fish diversity of Kashmir.

REFERENCES

Dehadrai, P.V. & Poniah, A.G. (1997). Conserving India's fish biodiversity. *Int. J. Ecol. Env. Sci.* 23:315-326.

Foteder, D.N. & Qadri, M.Y. (1974). Fish and fisheries and the impact of *Cyprinus carpio* L. on endemic fish. *J. Sci. Univ. Kash.* 2(1-2): 79-90.

Khan, M.A. (2000a). Status report on aquatic biodiversity in Kashmir Himalayan Valley. In: Proc. Second Natl. Workshop on Environ. Statistics. Hyderabad 6-7 April. pp. 96-103. Minst. Statis. Progrm. Impl. GOI, New Delhi.

Khan, M.A. (2000b) Anthropogenic eutrophication and red tide outbreak in lacustrine systems of the Kashmir Himalaya. *Acta hydrochem. hydrobiol* (Weinheim) 28: 95-101.

Nath, S. (1986). A checklist of fishes of Jammu and Kashmir state (India) with remarks on the Ichthyogeogrpahy of the state. *J. ZooL Soc. Ind.* 38: 83-98.

Sandhu, G.S., Tandon, K.K. & Johal, M.S.(1994). Growth studies of an endangered fishes, *Tor pottitora* (Hamilton), from Govindsagar (Himachal Pradesh) India. In: Threatened fishes of India.(eds) D.V. Dehadrai, P. Das,. & S.R. Verma pp. 137-147. Natcon Publ. 4. Soc. Nature Conserv. Muzzafarnagar, India.

Warren, M.L. & Burr, B.M. (1994). Status of coldwater species of United States. Overview of an imperiled fauna. *Fisheries* 19: 6-18.

Yousuf, A. R. (2000). Impact of man on fish diversity in Kashmir. In : Environment, Biodiversity and Conservation, (ed.) M. A. Khan. pp. 439-449. APH. Publ. Corp. New Delhi.

Zutshi, D.P. & Vass, K.K. (1970). High altitude lakes of Kashmir. *Ichthyologica* 10: 12- 15.

Zutshi, D.P. & Ticku, A. (1990). Impact of mechanical deweeding on Dal lake ecosystem. *Hydrobiologia.* 200/201: 419-426.

Zutshi, D.P: & Ticku, A. (1991). Effect of weed harvesting on the fish population of Dal lake. Int. *J. Ecol. Environ. Sci.* 17: 217-224.

Zutshi, D. P. & Gopal, B. (2000). Himalayan lake ecosystem : current issues and threat perceptions. Verh. *Int. Ver. Limnol.* 27: 2167-2170.

CHAPTER 24

OUTBREAK OF 'ICH' INFECTION IN AN ORNAMENTAL CATFISH, *PANGASIUS SUTCHI*

Sahar Masud*. Abhed Pandey and K. Paniprasad****

**Division of Livestock Production and Management*
Faculty of Veterinary Sciences & Animal Husbandry,
Sher-e-Kashmir University of Agriculture Sciences and Technology, Jammu, J & K
***Central Institute of Fisheries Education, Seven Bungalows, Versova, Mumbai-400061*

ABSTRACT

'Ich' *(Ichthyophthirius multifiliis)* infection was observed in laboratory reared ornamental catfish, *Pangasius sutchi* during the month of December, probably due to the introduction of some wild specimens of snowtrout *(Schizothorax* sp.) to the same aquarium without any quarantine measures. Head, gills and fins of the infected fish were covered by white spots in the form of small blisters. The fishes were showing some abnormal behavioural signs of irritation, flushing, sluggishness, and loss of appetite. The dorsal fin and barbels were frayed due to heavy infection. The parasites under light microscope were large and dark with horseshoe-shaped nucleus.

Key words: *Ich, white spot, fish*

'Ich' or 'white spot disease' caused by a parasite, *Ichthyophthirius multifiliis* is a ciliate protozoan placed in the class Oligohymenophora, subclass Hymenostomata, Order Hymenostomatida, suborder Ophryoglenina and the family Ichthyophthiridae. The parasite cyclically transforms between an obligate fish associated trophont, and a free-living reproductive stage, the tomont. The theront is the infective stage of the parasite. The theront swims rapidly, frequently changing directions until it contacts a fish (Mac Lennon, 1935a; Kozel, 1986). The theront attaches to the surface epithelium of the skin and gill and penetrates within few minutes in the basal layer (Ewing *et al.*, 1985).

I. multifiliis, a largest protozoan parasite of fish causes massive fatalities in intensive aquaculture system. It is easily introduced into a fishpond, tank, or home aquarium by new fish and its outbreak occurs when conditions are favourable for rapid multiplication of the parasite. Losses of fish due to this parasite are believed to total over one billion dollars per annum worldwide (Clayton, 1990).

Corresponding Author Dr. K. Pani Prasad, Central Institute of Fisheries Education 7-Bungalows, Versova, Mumbai-400061

'Ich' parasite are important opportunist pathogen in aquaculture and most likely to occur when fishes are in stress due to overcrowding, oxygen depletion, change in water temperature etc. It is an important disease of aquarium fish and food fish and creates a major problem to aquarists and commercial fish producers.

The healthy specimens of aquarium catfish *P. sutchi* of average size 8.5 cm (range 8-10 cm) and average weight 12 g (10-15 g) were obtained from the commercial aquarium shop at Jammu in the month of November. These were kept in 4mx1mx1m glass aquarium in the laboratory conditions without proper temperature control. They were fed with commercially available formulated feed. Five specimens of snow trout *(Schizothorax* sp.), a coldwater fish collected from wild during December were also stocked in the same aquarium without quarantine.

For diagnosis of disease, fish were first observed for their abnormal behavioural signs. The gills and skin scrapings with several white spots were taken on microscope slide with a few drops of water and cover glass for the observation under light microscope.

The infected fish were exhibited differential movement and activity including abundant discharge of mucus on gills and skin covering, irritation, flushing, sluggishness, loss of appetite. Body colour of infected fish was dull muddy in comparison to dark colour healthy fish. Fins and barbels become frayed. The body surface of the fish has large number of white spots with high intensity on head, gills and fins (Fig.24.1). Within the 48 hours of infection (first sign of disease) complete mortality were observed.

The disease was observed during the last week of December when temperature of atmosphere and water was going down. Nigrelli et al (1976) suggested that a critical condition for an outbreak is water temperature. The duration of developmental cycle of I. *multifiliis* is significantly influenced by the temperature and there are physiological races related to the temperature tolerance of the host fish. Thus, there are races of *I. multifiliis* that infect coldwater (7.2-10.6°C) fishes such as salmon and other that infect warm water (12.8-16.1°C) tropical fishes. Fishes with wide range of temperature tolerance such as carp and catfish may be susceptible to both cold and warm water parasite. The outbreak of infection in aquarium catfish probably due to the mixing of wild snowtrout without quarantine, which may have coldwater race. Infected fishes have numerous white spots on head, gills and fins with

Fig 24.1
***Pangasius sutchi* infected with *Ichthyophthirius multifiliis*.**

frayed fins and barbells (Fig.24.2). Hines and Spira, 1973a; Kozel, 1976 have also reported the same except in very severe infection; *I. multifiliis* is not uniformly distributed on the body of the host fish. The parasite occurs most frequently on the dorsal surface, particularly the head and fins. As the infection advances the ulcer develop in the skin and are often the sites of secondary bacterial and fungal infections. The fins become frayed due to loss of tissue between the fin rays. Hines and Spira (1974a) have reported thick lumpy covering of surface mucus in heavily infected carp due to excessive mucus secretion, which provides the protective covering to outer surface of fish body.

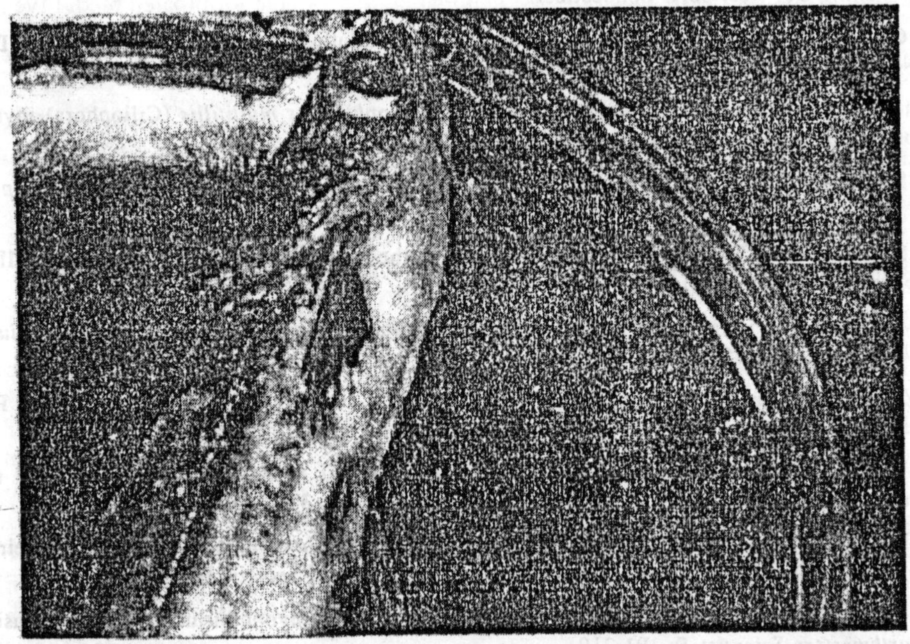

Fig 24.2
Part of fig. 24.1 magnified to show the trophonts above the surface of the fish

Under the stress due to presence of parasite in the medium, the fish exhibit variable shifting in their normal behaviour such as irritation, flashing, sluggishness and loss of appetite and dull body colour. Kabata (1985) noticed that in the early stages of disease fishes congregate near water inlet to reduce contact with free-swimming theronts. Fish also 'fash' or rub their bodies against objects in reaction to skin and gill irritation caused by theronts (Brown and Gratzek, 1980). With very heavy infection, fishes become lethargic and stop feeding (Hins and Spira, 1973a). The white spots in skin and gill scraping contains oval or round parasite that have horseshoe-shaped nucleus and the parasite move slowly in a tumbling manner. Peshkov and Tikhomirova (1968) and Houser (1972) reported that I. *multifiliis* is a holotrichous histophagous ciliate posses a large sausage or horseshoe-shaped nucleus. Maki *et al* (2001) reported unusual finding of *I. multifiliis* within the peritoneal cavities of experimentally infected channel catfish, *Ictalurus punctatus* and suggests that *A. multifiliis* may be found in tissue other than the skin and gills during the course of normal infection. It concludes that careful management should be taken when different species are kept in the aquarium and fish farm and prevention of infections is of the utmost importance.

ACKNOWLEDGEMENTS

We are grateful to Prof. H.U. Khan, Vice-chancellor, Sher-e-Kashmir University of Agriculture Science and Technology, Jammu for providing facilities necessary. Thanks are also due to the dean, Faculty of Veterinary Sciences and Animal Husbandry, Jammu and Director, CIFE, Versova, Mumbai.

REFERENCES

Brown, E.E. and Gratzek, J.B. (1980). *Fish Farming Handbook. Food, Bait, Tropicals and Goldfish.* AVI Publishing Co. west port, Connecticut.

Clayton, G.M. (1990). Ichthyophthiriasis in Fish: Genetic variation in resistance to infection. DISS-ABST-INT-PT-B-SCI and ENG. 51(1): 241 pp.

Ewing. M.S., Kocan, K.M. and Ewing, S.A. (1985). *Ichthyophthirius multifiliis* (Ciliophora) invasion of gill epithelium. *Journal of Protozoology,* 32: 305-310.

Hines, R.S. and Spira, D.T. (1973a). *Ichthyophthirius multifiliis* (Fouquet) in the mirror carp, *Cyprinus carpio L.*. I. Course of infection. *Journal of Fish Biology, 5* 385-392.

Hines, R.S. and Spira, D.T. (1974a). Ichthyophthiriasis in the mirror carp, *Cyprinus caroio* (L.) III. *Pathology. Journal of Fish Biology,* 6:189-196.

Houser, M. (1972). The intranuclear mitosis of the ciliates *Paracineta limbata* and *Ichthyophthirius multifiliis. Chromosoma,* 36:158-175.

Kabata, Z. (1985). Parasites and Diseases of fish cultured in the Tropics. Taylor and Francis, Philadelphia, Pennsylvania.

Kozel, T.R. (1976). The occurrence of *Ichthyophthirius multifiliis* (Ciliate: Hymenos-tomatida) in Kentucky waters. Transactions of the Kentucky Academy of Science, **37:** 41-43.

Kozel, T.R. (1986). Scanning electron microscopy of theronts of *Ichthyophthirius multifiliis:* their penetration into host tissue. *Transactions of the American Microscopical Society,* **105:** 357-364.

MacLennon, R.F. (1935a). Observations on the life cycle of Ichthyophthirius, a ciliate parasite on Fish. *Northwestern Scientist,* 9: 191-210.

Maki, J.L., Brown, C.C. and Dickerson, H.W. (2001). Occurrence of *Ichthyophthirius multifiliis* within the peritoneal cavities of infected channel catfish *Ichtalurus puntatus.* Diseases of Aquatic Organisms [Dis-Aquat-Org], 44(1): 41-45.

Nigrelli, R.F., Pokorny, K.S. and Ruggieri, G.D. (1976). Notes on *Ichthyophthirius multifiliis,* a ciliate parasite on Freshwater Fishes with some remark on possible physiological races and species. *Transactions of the American Microscopical Society,* **95:** 607-613.

Peshkov, M.A. and Tikhomirova, L.A. (1968). Ultrafine structure of nuclear apparatus in *Ichthyophthirius multifiliis* at the protomite cyst stage. *Doklady Akademiia Nauk SSSR, Novaia Serifa.* **183:** 451-452.

CHAPTER 25

A STUDY OF MATING AND BREEDING TACTICS OF *GAMBUSIA AFFINIS* (BAIRD & GIRARD 1853)

Vishal Tiwari, K.K Gaur and Rita Rani

Department of Zoology, R.B.S. College, Agra-282002.

ABSTRACT

The present investigation aimed to study the mating and breeding tactics of *Gambusia affinis* (Baird & Girard 1853). Rapid sexual maturity, prolific breeding habit, sex recognition, courtship and pair formation were some of the most important aspects of findings. Gonopodium as a secondary accessory sexual dimorphic feature in male along with gravid spot, ovoviviparity and superfaetation (producing number of broods from one mating) in female fish were promiscuous features of reproductive biology of *G.affinis*.

Key Words: *G.affinis: mating-breeding tactics.*

INTRODUCTION

The fish of the Poeciliidae family are the best known of all aquarium fish due to their relative ease of keep and the many colourful cultivated varieties available. *G.affnis* is one of the most renowned, cosmopolitan fish in terms of its larvicidal nature throughout the world. It is of great biological, ornamental, educational, commercial value, and thereby having its peculiar importance.

Perhaps this has certainly triggered off many hobbyist, commercial breeders, exporters to systematically search for attractive species and varieties through artificial breeding techniques. A good knowledge on the biology, feeding behavior and ambient environmental condition of the fish species is a prerequisite for culture techniques. Keeping this in view the present study on *G. affinis* has been under taken to work out detailed reporductive behaviour. Some important contributions on the said field of studies are from Baylis (1981); Wourms (1981); Emlen. *et al,* (1997); Evans *et al* (1999); and that of Pilastro *et al.* (1999).

MATERIALS AND METHODS

A set of 30 healthy adult male and female *of G. affinis* species of the same age group in the ratio 2:1 respectively were put in a glass aquarium of 24 X 12 X 12 inches filled with potable water. Netting, bagging, transporting and introducing fishes into an aquarium was done very carefully. Fishes that were healthy, having effortless swimming action and more active with clearly defined colour patterns on them, were chosen, eliminating diseased fishes and those with visible anatomical deformities in order to obtain the most vigorous fry.

The aquarium used was equipped with adequate illumination, meeting the requirements of the plants, filter, a thermostat, an aerator pump and the accessories. Another small sized aquarium of 12 X 6 X 6 inches was also prepared in the same manner possessing all the essential accessories as a breeding tank, to separate the females into it after the mating. Water quality parameters were monitored as they are considered to be the important factors influencing spawning and successful rearing of fry.

RESULT AND DISCUSSION

The males of *G. affinis* were noted comparatively smaller, more brighter, colourful and as a polygamous active members for mating preferences. The males were continuously in courtship display, always chasing after the females. The males were observed persistent drivers, so mating occurs frequently. Males sooner or later were send darting around a female, his fins spread and quivering. Acceptance of a male's advances is often signalled by head-shaking on the part of the female, followed by a dipping of the anterior part of her body (with a corresponding raising of the posterior half), which thus renders her genital aperture more accessible to the male once he adopts the correct mating position.

In order to fertilize the female, the anal fin of the male *G. affinis* developed into a structure known as "gonopodium", which inserts the packets of sperms into the female fish. It acts as the most significant external discriminatory feature between male and female and thereby, facilitating internal fertilization. The males were noted to court the female by approaching her and bending his body into 'S' shape. This type of behaviour was repeated. As the males wings the gonopodium forward to make contact, the female will often fold her anal fin out of the way helping him even further. When mating, advancing from behind, male brought his gonopodium to a forward or sideways position due to its flexibility and insert it into the vent of the female.

After successful mating the females transferred to the breeding tank showed their chest fills out almost upto the head and her middle gets quite distended, as the days of delivery approaches. Females about to give birth were recognized by their round tight stomach, which in late pregnancy become cuboidal in shape. In the mature gravid female, slightly forward to the normal shaped anal fin inside her body appears a dark area known as the gravid spot. It was infact considered to be the womb. As the eggs incubate, the eyes of fry are sometimes visible through the thin walls of the gravid spot.

About 4-8 weeks after mating, the females gives birth upto 10-30 youngs. At birth, the anal fin of the male *Gambusia* appears much like that of the female. As he grows, the fin becomes gradually modified so that in a resting stage, it appears narrow and somewhat longer than that of the female. At the age of 30-35 days the sexes can be distinguished by the development of larger body of the female and her gravid spot which males do not show. The fry were noticed gaining fast maturity and within

3-4 months they get ready for taking part in a successful mating. The highest state of evolution of parental care and breeding system is shown by viviparous teleosts (Baylis, 1981).

An interesting feature noticed in females that a fertile female have as many as half a dozen broods even if no males were present in the breeding tank. This behaviour may be attributed as a precautionary measure and to store the packets of sperms ejected by the male and uses them to fertilize a series of eggs and hatches as they ripen, following a single mating, females produces a number of fry at regular interval. This phenomenon is called superfaetation.

The wide range of tolerance to physico-chemical parameters of water (Table-25.1) and its adaptability and resilence along with superfaetation has made them suitable for aquarium life.

Tale 25.1
Reproductive biology of *Gambusia affinis*

Sexual Characteristics	Gestation period	Size	Temp	pH	Hardness	Number of
Males (i) Smaller in size (ii) Brighter, Colourful (iii) Gonopodium (iv) Attaining fast sexual maturity		3-4 cm				
Females (i) Comparatively larger (ii) Dull coloured (iii) Pot-bellied (iv) Gravid spot (v) Prolific breeders (vi) Ovoviviparity & Superfaetation (v) Attaining fast sexual maturity	4-6 weeks	5-7 cm	20-28°C	Slightly acidic 6-7	179 ppm	10-30

ACKNOWLEDGEMENT

The authors are thankful to the Principal, R.B.S. College, Agra for providing the essential research facilities during this investigation.

REFERENCES

A. P. H. A.; A. W. W. A. and W. E. F. 1998. *Standard methods for the examination of water and waste water.* A. D. Eatson, L. S. Clesceri and A. E. Greenberg (20th edition). Amer. Pub. Health. Assoc., Washington, D. C.

Baylis, J.R. 1981. The evolution of parental care in fishes, with reference to Darwin's rule of male sexual selection. *Environ. Biol. Fish.* 6 : 223-257.

Emlen, S. T. and Orign, L. W. 1997. Ecology, sexual selection, and the evolution of mating systems. *Sci.* 197 : 215-223.

Evans, J. P. and Magurran, A. E. 1999. Male mating behaviour and sperm production characteristics under varying sperm competitions risk in guppies. *Anim. Behav.* 58 (5): 1001-1006.

Pilastro, A. and Bisazza, A. 1999. Insemination efficiency of two alternative male mating tactics in the guppy, *Poecilia reticulata. Proc. Royal. Soc. Biol. Sci. Series.* 266(1431): 1887-1891.

Wourms, J. P. 1981. Viviparity : the maternal fetal relationship in fishes. *Am. Zool.* 21 : 473-515.

CHAPTER 26

IMPACT OF TEMPERATURE ON OUTBREAK OF EPIZOOTIC ULCERATIVE SYNDROME IN FISH

Tun-Tun Singh, Shambhu Prasad and N.K. Dubey

University Department of Zoology, L.N. Mithila University, Darbhanga 846004 - Bihar, India.

ABSTRACT

The present paper discuss about the influence of temperature an outbreak of epizootic ulcerative syndrome (EUS) in fish in association with seasonal variation. The study revealed that the gradual decline in temperature directly affect growth and survival of fish might be one of the important predisposing stress factor for out break of EUS in Fish.

INTRODUCTION

Aquaculture is the most promising and fast growing food farming sector of India contributing 70% percent of total fish production and having an annual growth rate of 9 percent. Inadequate knowledge of fish health management among the fish farmers is one of the major constrants in popularizing large scale development of rural aquaculture entrepreneurship. Fish are always in equilibrium between potential disease organism and their environment. Change in equilibrium due to deterioration of water quality can lead to stress and consequently make fish prone to diseases (Boyd 1982 Boyd and Pillay - 1984). The outbreak of epizootic ulcerative syndrome in fish is one of the major threats and challenge to aquaculturists not only in India but throughout the world. The earlier work suggested heavy rainfall, decline in temperature and poor dissolved oxygen to be the important pre-disposing stressed factors for out break of epizootic ulcerative syndrome (Patliratne, et.al. 1992, 2001. Lio-po-1999, Chinabut et al 1995)

MATERIALS & METHODS

The experiments were conducted during winter season (November 2002 to March 2003) to know the impact of temperature on outbreak of epizootic ulcerative syndrome (EUS) The experiments were conducted on a perennial pond having total water spread area of 5 has and average depth of 1.5 meter. The pond is situated near Baghmor of Darbhanga Town in Bihar state having 25 5' to 26 27' (North) latitude and 85.45'to 86.25' (East) longitude.

Table 26.1

Average weekly fluctuation of temperature and dissolved oxygen in association with average infected fish and average mortality recorded during 19 Nov. 02 to 3 March 03.

No. of week	Average Temperautre (°C)	Dissolved Oxygen (in ppm)	Average Infected Fish (No)	Average Mortality (No.)
1st (19-25 Nov. 02)	21.45	6.5	50	32
2nd (26 Nov - 2 Dec 02)	20.8	6.5	58	36
3rd (3-9 Dec 02)	18.5	6.3	62	40
4th (10-16 Dec. 02)	18.05	6.1	70	41
5th (17-23 Dec. 02)	16.0	5.8	76	52
6th (24-.31 Dec. 02)	14.7	5.6	75	58
7th (1-7 Jan. 03)	13.85	5.3	74	63
8th (8-14 Jan. 03)	9.8	4.1	150	122
9th (15-21 Jan. 03)	9.4	3.6	170	151
10th (22-28 Jan. 03)	11.85	4.2	140	125
11th (29 Jan - 4 Feb.03)	13.45	4.5	150	84
12th (5-11 Feb. 03)	16.75	5.3	140	63
13th (12-18 Feb. 03)	18.00	5.8	170	44
14th (19-25 Feb. 03)	20.5	6.1	160	33
15th (26 Feb.-3 March 03)	22.9	6.8	150	20

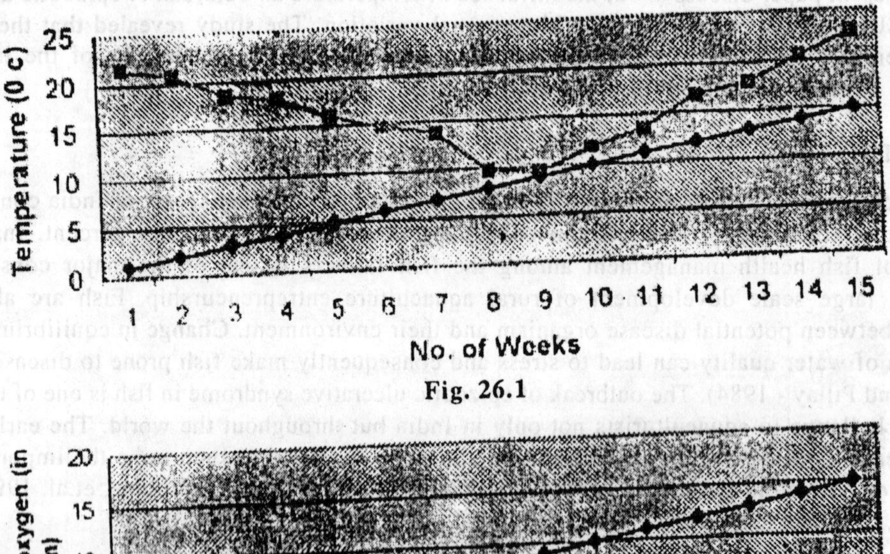

Fig. 26.1

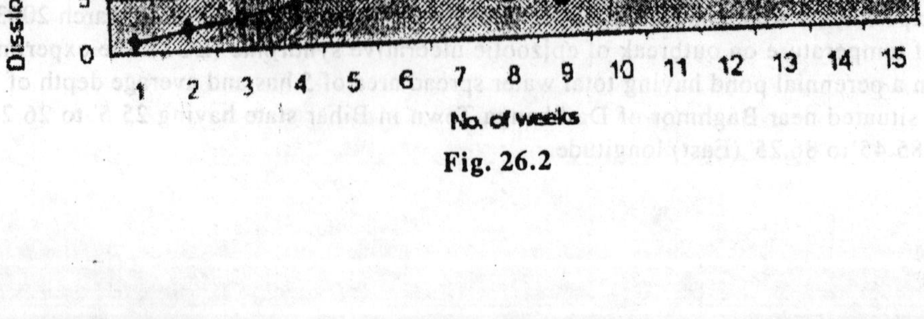

Fig. 26.2

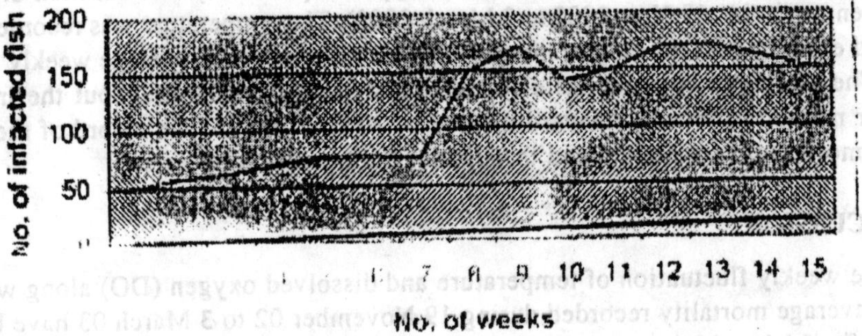

Fig. 26.3

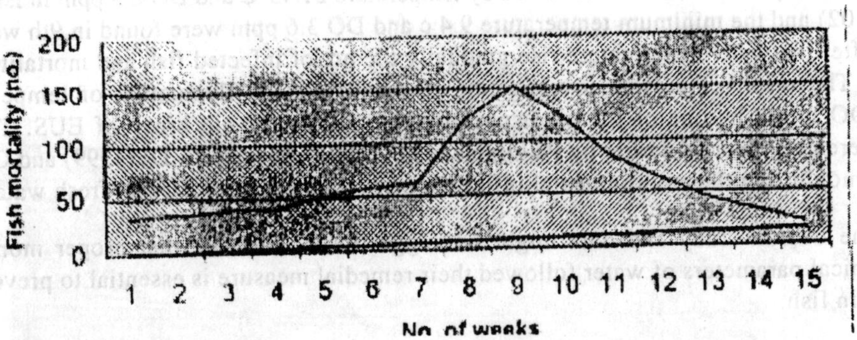

Fig. 26.4

The species stocked in pond were catla *(Catla Catla)*, Rohu *(Labeo, rohita)*. Mrigal (*cirrhinus mrigala*) Grass carp (*Ctenophraryangodon idella*), Common carp *(Cyprnus carpio)*, Silver carp *(Hypophthalmicthys molitrix)* and Calbasu *(Lobeo calbasu)* having stocking density @ 10000 fingerlings per hectare. The fish were led ricebran and mustared oil cake @ 2 to 5 percent of their body weight. The seed were stocked in the month of June- July 2002. The water temperature and dissolved oxygen value at the time of stocking were 28°c. and 6.8 ppm respectitvely The duration of experiment were fixed as fifteen week (i.e. 19 Nov. 2002 to 3 March 2003). The temperature was recorded every day in the morning and evening with the help of mercury thermometer and then average weekly temperature was calculated. The dissolved oxygen was also calculated every week tnroughout the investigation period by Winkler method following standard method of APHA (1989). The record of the number of infected fish and mortality were maintained throughout the investigation period.

RESULTS & DISCUSSION

The average weekly fluctuation of temperature and dissolved oxygen (DO) along with average infected fish and average mortality recorded during 19 November 02 to 3 March 03 have been shown in table-26.1, and fig. 26.1 to 26.1. It was observed that DO decreases with the decline in temperature. However, the minimum temperature 9.4 and minimum DO 3.6 ppm were recorded in 9th week (i.e 15 to 21 January 03). The result revealed positive correlation between temperature and DO.

The experiment was conducted in winter (November 2002 to Match 2003) the water temperature declined significantly upto 15 January 2003 and afterward started using gradually. It was observed that the number of infected fish and consequently the mortality increased with the gradual decrease of temperature and DO.When temperature declined from November 2002 the number of infected fish and mortality increased significantly and maximum no. of infected fish (170) and mortality (151) were recorded in the month of January. The average infection and the average mortality increased with the decease in temperature

The Physico-chemical parameter like temperature and dissolved oxygen were recorded through out the investigation period revealed that there is direct correlation between temperature and DO. During the investigation period the maximum temperature 22.9°C. and DO 6.8 ppm recorded in 15th week (i.e. 26 February 03 to March 03 followed by temperature 21.45°C and DO 6.5 ppm in Ist week (19-25 November 02) and the minimum temperature 9.4 c and DO 3.6 ppm were found in 9th week (15-21 January 03) after that when temperature rise gradually the no. of infected fish and mortality decrease considerably. The result of present investigation revealed that the fluctuation of temperature and variation of DO might be one of the predisposing stress factor for out break of EUS. The similar observation were also reported earlier by Pathiratne et al., (2001), Lio- Po G. D) (1999) and Chinabut et al. (1995) by monitoring the environmental influence on the out break of EUS in fresh water fish.

Thus the overall study and data might be suggestive of the fact that proper monitoring of Physico- chemical parameters of water followed their remedial measure is essential to prevent the out break of EUS in fish.

ACKNOWLEDGMENTS

The author wish to express their gratitude to Dr. Shishir Kumar Verma, Principal Investigatoi Department of Zoology, L.N.Mithila University, Darhhanga and Sri Ram Lakhan Sahani, for their co-operation and providing necessary facilities.

REFERENCES

APHA (1985), Standard methods for the examination of water and waste Water. American Public Health Association, New York.

Boyd, CE, and Pillay. V.C. (1984), Water Quality Management in Aquaculture. Country' Special Publication No. 22: 5-11/22.48.

Boyd, C.E. (1984) Water Quality Management for Pond Fish Culture, Elseviey Scientific Publishing Company P.44-46.

Chinabute. S. Roberts. R.J., Willoughby, U.R., Pearson, M.D., (1995). Histopathology of Snake Head *Channa straitus* (Bloch), Experimentally infected with aphanomyces fungus associated with Epizootic ulerative Syndrome at different temperature, J. of fish discases.V-18 (1) p. 91-97.

Lio-Po-G.D. (1999) Epizootic Alternative syndrome (EUS) recent development forth symp. on disease in Asian Aquaculture (I)AAN) Cebu City Phillipinss;22-26NOV.

Pathiratune. A., Jaysinghe R. P. K. 2001 Environmental factures influence on the occurrence of EUS in freshwater fish. The Bellanwilla.

Attidiya Wetland, Sri Lanka. J. of applied Ichthyology Vol. 17 (1) P. 30-34. Pathname A Jaysinghe, R.P.K., and Epa, U, P.K. 1999 Environmental factors Associated with the occurrence of EUS in freshwater fishes in Sri Lanka. Symp on Diseases in Asian Aquacultuer (DAAN) Cevucity Phillipins 22-26 Nov. 1999.

CHAPTER 27

Need for Multi-Media Strategy for Accelerating Inland Fish Production

Utpal Bhaumik, T. Paria, S. K. Saha and A. Mitra

Central Inland Fisheries Research Institute, Barrackpore. Kolkata. 700120. West Bengal

ABSTRACT

The innovation of scientific technologies in inland fisheries sector not only has rendered support to increase fish production hut also has imparted new dynamism to Indian fisheries and new hopes and confidence to million of fish farmers/fishermen. This has provided both a challenge and opportunity to the communication of information to the clientele, Experience in communication research showed that no single medium holds good in all situations. Research studies in communication indicated that the same message if transmitted through a variety of channels/media would leave and imprint on the memory of a farmer. In order to reach the fish farmers/fishermen at different stages of innovation decision process, the fishery extension personnel are required to use several media in combination. The differential sources credibility accorded by fish farmers demand the need to use multimedia for effective communication for technology. The study conducted at CIFRI depicted that three media combination viz Group discussion + Slide show + Demonstration contributed the most in terms of overall effectiveness of communication while disseminating the technology of *Pen* and *Cage culture* of fish and prawn in the floodplain wetlands of West Bengal. Significant increase in the adoption of technologies has been observed with the increase in number of combination of media. Thus, concerted effort need be taken by the inland fishery extension personnel to adopt proper combination media for accelerating inland fish production in the country.

Key words: *Communication: Clientele: Message; Media: Dissemination*

INTRODUCTION

The breakthrough in inland fisheries production triggered off by the new seientifie innovations in inland fisheries sector and their use on commercial scale particulary adoption of the high productive technologies and strategy has imparted a new dynamism to Indian fisheries and has given new hopes and confidence to millions fish farmers/fishermen. The Indian fisheries sector has heaved itself out of the ruts of traditional inland fish farming based on customs and traditions and is increasingly assimilating the modern and most scientific techniques for stepping up the productivity. This whole process of transformation of traditional inland fish farming into a modern commercial enterprise has given strong support towards *Blue revolution.* All these have created unprecedented upsurge in the demand for appropriate, authentic and timely information on new technologies in inland fishery sector including effective use of fish health management, environmental impact assessment, fish conservation, inputs requirement and post-harvest technology and above all information on cost benefit ratio. This has provided both a challenge and an opportunity to the communication of information on inland fishery specially the research and extension works to the needs of rapidly changing and developing Indian fisheries and fish farming community. Experience in communication research showed that no single media holds good in all situations. The complexity of technology, various categories of farmers i.e., illiterate to highly educated and the variance in communication skills of extension personnel demand the need to have a combination of media for effective communication of technology. To find out the effectiveness of different combinations of media a study was conducted in 3 flood plain lakes of West Bengal during dissemination of pen & cage culture of fish and prawn.

MATERIAL AND METHODS

The investigation was carried out amongst the member fishermen of 3 Fishermen Cooperative Societies managing Saguna & Mathura and Amda beels of North 24 Parganas and Nadia districts of West Bengal respectively. Initially list of member fishermen of above 3 Fishermen Cooperative Societies were prepared and 25 members from each society were selected at random which constituted the sample of the study. Six different communication media *viz* field trip, group discussion, demonstration, fish farmers day, film show. poster exhibition were utilized individually and in different combinations during dissemination of pen and cage culture operation of fish and prawn to assess their effectiveness. The effectiveness was judged on the basis of fishermen's score given to the extension techniques.

RESULT AND DISCUSSION

Communication researches on source credibility at CIFRI conclusively showed that different categories of fish farmers/fishermen attached differential credibility to various sources of information. Bhaumik and Saha (1998) reported that demonstration was ranked as 1st in sources credibility in rural areas whereas 3rd rank in suburban areas. Thus, the differential sources credibility accorded by fish farmers demand the need to use multimedia for effective communication of technology.

The modern inland fishery technologies with its complex nature demand a variety of media for effective dissemination of information to the fish farmers/fishermen. Bhaumik and Saha (1998) highlighted the need of using different communication media by the fishery extension personnel on the basis of the nature of technology.

EFFECTIVENESS OF DIFFERENT COMBINATIONS OF MEDIA

The media commonly used for inland fisheries extension work are: Fieldtrip. group discussion, demonstration, films exhibition, radio, fish farmers' day poster show etc..The relative effectiveness of each medium of communication has been experimentally found to differ.

It is clear from the Table 27.1 that Field trip was ranked as First in terms of overall effectiveness followed by Demonstration. On the other hand poster had lowest scores of in terms of awareness, knowledge and adoption. Chin-wen chen (1981) reported a relative influence of extension methods, personal_contact (38.23%) was ranked as first in terms of number of farm families adopted the innovation. It was followed by indirect influence (25.11%).Group discussion (21.95%). Mass media (9.24%) and other influences (5.47%).

Table 27.1
Effectiveness of single medium

Rank in descending order of overall effectiveness	Media	Score
1	Field trip	7.32
2	Demonstration	6.67
3	Fish farmers' day	4.66
4	Group discussion	3.05
6	Film/show	2.82
7	Poster/exhibition	0.62

EFFECTIVENESS OF COMBINATION OF TWO MEDIA

Two media combinations were found to be statistically similar in effect. The combinations of poster slide show and group discussion + slide show were ranked lowest in terms of overall effectiveness. Other media combinations are indicated in Table 27.2 in descending order of effectiveness. While studying the two media combinations, Singh and Singh (1976) reported that the combination of Field trip + Slide show emerged as the best in terms of overall effectiveness.

Table 27.2
Effectiveness of combination of two media

Rank in descending order of overall effectiveness	Media	Score
1.	Demonstration + Slide show	12.45
2	Demonstration+ Group discussion	10.62
3	Field trip + Group discussion	8.06
4	Exhibition + Group discussion	7.18
5	Radio + Demostration	6.90
6	Film + Group discussion	6.62
7	Poster exhibition + Slide show	5.02
8	Group discussion + slide show	4.13

Mishra (1970) observed that the result and method demonstrations together have influenced 64 percent of farmers in changing their attitudes towards adoption of improved practices. Jalihal (1965) found that the effectiveness of the village meeting was increased due use of movie films. From the study mentioned above, it is clear that demonstration +Slide show or Demonstration Group discussion are the best two media combinations for motivate the fish farmers/fishermen to adopt the new technology.

EFFECTIVENESS OF COMBINATIONS OF THREE MEDIA

Among the three media combinations demonstration along with other two combinations contributed the most in terms of overall effectiveness as reported in Table 27.3. The combination of Group discussion Slide show + demonstration. Flip chart + Group discussion demonstration and Field trip + Group discussion + demonstration were found to he statistically similar to the top ranking combination to change the attitude of the fish farmer/fishermen towards adoption of Pen & Cage culture technology in open water system. However. Mankar (1966) (blind that the most effective combination in brining about a change in knowledge, attitude and skill of farmers was personal contact method demonstration + leaflet. Rao (1958) observed that to popularise most improved agricultural practices the combination of Field trip +Movie + Result demonstration was the best.

Table 27.3

Effectiveness of combination of two media

Rank in descending order of overall effectiveness	Media	Score
1	Group discussion + Slide show + demonstration	16.14
1	Poster exhibition + Group discussion + demonstration	10.62
2	Field trip + Group discussion + demonstration	14.56
3	Fish farmer's day + Group discussion + demonstration	12.76
4	Field trip + Slide show + Group discussion	9.07
5	Poster exhibition + Group discussion + Slide show	8.70
6	Field trip + Group discussion + Slide show	7.48

CONCLUSION

No single communication medium is suitable under any situation of transfer of technology system. The proper combination of media, which increases the effectiveness of communication is always warranted. In addition to this, the probability or changes of reaching message to the target: ownership of information by certain individuals, the complexity of technology, stage of the innovation decision process of an individual fish farmer/fishermen and differential credibility accorded by fish farmers to various sources of information demand the need to have multi-media for effective transfer of technology. Significant increase in the adoption of inland fisheries technologies has been observed with the increase in number of combination media used. Thus, concerted effort need be taken by the inland fisheries extension personnel to adopt proper combination of media for accelerating inland fish production in the country.

ACKNOWLEDGEMENTS

The authors are grateful to the Director of CIFRI, for providing facilities for the research. Authors also thankful the Fishermen Cooperative Societies of North 24 Parganas and Nadia districts of West Bengal for rendering help for these study.

REFERENCES

Bhaumik. utpal and Saha. S. K. 1998. Need for composite fish culture technology in West Bengal as perceived by the fish farmers. *Ed.* Thomas. P. C. *In* Current and Emerging trends in aquaculture. Daya Publishing House, New Delhi.

Chin-wen Chen. 1981. Effective extension approaches applied to Taiwan. Republic of Chaina. In extension education and rural development. Vol. 2, John Wiley & Sons Ltd.

Jalihal. K. A. 1965. The procedure followed in effecting a break through from demonstration to adoption stages in fertilizing in dry land Ragi in Mysore slate. Alumni Association Bulletin. Agricultural College, Hebbal. Bangalore.

Mankar. S. S. 1966. Relative effectiveness of selected extension methods and their combinations in introducing chemical control of weeds. Unpublished Ph. D. Thesis, University of Nagpur.

Mishra. R. S. 1970. Quoted in agricultural communication at extension teaching methods including audio visuals aids. Sahay, B. N. Research in Extension Education of accelerating development process. Indian Agricultural Research Institute, New Delhi. 165-183.

Rao. C. S. S. 1958. Evaluation of extension methods for successful introduction of improved farming practices. Ph. D. Thesis, Indian Agricultural Research Institute. New Dehli.

Singh. K. N. and Singh. S. N.. 1976. Effective communication media for rural audiences (An experimental study). The Dharamsi Morarji Chemical Corporation Ltd. 317/21, Dr. Dadabhai Naoroji Road. Bobbay.

CHAPTER 28

MALATHION AND TEMPERATURE INDUCED ALTERATIONS ON SOME BLOOD PARAMETERS OF FRESHWATER GOBIID FISH, *GLOSSOGOBIUS GIURIS* (HAM).

G.V. Venkataramana, P.N. Sandhya Rani and P.S. Murthy

Department of Zoology, Bangalore University, Bangalore- 560056, Karnataka.

ABSTRACT

Glossogobius giuris exposed to different sub-lethal concentrations of malathion for 24,48,72, and 96 h in constant low and high temperature (20, 27 and 30 ± 1°C) altered the blood parameters such as red blood corpuscle (RBC) and white blood corpuscle (WBC) counts, haemoglobin concentration (Hb), mean corpuscular haemoglobin concentration (MCHC), Colour index (CI), mean corpuscular haemoglobin (MCH), mean corpuscular volume (MCV) and haematocrit (Ht). A decreasing tendency in RBC and WBC counts, Hb content, Ht and MCV were noticed in the treated fish with malathion at 30°C and variations in colour index and MCHC concentration. The RBC, Hb, Ht was found to be decreased when exposed for 24 and 48 h with malathion at 20°C. However WBC, MCV, CI and MCH increased. The fishes treated with 0.05, 0.25 and 0.5 ppm at 20°C for 72 and 96 h the RBC, Hb and Ht was found to be increased whereas WBC, MCHC declined with variations in colour index, MCV, MCH values. In the laboratory temperature (27 ± 1°C) the RBC, Ht, Hb and MCV decreased.

On the other hand WBC, MCHC and MCH increased at 24, 48, 72 and 96 h exposure in different sub-lethal concentrations.

Key words: *G. giuris, malathion, temperature, heamatology.*

INTRODUCTION

The extensive use of Pesticides, Insecticides, Herbicides / Seedicides and Fungicides are being promoted by Government of India to enhance the crop production to meet the demand of the growing population. The rationale behind this is the eradication to meet the demand of the growing population.

The rationale behind this is the eradication of undesirable insects, pests, weeds and herbs for increased yield but on the other hand, these lead to large scale mortality of one of the most important aquatic fauna, the fishes forming stable food to the mankind. All these intern cause severe ecological imbalance. These chemicals reach aquatic systems by direct application, spray drift from ground or aerial, atmospheric fallout, runoff from agricultural land and discharge of effluents. The concentration of pesticides can produce chronic histopathological changes in aquatic organisms. Survival and dispersal of organisms therefore depend upon their successful adaptation in the changes of environmental conditions in aquatic media Mount and Stephen (1967) and Ramalingam (1982).

These pesticides persist in the environment and are found to be toxic to various aquatic animals including the fishes (Kumar, 1994). The physiological changes occurring as a result of pesticide intoxication may be due to alterations in pH, temperature, dissolved oxygen and hardness of water and temperature variations influence the degree of toxicity of various pesticides in aquatic organisms. (Pascoe *et al,* 1986 and Nemcosk *et al.*,1987). Haematological parameters have been used for detecting diseases, dietary deficiencies and environmental stress (George and Alexender, 1980 and Srivastava, 1989). Such information are important when interpreting the changes which are induced by toxicants and other environmental factors (Basile *et al.*, 1976; Christensen *et al.*, 1978). Studies on haematology of fishes are of physiological importance. Such study provides valuable information in the health management of fishes. Hence in the present investigation an attempt has been made to study the haematological changes in response to malathion-temperature combination.

MATERIALS AND METHODS

The freshwater gobiid fish, *Glossogobius giuris* were collected from Kelaverapalli Dam (located near Bangalore) by using cast and gill nets (mesh: 10mm), brought alive to the laboratory and were acclimatized for 15 days in 15 liters glass aquarium (60" x 30" x 20")containing aerated tap water prior to use for experiments. They were maintained at 27°C + 1°C temperature and were fed daily with earthworms. Sexually mature fishes were selected for the study.

The pesticide malathion [0.0-dimethyl, S(l-2 dicarbethoxyethyl) phodophorodithioate] was dissolved in acetone and added to the test water to obtain the desired concentrations. The stock solution, 1mg / litre was prepared separately and desired concentrations were made by adapting the dilution techniques as outlined in APHA AWWA WPCF (1985).

The acclimatised fishes were divided into four experimental groups of eight fishes each. The first three groups of fishes were placed in sublethal concentrations (0.05, 0.25 and 0.5ppm) of malathion and the fourth group in freshwater served as control. The second set was divided into four groups and were subjected to different concentrations of malathion (0.05,0.25 and 0.5 ppm) and temperature variation (20°C and 30°C). Eight fishes were used for each concentration in 15 litres glass trough in all experiments. The acclimated fishes were starved for 24 h prior to their use in the experiment and were not fed during the course of experiments (Dalela *et al,* 1978). The tap water was changed on alternate days and the concentration of pesticide was maintained. In the experimental and control fishes, the blood was collected by severing the caudal peduncle in the intervals of 24,48,72 and 96 h, causing minimum stress to the fish, and the blood was drawn with the help of 1 ml graduated syringe fitted with 2.4 Gxl" needle coated with sodium citrate as anticoagulant.

Total counts of erythrocytes (TEC), leucocytes (TLC), haemoglobin content (Hb), haematocrite (Ht) and blood indices like, Mean corpuscular volume (MCV), Mean corpuscular haemoglobin (MCH),

Mean corpuscular haemoglobin concentration (MCHC) and Colour index (CI) were determined following the methods of Donald Hunter and Bomford (1963); Blaxhall and Daisley (1973) and Wintrobe (1977). From the total count of erythrocyte and leucocyte, the erythrocyte and leucocyte ratio (RBC/WBC ratio) was calculated.

RESULTS

The observations revealed a significant reduction in RBC, Hb, PCV, MCV with an increase in WBC, MCH, MCHC. The values of WBC, MCH. MCHC exhibited in inverse proportion with the changes in RBC. Hb and PCV profile.

The decrease in RBC count and Hb concentration were more conspicuous in 0.5 ppm of malathion when compared to control. Similarly a significant decrease of PCV was observed in higher concentration of malathion (0.5 ppm Table 28.1). MCV content was found to have decreased in the higher concentration of malathion (0.5ppm). On the other hand, the WBC count was found to have increased significantly in the 0.05 ppm of malathion. However in 0.5 ppm the WBC count had increased from 20.58 ± 0.13 x 10^3 cmm to 39.89 ± 0.38 x 10^3 cmm for 24 to 96 h respectively (Table 28.1).

After 24 and 48 h of exposure to sublethal concentrations of malathion (0.05,0.25 and 0.5 ppm) at 20°C temperature the RBC count significantly decreased, and was more conspicuous in higher concentration (0.5 ppm) (Table 28.2) when compared to control (RBC: 2.85 ± 0.04 x 10^6 cmm : 2.51 ± 0.60 x 10^6 cmm for 24 and 48 h). However at 20°C temperature combination with malathion (0.5 ppm), RBC count significantly increased during 72 and 96 h (RBC: 2.40 ± 0.04 x 10^6 cmm and 2.45 ± 0.08x10^6 cmm) when compared to control (RBC: 2.22 ± 0.06 x 10^6 cmm and 1.93 ± 0.13x10^6 cmm).

At 30° C temperature with malathion combination a decrease in RBC count was observed in 0.05 ppm of malathion (RBC: 2.07 ± 0.03x 10^6 cmm; 1.52 ± 0.05 x 10^6cmm; 1.50 ± 0.10 x 10^6cmm; 1.46 ± 0.40 x 10^6 cmm) for 24, 48, 72 and 96 h respectively (Table 28.3). The significant decrease was also observed in 0.5 ppm malathion (RBC: 1.66 ± 0.06 x 10^6 cmm for 24 h: 1.43 ± 0.10 x 10^6 cmm for 48 h; 1.36 ± 0.08x 10^6 cmm for 72 h and 1.37 ± 0.09x 10^6cmm for 96 h) when compared to control (30°C: RBC: 2.32 ± 0.04x 10^6; 1.78 ± 0.05 x 10^6;1.77 ± 0.07 x 10^6cmm, 1.69 ± 0.09 x 10^6 cmm) for 24, 48, 72 and 96 h respectively.

The haemoglobin content (Hb) was found to be decreased after exposure to different concentrations of malathion at 20 °C for 24 and 48 h (Table 28.2). At 20°C with higher concentration of malathion (0.5 ppm) the Hb content had further decreased (Hb: 6.52 ± 0.19 g% and 6.16 ± 0.72g %). On the other hand an increase in Hb content in 0.5 ppm malathion at 20°C was recorded (Hb: 7.88 ± 0.17g% and 9.20 ± 0.84g%) for 72 and 96 h.

At 30° C the Hb concentration had decreased from 24 to 96 h 6:88 ± 0.19g%; 6.24 ± 0.17g%; 5.96 ± 0.48g% and 5.44 ± 0.78g% for 24, 48, 72 and 96 h respectively after exposing to 0.05 ppm malathion. The maximum decrease in Hb content was observed at 72 and 96 h of treatment (5.16 ± 1.28g% and 3.92 ± 1.23g %) with 0.5ppm of malathion at 30°C (Table 28.3).

A significant decrease of PCV (Ht %) was observed in higher concentration (0.5ppm) of malathion (PCV:4.68 ± 0.22% for 24 and 4.28 ± 0.23 for 48 h) when compared to control (PCV : 6.16 ± 0.09 and 5.28 ± 0.83%) during 24 and 48 h respectively at 20°C. The significant increase of PCV value was observed at 20 °C (PCV: 6.86 ± 0.15% and 8.00 ± 0.51%) for 72 and 96 h of exposure in 0.5 ppm of malathion than the control (5.20 ± 0.77% and 5.94 ± 0.71 % for 72 and 96 h) (Table 28.2). The PCV

Table 28.1: Effect of sublethal concentrations of malathion on haematological parameters of *Glossogobius giuris*. (mean ± S.E.)

Parameters	RBC ($X10^6$/mm3)	WBC ($X10^3$/mm3)	RBC/WBC ratio	Hb (g%)	Ht (%)	CI	MCV (μm^3)	MCH (pg)	MCHC(%)
24 Hours									
Control	3.33 ±0.07	14.08 ±0.45	23.65 ±0.38	11.21±0.08	10.0 ±0.28	0.200 ±0.01	25.06 ± 0.40	6.01 ±0.28	24.00± 1.22
0.05	2.44 ±0.02	16.24 ± 0.56	15.02 ±0.91	10.53 ±0.04	7.93 ± 0.08	0.196 ±0.02	23.06 ±0.28	5.91 ±0.10	25.41 ±0.94
0.25	2.23 ±0.03	16.96 ± 0.67	13.14±0.31	10.02 ± 0.07	7.39 ±0.35	0.186 ±0.02	15.58 ±0.95	5.58±0.17	35.82 ± 1.21
0.5	2.04 ± 0.03	20.58 ±0.13	10.12 ±0.48	9.40 ±0.07	4.26 ±0.04	0.243 ±0.01	18.20 ±0.42	7.30 ±0.11	40.14± 1.32
48 Hours									
Control	3.11±0.19	15.08 ±0.35	20.62 ± 0.68	10.18 ±0.09	9.43 ±3.5	0.220 ± 0.09	26.08 ± 0.48	6.21+0.38	25.55±2.11
0.05	2.76 ± 0.06	18.88 ±0.22	14.61 ± 1.00	10.42 ± 0.04	7.39±0.13	0.189±0.12	22.53 ±0.85	5.67 + 0.03	25.20 ± 1.00
0.25	2.28 ±0.05	27.41 ±0.22	8.31 ±1.31	9.86 ±0.07	4.47 ± 0.07	0.200 ± 0.02	16.19 ±0.41	6.01+0.36	31.13±2.27
0.5	2.07 ± 0.04	29.33 ±0.34	7.05 ±0.97	9.06 ±0.08	4.06 ± 0.03	0.262 ± 0.02	19.61 ±0.25	7.87 ±0.08	40.14± 1.31
72 Hours									
Control	3.20 ±0.29	15.43 ±0.41	20.73 ±0.40	9.89 ±0.06	9.38 ±3.00	0.241 ±0.02	25.40 ±0.40	6.91+0.27	25.71 ±2.66
0.05	2.45 ± 0.06	27.76 ± 0.36	9.81 ±1.12	9.48 ±0.02	6.71 ±0.07	0.186 ±0.02	21.78 ±0.58	5.59 ±0.07	25.67 ±0.39
0.25	2.08 ±0.90	34.45 ±0.34	6.03 ±0.71	8.53 ±0.01	3.90 ±0.02	0.212±0.01	15.91 ±0.73	6.44 ±0.07	40.51 ±2.06
0.5	1.98 ±0.05	36.58 ±0.35	5.41 ±0.41	8.17 ±0.05	3.80 ±0.12	0.256 ± 0.02	19.69 ±0.47	7.70 ±0.12	40.15 ±0.01
96 Hours									
Control	3.00 ±0.46	14.48 ±0.68	20.71 ±0.65	10.41±0.09	10.00 ±1.2	0.250 ± 0.02	24.48 ±0.48	6.48 + 0.33	26.78 ± 2.06
0.05	2.38 ±0.05	30.93 ± 0.05	7.69 ± 0.79	7.70 ± 0.02	5.82 ±0.08	0.173 ±0.02	20.93 ± 0.66	5.23 ±0.19	25.01 ±0.41
0.25	2.24 ±0.09	39.50 ±0.65	5.67 ±0.51	8.50 ±0.04	3.69 ±0.05	0.189 ±0.02	16.47 ±0.52	5.69 ±0.24	34.57 ± 1.52
0.5	1.65 ±0.07	39.89 ±0.38	4.23 ±0.38	7.06 ± 0.03	2.58 ±0.18	0.226 ±0.01	15.63 ±0.29	6.42 ±0.33	41.08 ±0.98

Table 28.2 : Effect of sublethal concentrations of malathion on haematological parameters of *Glossogobius giuris* at 20 °C (mean ± S.E.)

Parameters	RBC ($X10^6/mm^3$)	WBC ($X10^3/mm^3$)	RBCAVBC ratio	Hb (g%)	Ht (%)	CI	MCV μm^3)	MCH (pg)	MCHC(%)
24 Hours									
Control	2.85 ±0.04	22.40 ±0.88	12.72 ± 1.83	8.52 ± 0.90	6.16 ± 0.09	0.996 ± 0.07	21.61±1.78	29.89±0.81	138.31±0.31
0.05	2. 16±0.22	17.90 ±1.64	12.06 ± 2.40	8.24 ± 0.25	4.20 ± 0.99	1.271 ± 0.09	19.44±2. 11	38.14±0.91	196.10±0.41
0.25	1.72 ±1.50	27.50±4.1	6.25 ±1.71	8.00 ±0.80	4.40 ± 0.75	1.425 ± 0.08	25.58±2.61	46.51±1.20	181.81±0.91
0.5	1.56±0.60	29.40 ±1.93	5.30±2.79	6.52±0.19	4.68 ± 0.22	1.392 ± 0.02	30.00±1.99	41.79±0.41	139.31±0.81
48 Hours									
Control	2.51±0.60	25.28 ± 1 .60	9.92±1.70	8.00 ±0.06	5.28 ± 0.83	1.062 ± 0.07	21.03±2.41	31.87±0.91^	151.51±1.11
0.05	1.83±0.06	23. 72 ±1.26	7.71 ±2.00	7.60 ±0.34	4.08 ± 0.76	1.384±0.05	22.29±3.11	41.53±0.68	186.27±1.33
0.25	1.53±0.02	30:00 ±2.00	5. 10 ±3.22	6.52 ±0.20	4.00 ± 0.40	1.420±0.03	26.14±3.67	42.61±0.81	163.00 ±0.17
0.5	1.34±0.06	30.85 ±0.60	4.34 ± 4.20	6.16 ±0.72	4.28 ± 0.23	1.532±0.06	3 1.94±2.66	45.97±1.11	143.92±1.11
72 Hours									
Control	2.22 ± 0.06	22.23 ±0.57	9.98 ±2.82	7.56 ±0.73	5.20 ± 0.77	1.198±0.09	23.42 ± 1.77	35.94±0.71	153.46±0.91
0.05	2.38±0.32	27.84 ± 1.80	8.54 ±3.33	7.98 ± 0.20	6.40 ± 0.18	1.058±0.04	26.89 ± 2.77	31.76±0.86	118.12± 1.60
0.25	2.24±0.25	22.72±0.41	9.85 ±3.63	6.94 ±1.36	5.80 ± 0.26	1.032±0.07	25.89 ± 3.11	30.98±1.12	119.65± 1.32
0.5	2.40±0:04	18:06 ±0.83	13.28 ± 2.46	7.88 ±0.17	6.86±0.15	1.063± 0.04	27.77 ± 3.00	31.90±1.25	114.86 ± 2.00
96 Hours									
Control	1.93 ±0.13	20.64 ±0.88	9.35 ±3.31	6.36 ±0.73	5.94 ± 0.71	1.098±0.02	30.77 ± 1.97	32.95 ±1.33	117.07±0.71
0.05	2. 15 ±0.07	22.72 ± 1.36	9.46 ±2. 10	8.28 ± 0.31	7.28±0.33	1.283±0.04	33.86 ± 1.66	38.51 ±0.95	113.73 ±1.21
0.25	2.20±0!05	18.88 ±0.76	11.65 ± 3.70	7.96 ±0.38	6.96 ± 0.44	1.205±0.05	31.63 ± 1.81	36. 18 ±0.71	112.93 ±1.61
0.5	2.45 ±0:08	16.80 ±1.78	14.58 ±2.80	9.20 ± 0.84	8.00 ± 0.51	1.251±0.06	32.65 ±2.22	37.55 ±1.91	115.00±1.51

Table 28.3: Effect of sublethal concentrations of malathion on haematological parameters of *Glossogobius giuris* at 30 °C (mean ± S.E.)

Parameters	RBC ($X10^6$/mm3)	WBC ($X10^3$/mm3)	RBCAVBC ratio	Hb (g%)	Ht (%)	CI	MCV mm³)	MCH (pg)	MCHC(%)
24 Hours									
Control	2.32 ±0.04	15.92 ± 0.48	14.57±1.34	8.04 ± 0.17	8.56±1.78	1.150 ± 0.02	36.89±2.16	34.65 ±0.45	93.92 ±0.39
0.05	2.07 ±0.03	24-.08 ± 0.72	8.59±2.08	6.88±0.19	6.16±1.37	1.107 ±0.07	29.75±2.27	33.23 ±0.67	111.68±0.86
0.25	1.82 ±0.02	24.44 ±1.42	7.44±11.27	6.40 ± 1.15	5.84±1.49	1.170 ±0.09	32.08 ±3. 05	35. 16 ±0.29 1	109.58±0.32
0.5	1.66 ±0.06	17. 12 ±1.72	9.69 ±2.79	5.92±0.23	5.04 ± 0.97	1.188 ±0.07	30.36 ±2.86	35.66 ±1.00	117.46±0.83
48 Hours									
Control	1.78 ±0.05	24.24 ± 2.0	7.34 ±1.80	6.64±0.21	4.72±1.42	1.243 ±0.02	26.51 ± 1.86	37.30 ±0.47	140.67±0.63
0.05	1.52 ±0:05	13.68 ±1.54 1	11.11±1.70	6.24±0.17	4.76 ±2.00	1.368 ±0.22	31.31±2.60	41.05±0.38	13 1.09 ±0.26
0.25	J .48 ± 0.09	12. 80 ±2.40	11.56±3.27	5.84±0.29	3.80 ± 2.11	1.315±0.11	25.67±2.63	39.45±1.31	153.68±0.66
0.5	1.43 ±0.10	10.40 ±3. 40	13.75±14.30	4.96±0.49	3.60±2.20	1.155 ±0.07	25. 17±2.65	34.68 ± 0.71	137.77± 1.20
72 Hours									
Control	1.77 ±0.07	22.32 ± 2.20	7.93 ±12.80	6.60±0.40	4.52±1.20	1.242±0.09	25.53±1.66	37.28 ± 0.78	146.01 ±0.91
0.05	1.50±10.10	16.56±13.12	9:05 ±3.86	5.96±0.48	4.00 ± 0.94	1.324 ±0.19	26.66±2. 11	39.73 ±0.73	149.00±0.94
0.25	1.38 ±0.07	14.08±12.40	9.80 ± 3.08	5.24±0.48	3.44 ±0.96	1.265 ±0.07	24.92±4.13	37.97 ±0.92	152.32±0.66
0.5	1.36 ±0.08	11.60±0.09	11.72 ±2.44	5.16±0.28	3.44±1.20	1.264 ±0.09	25.29 ±5.55	37.94 ± 0.69	150.00±1.05
96 Hours									
Control	1.69 ±0.09	24.72 ± 0.31	6.83 ±2.14	5.92±0.88	4.40±1.72	1.167±0.06	26.03 ± 5.93	35.02 ±0.71	134.54±0.62
0.05	1.46 ±0.40	18.00 ±0.21	8.11 ±2.72	5.44±0.78	3. 88±0.99	1.241±0.05	26.57±6.11	37.26±1.10	140.20 ± 0.72
0.25	1.40 ±030	14.80 ±0.31	9.45 ±1.81	4.28±1.22	3.64±0.71	0.907 ± 0.05	24.97±3.95	29.31±1.31	117.58±1.40
0.5	1.37 ±0.09	12.48±10.41	10.97 ±1.91	3.92±1.23	3.32±0.91	0.953 ±0.04	24.23±2.85	38.61 ±0.89	118.07±1.61

values were found to have significantly declined in the malathion (0.05 ppm) and temperature (30 °EC) combination, after exposure for 24 to 96 h (PCV: 6.16 ± 1.37%; 4.76 ± 2.00% ; 4.00 ± 0.94% and 3.88 ± 0.99%) than the controls. The decrease in the values were more conspicuous in 0.5 ppm (PCV: 5.04 ± 0.97% to 3.32 ± 0.91%) for 24 to 96 h (Table 28.3).

When the fishes were exposed to malathion (0.5 ppm) at 20°C temperature for 24 and 48 h the WBC count was found to have increased (WBC: 29.40 ± 1.93 x 10^3 cmm and 30.85 ± 0.60 x 10^3 cmm) and decreased in the higher concentration (0.5 ppm) of malathion and temperature combination (20°C) for 72 and 96 h (WBC : 18.06 ± 0.83 x10^3 cmm and 16.80 ± 1.78 x 10^3 cmm) than the control (Table 28.2).

Significant decrease of WBC count was observed in higher concentration of malathion after 24 to 96 h of exposure at 30°C (WBC: 17.12 ± 1.72 x10^3 cmm to 12.48 ± 0.41x 10^3 cmm) for 24, 48, 72 and 96 h respectively (Table 28.3).There were variations in values of MCV, MCH and MCHC when the fishes were exposed to sublethal concentrations (0.05, 0.25 and 0.5 ppm) of malathion for 24 to 96 h at 20°C and 30°C(Table 28.2 and 28.3).

DISCUSSION

Blood analysis forms an useful tool for determining pathological conditions of the fishes. The pathological changes were noticed mostly in the peripheral blood of erythrocytes and leucocytes as observed by Reichenbach-Klinke, (1967). In the present study changes in erythrocyte count was observed after the fish was treated for longer duration. Wlasow and Dabrowska (1989) have reported inhibition of erythroblastosis in circulating blood due to ammonia intoxication in carp. In the present study similar changes of erythrocyte formation was noted in treated fishes. The haematological parameters like Hb content, RBC and WBC count showed significant decrease in treated fishes. The decrease in Hb content and RBC number may result in hypochromic microcytic anaemia, which could be due to deficiency of iron content (Bhai *et al,* 1971). Since Hb and RBCs are oxygen carrying devices, the quantitative decrease in their levels might have led to the dearrangement of the oxidative metabolism with a concomitant decrease in the tissues of respiratory potential.

G. giuris exposed to sublethal concentrations of malathion resulted in a significant decrease in RBCs count leading to anaemia as a result of inhibition of erythropoiesis, haemosynthesis and increase in the rate of erythrocyte destruction in haemopoietic organs. Similar reports have been made by Goel and Kalpana (1985) and Sampath *et al,* (1998) on the Hb content, RBC count and PCV values resulting in macrocytic anaemia in *Heteropneustes fossilis* and *Oreochromis mossambicus* alter exposure to zinc and copper. Srivastava and Shashikala (1979) have shown that exposure to sublethal concentration of Pb produced haemolytic anaemia leading to the lysis of erythrocytes, decrease in Hb content and PCV value in *Colisa fasciatus.* The significant reduction of RBC and Hb content are also reported in fishes exposed to different heavy metals (Goel and Sharma, 1987). Natarajan (1981) reported a reduction in Hb content, RBC count and PCV values, resulting in hypochronic anaemia due to deficiency of iron and decreased utilization for Hb synthesis. In the present study on *G.giuris* sublethal concentration of malathion might have impaired the synthesis of iron as a result destruction of mature RBC and reduction in RBC count causing macrocytic anaemia. Sampath *et al,* (1998) have also reported such macrocytic anaemia in *O. mossambicus* after exposure to copper and zinc. Joseph *et al,* (1993) have showed an increase in the relative proportion of immature RBC and decrease maturation of RBC in *Cyprinus carpio* after exposure to sublethal concentration of ammonia. The increased WBC count in treated fishes may be due to the inclusion of thrombocytes in the leucocyte population.

The anaemic condition recorded in the present study could be due to the destruction of mature RBC or inhibition of erythropyte production. Such a decrease in RBC and anaemic response have been observed by Koundinya and Ramamurthy (1979) in *Sarotherodon mossambicus* and *Tilapia mossambicus* after exposure to lethal concentrations of sumithion; Chouhan *et al*, (1983) in *Puntius ticto* treated with herbicide; Lal *et al*., (1986) in *H. fossilis* exposed to malathion; and Sakthivel and Sampath (1989) in *C. carpio* treated with sublethal levels of tannery Natarajan (1981) reported that impairment in iron synthesizing mechanism corresponds to drastic decrease in haemoglobin content in the fish *Channa striatus* exposed to metasystox. It is evident that the reduction in number of RBC and Hb content and decrease in MCV values might have caused microcytic anaemia in *G. giuris*. Varadharaj *et al*, (1993) have opined that a reduction in the number of RBC and Hb content with an increase in MCV and MCHb values might cause macrocytic anaemic in *O. mossambicus*. Further Pandey *et al*., (1979) and Subramanian *et al*, (1988) have observed a decrease in the rate of oxygen consumption in fishes due to fall in RBC count and Hb content.

An increase in the total WBC count and MCHC content in *G. giuris* after exposure to different sublethal concentrations of malathion coincides with the observations made by Joshi (1987); Radhaknshnan and Prasad (1994). In the present investigation MCV significantly decreased which coincide with the findings of Rai and Quayyam (1981); Dhanekar *et al*, (1985) and Surendra Singh and Bhati (1994). *G. giuris* exposed to different temperatures (20°C and 30°C) and sublethal concentrations (0.05, 0.25 and 0.5 ppm) of malathion the heamatological paramcters exhibited variations. Among the various external factors, temperature considered to be as controlling factor and regulating the metabolism (energy utilization) in fishes (Prosscr and Brown, 1961 and Kasim, 1979).

The decrease in RBC count Hb content and PCV in *G. giuris* subjected to temperature (20°C) and malathion toxicity are in agreement with the observations made by Slicher and Pickford (1968) in *Fundulus heteroclitus* and Anthony (1961) in *Carassius auratus*. The results in the present study indicate that the fishes exposed to malathion at 20°C the total number of leucocytes was significantly increased and RBC count, Hb concentration and PCV gradually declined during 24 and 48 h. On the other hand RBC count and PCV increased gradually when the fishes were exposed to malathion with the increased period of exposure at 20°C. However, the total WBC, MCHC and MCH significantly decreased when compared to those of controls. The significant decrease in MCV with simultaneous increase in MCHC might have led to microcytosis in *G.giuris* as pointed out by Haws and Goodnight (1962) and Joshi *et al*, (1980).

The fishes exposed to malathion at a temperature of (30°C) and malathion-temperature combination, the haematological parameters except MCHC were found to have decreased with the increase in the exposure period. This phenomenon suggests that a high temperature and malathion toxicity may cause anaemic and macrocytic changes in *G.giuris* when compared to low temperature and malathion stress. Similar results have been recorded by Joshi *et al*, (1980) in *C.batrachus*. In *G. giuris* a decrease in WBC count could have led to a fall in the resistance for malathion and temperature (at 30°C) stress, and this suggests that malathion in combination with high temperature has an adverse effect resulting in anemic and macrocytic conditions.

ACKNOWLEDGEMENTS

The author is grateful to Dept. of Zoology Bangalore University, Bangalore for providing necessary research facilities and sincere thanks to Dr.S.Krishnan, professor, Dept. of Zoology, Bangalore University, Bangalore for their kind help.

REFERENCES

Anthony, E.H. 1961. The oxygen capacity of gold fish, *Carassius auratus* in relation to the thermal environment. J. Exp. Biol. 38: 93-107.

Apha Awwa and Wpcf 1985. Standard methods for the examination of water and wastewater, 16th Edition, American Public Health Association, Washington, D.C.

Basile, C, Goldspink, G. Modigh, M and Tota, B. 1976. Morphological and biochemical characterization of the inner and outer ventricular myocardial layer of adult Tuna fish (Thymus thymus) biochem. Physio., 854:279-283.

Bhai, L., Nath, N. and Nath, M.C. 1971. Haematological changes in rats injected with acetoacetate, Proc, Soc. Exp. Biol. Med., **138:** 597-599.

Blaxhall, P.C. and Daisley, K.W. 1973. Routine haematological methods for use with fish blood, *J. Fish Biol,* 5: 771-782.

Chouhan, M.S., Verma, D. and Pandey, A.K. 1983. Herbicide induced haematological changes and their recovery in a fresh water fish, *Puntius ticto* (Ham.), Comp. Physiol. Ecol., 8(4): 249-251.

Christensen, G.M., MC kim., J.M.. Brungs, W.A. and Hunt, E.P. 1978. Changes in the blood of the brown bull head, *Ictahurus nebulosus* following short and long term exposure to copper (II), Toxicol, Appl. Pharmacol., **23:** 477-482.

Dalela, R.C., Bhatnagar, M.C., Tyagi, A.K. and Verma, S.R. 1978. Adenosine triphosphatase activity in few tissues of a freshwater teleost, *Channa gachua,* following *in vivo* exposure to endosulfan, Toxicology, **11:** 361-368.

Dhanekar. S., Srivastava, S. and Rao. K.S. 1985. Haematological studies on some fresh water fishes with references to zinc toxicity. J. Hydrobiol., **1(2):** 105-107.

George and Alexander, K.M. 1980. Studies of physiological of mammalian cardiac muscle L. A report on differential distribution of some metabolities in myocardia of ox: Boss Sp., Abstr., Proc., 4th All Indian Symposium on comparative physiology and endocrinology. S.V. Unive., Tirupati, Dec.

Goel, K.A.and Kalpana,G.1985. Haematological characteristics of *Heteropneutes fossilis* under the stress of zinc. Indian. J. Fish., **36:** 256-259.

Goel, K.A., and Sharma, S.D., 1987. Some haematological characteristics of *Clarias batrachus* under metallic stress of arsenic. Comp. Physiol. Ecol., **12:** 63-66.

Haws, T.G. and Goodnight, C.J. 1962. Some aspects of the haematology of two species of catfish in relation to their habitats. Physiol. Zool., **35:** 8-17.

Joseph, K. Manissery and Madhyastha, M.N. 1993. Haematological and histopathological effect of ammonia at sublethal levels of fingerlings of common carp *Cyprinus carpio.* The science of the total environment, supplement, Elsevier Science Publishers B.V. Amsterdam. 913-920.

Joshi, B.D., Chaturvedi, L.D. and Dabral, R. 1980. Some haematologic values of *Clarias batrachus* following its sudden transfer to varying temperature. Indian J. Exp. Biol., **18:** 76-77.

Joshi, B.D. 1987. Changes in some blood values of *Clarias batmchus* exposed to lead nitrate. Him. J. Env. Zool., 1: 33-36.

Kasim, H.M. 1979. Ecophysiological studies on fry and fingerlings of some freshwater fishes with special reference to temperature tolerance. Ph.D., thesis. Madurai Kamaraj Univ., India.

Koundinya, P.R and Ramamurthy, R. 1979. Effect of Organophosphate sumithion (Fenthion) on some aspects of carbohydrate metabolism in a freshwater fishes *Sarotherodon, mossambicus* and *Tilapia mossambicus* (peters), J.Expt., 15:1632-1633.

Kumar. A.V. 1994. Endosulfan induced biochemical and pathophysiological changes in freshwater fish. *Clarias batractus* (Linn.) Ph.D. Thesis. Osmania University, Hydrabad, India.

Lal, A.S.B., Anitakumari, S. and Sinha, R.N. 1986. Biochemical and haemotological changes following malathion treatment in the freshwater catfish *Heteropneustes fossilis* (Bloch). Environ. Pollut, 42(A): 151-156.

Mount, D.I and Stephan. C.E. 1967. A method for establishing acceptable toxicant limits for fish malathion and the butoxy-ethanol ester of 2,4-D. Trans. Amer. Fish. Soc. **21:** 185-193.

Natarajan, G.M. 1981. Changes in the biomodal gas exchange and some blood parameters in the air-breathing fish *Channa striatus* (Bloch) following lethal exposure to metasystox (Dimeton). Curr. Sci., **50:** 40-41.

Nemcosk, J., Orban, L., Asztalos, B. and Vig, F. 1987. Accumulation of pesticides in the organs of carp *Cyprinus carpio* L. at 4 and 20°C. Bull. Enviorn. Contam, Toxicol., **39:** 370-378.

Pandey, B.N., Chanchai, A.K., Singh, S.B., Prasad, S. and Singh, M.P. 1979. Effect of some biocides on the blood and oxygen consumption in *Channa punctatus. The* academy of environmental Biology Muzaffarnagar, 343-348.

Pascoe, D., Evans, S. and Woodworth, J. 1986.Heavy metal toxicity to fish and the influence of water hardness. *Arch. Environ. Contam. Toxicol.*, **15:** 481-485.

Prosser, C.L. and Brown, F.W. 1961.Comparative animal physiology (2nd Edn.) Saunders, Philadelphia.

Radhakrishnan, S. and Prasad, G. 1994. A unique haematological response of *Oreochromis mossambicus* to sublethal concentrations of the insecticide, Ekaluk EC-25. Proceedings of the Indian National Science Academy Part B. Biological Sciences 60(6): 499-503.

Rai, R. and Quayyam, M.A. 1981. Haematological studies of mercury intoxicated fleets fish *Catla Catla* (Hamilton). Ind. J. Zool, 9: 87-91

Ramalingam, K. and Ramalingam, K. 1982. Effect of sublethal levels of DDT, malathion and mercury on tissue proteins of *Sarotherodon mossambicus* (Peters). Proc. Indian. Acad, Sci. (Anim. Sci) **91:** 501-503.

Reichenbach-klinke, H.H. 1967. Untersuchungen uber die Einwirking des Ammoniakgchalts aufden fish organisums, Arch. Fisch, Wise, 17: 122-130.

Sakthivel, M. and Sampath, K. 1989. Haematological responses of *Cyprinus carpio* in relation to starvation, Geobios, 16: 61-65.

Sampath, K., James, R. and Akbar Ali, K.M. 1998. Effects of copper and zinc on blood parameters and prediction of their recovery in *Oreochromis mossambicus*. Indian. J. Fish., 45(2): 129-139.

Slicher, A.M. and Pickford, G.E., 1968. Temperature controlled stimulation of haemopoiesis in a hypophysectomized cyprinodont fish *Fundulus heteroclitus*. Physiol. Zool., 41: 293-297.

Srivastava, A.K. and Shashikala, M. 1979. Blood dyscrasia in a teleost, *Colisa fasciatus* after acute exposure of sublethal concentration of lead. J. Fish Biol., 14: 199-203..

Srivastava. A.K. 1989.A review of landmarks in teleostean haematology. Proc. Natl., Symp. Emerg. Tr. Anim. Haematol., 165-177.

Subramanian, M.A., Shahul, H and Veradaraj. G. 1988. Haematological changes in the fish *Marcones keletius*

(Gunther) after exposure to the tannary effluent the Indian J.zoo. 12:71-74.

Surendra Singh and Bhati, D.P.S. 1994. Effect ot zinc chloride on certain morphological parameters of the blood in *Channa punctatus* (Bloch.). Poll Res. 13(14): 381-384.

Varadharaj. G., M.A. Subramanian and B. Nagarajan, 1993. The effect of sublethal concentration of paper and pulp mill effluent on the haematological parameters of *Oreochromis mossambicus* (Peterns) J. Environ, Biol., 14(4): 321-325.

Wintrobe, M.M. 1977. Clinical haematology. Henry Kimpton. London, 448.

Wlasow, T. and Dabrowska, H. 1989. Cellular changes in the blood and haemopoietic tissues of common carp exposed to sublethal concentration of ammonia. Aquat. Living Resour., 2(3): 169-174.

CHAPTER 29

EFFECT OF MALATHION AND MALATHION-TEMPERATURE COMBINATION ON HAEMATOLOGICAL PARAMETERS OF GOBIID FISH, *GLOSSOGOBIUS GIURIS* (HAM).

G.V.Venkataramana, P.N.Sandhya Rani, M.B.Nadoni and P.S.Murthy

Department of Zoology. Bangalore University, Bangalore-560056, Karnataka.

ABSTRACT

Temperature variations influence the degree of toxicity of various pesticides in aquatic organisms. *Glossogobius giuris* exposed to various sub-lethal concentrations of malathion for 24,48, 72, and 96 h. in constant low and high temperature (20. 27 and 30 ±1°C) altered the blood parameters such as red blood corpuscle (RBC) and white blood corpuscle (WBC) counts, haemoglobin concentration (Hb). mean corpuscular haemoglobin concentration (MCHC), colour index (CI), mean corpuscular hemoglobin (MCH), mean corpuscular volume (MCV) and haematocrit (Ht). A decreasing tendency in RBC and WBC counts. Hb content Ht and MCV were noticed in the treated fish with malathion at 30°C and variations in colour index and MCHC concentration. The RBC, Hb. Ht was found to be decreased when exposed for 24 and 48 h with malathion at 20°C. However WBC, MCV, CI and MCH increased. The fishes treated with 0.05, 0,25 and 0.5 ppm at 20°C for 72 and 96 h the RBC, Hb. Ht was found to be increased whereas WBC. MCHC declined with variations in colour index, MCV, MCH values. In the laboratory temperature (27±1 °C) the RBC, Ht, Hb and MCV decreased. On the other hand WBC, MCHC and MCH increased at 24. 48. 72 and 96 h exposure in different sub-lethal concentrations.

Key words: *G.giuris, Malathion, Temperature, Heamatology.*

INTRODUCTION

The rapid use of pesticides to control pests poses serious hazards to non-target organisms including fishes in the aquatic bodies. Pesticides reach aquatic systems by direct application, spray drift from ground or aerial, atmospheric fallout, run-off from agricultural land and discharge of effluents.

The concentration of pesticides can produce chronic histopathological changes in aquatic organisms. Survival and dispersal of organisms therefore depend upon their successful adaptation in the changes of environmental conditions in aquatic media Mount and Stephen (1967) and Ramalingam (1982).

These pesticides persist in the environment and are found to be toxic to various aquatic animals including the fishes (Kumar, 1994). The physiological changes occurring as a result of pesticide intoxication may be due to alterations in pH. temperature, dissolved oxygen and hardness of water (Pascoe *et al*,. 1986 and Nemcosk *et al*, 1987) Studies on haematology of fishes are of physiological importance. Such study provides valuable information in the health management of fishes. Hence in the present investigation an attempt has been made to study the haematological changes in response to malathion, malathion temperature combination.

MATERIALS AND METHODS

The freshwater gobiid fish, *Glossogobius giuris* were collected from Kelaverapalli Dam (located near Bangalore) by using cast and gill nets (mesh: 10 mm), brought alive to the laboratory- and were acclimatized for 15 days in 15 liters glass aquarium (60" x 30" x 20") containing aerated tap water prior to use for experiments. They were maintained at 27°C ± 1 °C temperature and were fed daily with earthworms. Sexually mature fishes were selected for the study.

The pesticide malathion [0.0-dimethyl. S (1-2 dicarbethoxyethyl) phodophorodithioate was dissolved in acetone and added to the test water to obtain the desired concentrations. The stock solution, 1mg / litre was prepared separatelv and desired concentrations were made by adapting the dilution techniques as outlined in APHA AWWA WPCF (1985).

The acclimatised fishes were divided into four experimental groups of eight fishes each. The first three groups of fishes were placed in sublethal concentrations (0.05. 0.25 and 0.5ppm) of malathion and the fourth group in freshwater served as control. The second set was divided into four groups and were subjected to different concentrations of malathion (0.05, 0.25 and 0.5 ppm) and temperature variation (20 °C and 30 °C). Eight fishes were used for each concentration in 15 litres glass trough in all experiments. The acclimated fishes were starved for 24 h prior to their use in the experiment and were not led during the course of experiments (Dalela *et al*, 1978). The tap water was changed on alternate days and the concentration of pesticide was maintained. In the experimental and control fishes, the blood was collected by severing the caudal peduncle in the intervals of 24,48,72 and 96 h, causing minimum stress to the fish, and the blood was drawn with the help of 1 ml graduated syringe filled with 2.4 Gxl" needle coated with sodium citrate as anticoagulant.

Total counts of erythrocytes (TEC), leucocytes (TLC), haemoglobin content (Hb), haematocrite (Ht) and blood indices like. Mean corpuscular volume (MCV), Mean corpuscular haemoglobin (MCH), Mean corpuscular haemoglobin concentration (MCHC) and Colour index (CI) were determined following the methods of Blaxhall and Daisley (1973) and Wintrobe (1977). From the total count of erythrocyte and leucocyte, the erythrocyte and leucocyte ratio (RBC/WBC ratio) was calculated.

RESULTS

The observations revealed a significant reduction in RBC, Hb, PCV, and MCV with an increase in WBC, MCH, and MCHC. The decrease in RBC count and Hb concentration were more conspicuous in 0.5 ppm of malathion when compared to control. Similarly a significant decrease of PCV was observed in higher concentration of malathion (0.5 ppm Table 29.1). MCV content was found to have decreased in the higher concentration of malathion (0.5ppm). On the other hand, the WBC count was found to have increased significantly in the 0.05 ppm of malathion. However in 0.5 ppm the WBC count had increased from 20.58 ± 0.13 x 10^3 cmm to 39.89 ± 0.38 x 10^3 cmm for 24 to 96 h respectively (Table 29.1).

After 24 and 48 h of exposure to sublethal concentrations of malathion (0.05,0.25 and 0.5 ppm) at 20°C temperature the RBC count significantly decreased, and was more conspicuous in higher concentration (0.5 ppm) (Table 29.2) when compared to control. However at 20°C temperature combination with malathion (0.5 ppm). RBC count significantly increased during 12 and 96 h (RBC: 2040 ± 0.04 x 10^6 cmm and 2,45 ± 0.08x10^6 cmm) when compared to control (RBC: 2.22 ± 0.06 x 10^6 cmm and 1.93 ±0.13x 10^6 cmm).

At 30°C temperature with malathion combination a decrease in RBC count was observed in 0.05 ppm of malathion for 24, 48, 72 and 96 h (Table 29.3). The significant decrease was also abserved in 0.5 ppm malathion (RBC: 1.66 ± 0.06 x 10^6 cmm for 24 h; 1.43 ± 0.10 x 10^6 cmm for 48 h; 1.36 ±0.08x 10^6 cmm for 72 h and 1.37 ± 0.09x 10^6cmm for 96 h) when compared to control (30 °C: RBC: 2.32 ±0.04x 10^6; 1.78 ± 0.05 x 10^6:1.77 ±0.07 x 10^6cmm, 1.69 ± 0.09 x 10^6 cmm) for 24, 48, 72 and 96 h respectively.

The haemoglobin content (Hb) was found to be decreased after exposure to different concentrations of malathion at 20°C for 24 and 48 h (Table 29.2). At 20 °C with higher concentration of malathion (0.5 ppm) the Hb content had further decreased. On the other hand an increase in Hb content in 0.5 ppm malathion at 20 °C was recorded for 72 and 96 h.

At 30 °C the Hb concentration had decreased from 24 to 96 h 6:88 ±0.19g%: 6.24 ± 0.17g%; 5.96 ± 0.48g% and 5.44 ± 0,78g% for 24, 48, 72 and 96 h respectively after exposing to 0.05 ppm malathion. The maximum decrease in Hb content was observed at 72 and 96 h of treatment with 0.5ppm of malathion at 30 °C (Table 29.3).

A significant decrease of PCV (Ht %) was observed in higher concentration (0.5ppm) of malathion. when compared to control during 24 and 48 h respectively at 20 °C. The significant increase of PCV value was observed at 20 °C (PCV: 6.86 ± 0.15% and 8.00 ± 0.51%) for 72 and 96 h of exposure in 0.5 ppm of malathion than the control (Table 29.2). The PCV values were found to have significantly declined in the malathion (0.05 ppm) and temperature (30 °C) combination, after exposure for 24 to 96 h than the controls. The decreases in the values were more conspicuous in 0.5 ppm for 24 to 96 h (Table 29.3).

When the fishes were exposed to malathion (0.5 ppm) at 20 °C temperature for 24 and 48 h the WBC count was found to have increased and decreased in the higher concentration (0.5 ppm) of malathion and temperature combination (20°C) for 72 and 96 h than the control (Table 29.2).

Significant decrease of WBC count was observed in higher concentration of malathion after 24 to 96 h of exposure at 30°C for 24, 48, 72 and 96 h (Table 29.3).There were variations in values of MCV, MCH and MCHC when the fishes were exposed to sublethal concentrations (0.05, 0.25 and 0.5 ppm) of malathion for 24 to 96 h at 20 °C and 30 °C (Table 29.2 and 29.3).

Table 29.1: Effect of sublethal concentrations of malathion on haematological parameters of *Glossogobius giuris*. (mean ± S.E.)

Parameters	RBC ($X10^6$/cmm)	WBC ($X10^3$/cmm)	RBCAVBC ratio	Hb (g%)	Ht (%)	CI	MCV (μm^3)	MCH (pg)	MCHC(%)
24 HOURS									
Control	3.33±0.07	14.08±0.45	23.65±0.38	11.21±0.08	10.0±0.28	0.200±0.01	25.06±0.40	6.01±0.28	24.00±1.22
0.05	2.44±0.02	16.24±0.56	15.02±0.91	10.53±0.04	7.93±0.08	0.1 96±0.02	23.06±0.28	5.91±0.10	25.41±0.94
0.25	2.23±0.03	16.96±0.67	13.14±0.31	10.02±0.07	7.39±0.35	0.1 86±0.02	15.58±0.95	5.58±0.17	35.82±1.21
0.5	2.04±0.03	20.58±0.13	10.12±0.48	9.40±0.07	4.26±0.04	0.243±0.01	18.20±0.42	7.30±0.11	40.14±1.32
48 HOURS									
Control	3.11±0.19	15.08±0.35	20.62±0.68	10. 18±0.09	9.43±3.5	0.220±0.09	26.08±0.48	6.21±0.38	25.55±2.11
0.05	2.76±0.06	18.88±0.22	14.61±1.00	10.42±0.04	7.39±0.13	0.189±0.12	22.53±0.85	5.67±0.03	25. 20±1.00
0.25	2.28±0.05	27.41±0.22	8.31±1.31	9.86±0.07	4.47±0.07	0.200±0.02	16.19±0.41	6.01±0.36	31. 13±2.27
0.5	2.07±0.04	29.33±0.34	7.05±0.97	9.06±0.08	4.06±0.03	0.262±0.02	19.61±0.25	7.87±0.08	40.14 ±1.31
72 HOURS									
Control	3.20±0.29	15.43±0.41	20.73±0.40	9.89±0.06	9.38±3.00	0.241±0.02	25.40±0.40	6.91±0.27	25.71±2.66
0.05	2.45±0.06	27.76±0.36	9.8i±1.12	9.48±0.02	6.71±0.07	0.1 86±0.02	2 1.78±0.58	5.59±0.07	25.67±0.39
0.25	2.08±0.90	34.45±0.34	6.03±0.71	8.53±0.01	3.90±0.02	0.212±0.01	15.91±0.73	6.44±0.07	40.51±2.06
0.5	1.98±0.05	36.58±0.35	5.41±0.41	8.17±0.05	3.80±0.12	0.256±0.02	19.69±0.47	7.70±0.12	40.15±0.01
96 HOURS									
Control	3. 00±0.46	14.48±C.68	20.71±0.65	10.41±0.09	10.00±1.2	0.250±0.02	24.48±0.48	6.48±0.33	26.78±2.06
0.05	2.38±0.05	30.93±0.05	7.69±0.79	7.70±0.02	5.82±0.08	0.173±0.02	20.93±0.66	5.23±0.19	25.01±0.41
0.25	2.24±0.09	39.50±0.65	5.67±0.51	8.50±0.04	3.69±0.05	0.189±0.02	16.47±0.52	5.69±0.24	34.57±1.52
0.5	1.65±0.07	39.89±0.38	4.23±0.38	7.06±0.03	2.58±0.18	0.226±0.01	15.63±0.29	6.42±0.33	41.08±0.98

Table 29.2: Effect of sublethal concentrations of malathion on haematological parameters of *Glossogobius giuris* at 20°C (mean ± S.E.)

Parameters	RBC ($X10^6$/mm3)	WBC ($X10^3$/mm3)	RBCAVBC ratio	Hb (g%)	Ht (%)	CI	MCV (μm^3)	MCH (pg)	MCHC(%)
24 HOURS									
Control	2.85±0.04	22.40±0.88	12.72±1.83	8.52±0.90	6.1.6±0.09	0.996±0.07	21.61±1.78	29.89±0.81	138.31±0.31
0.05	2.16±0.22	17.90±1.64	12.06±2.40	8.24±0.25	4.20±0.99	1.271±0.09	19.44±2. 11	38.14±0.91	196.10±0.41
0.25	1.72±1.50	27.50±4.1	6.25±1.71	8.00±0.80	4.40±0.75	1.425±0.08	25.58±2.61	46.51±1.20	181.81±0.91
0.5	1.56±0.60	29.40±1.93	5.30 ± 2.79	6.52±0.19	4.68±0.22	1.392±0.02	30.00±1.99	41.79±0.41	139.31±0.81
48 HOURS									
Control	2.51±0.60	25.28±1.60	9.92±1.70	8.00±0.06	5.28±0.83	1.062±0.07	21.03±2.41	31.87±0.91	151.51±1.11
0.05	1.83±0:06	23.72±1.26	7.71±2.00	7.60±0.34	4.08±0.76	1.384±0.05	22.29±3. 11	41.53±0.68	186.27±1.33
0.25	1.53±0.02	3000±2.00	5.10±3.22	6.52±0.20	4.00±0.40	1.420±0.03	26.14±3.67	42.61±0.81	163.00±0.17
0.5	1.34±0.06	30.85±0.60	4.34±4.20	6.16±0.72	4.28±0.23	1.532±0.06	31.94±2.66	45.97±1.11	143.92±1.11
72 HOURS									
Control	2.22±0.06	22.23±0.57	9.98±2.82	7.56±0.73	5.20±0.77	1.198±0.09	23.42±1.77	35.94±0.71	153. 46±0.91
0.05	2.38±0.32	27.84±1.80	8.54±3.33	7.98±0.20	6.40±0.18	1.058±0.04	26.89±2.77	31.76±0.86	118.12±1.60
0.25	2.24±0.25	22.72±0.41	9.85±3.63	6.94±1.36	5.80±0.26	1.032±0.07	25.89±3. 11	30.98±1.12	11 9.65±1.32
0.5	2.40±0.04	18.06±0.83	13.28±2.46	7.88±0.17	6.86±0.1 5	1.063±0.04	27.77±3. 00	31.90±1.25	11 4.86±2.00
96 HOURS									
Control	1.93±0.13	20.64±0.88	9.35±3.31	6.36±0.73	5.94±0.71	1.0981±0.02	30.77±1.97	32.95±1.33	117.07±0.71
0.05	2.15±0.07	22.72±1.36	9.46±2.10	8.28±0.31	7.28±0.33	1.283±10.04	33.86±1.66	38.51±0.95	113.73±1.21
0.25	2.20±0,05	18.88±0.76	11.65±3.70	7.%±0.38	6.96±0.44	1.2051±.05	31.63±1.81	36.18±0.71	112.93±1.61
0.5	2.45±0.08	16.80±1.78	14.58±2.80	9.20±0.84	8.00±0.51	1.2511±0.06	32.65±2.22	37.55±1 .91	115.00±1.51

Table 29.3 : Effect of sublethal concentrations of malathion on haematological parameters of *Glossogobius giuris* at 30°C (mean± S.E.)

Parameters	RBC ($X10^6/mm^3$)	WBC ($X10^3/mm^3$)	RBC/WBC ratio	Hb (g%)	Ht (%)	CI	MCV (μm^3)	MCH (pg)	MCHC (%)
24 HOU RS									
Control	2.32±0.04	15.92±0.48	14.57±1.34	8.04±0.17	8.56±1.78	1.150±0.02	36.89±2.16	34.65±10.45	93.92±0.39
0.05	2.07±0.03	24.08±0.72	8.59±2.08	6.88±0.19	6.16±1.37	1.107±0.07	29.75±2.27	33.23±0.67	111.68±0.86
0.25	1.82±0.02	24.44±1.42	7.44±.27	6.40±1.15	5.84±1.49	1.170±0.09	32.08±3.05	35. 16 ±0.29	109.58±0.32
0.5	1.66±0.06	17.12±1.72	9.69±2.79	5.92±0.23	5.04±0.97	1.188±0.07	30.36±2.86	35.66 ±1.00	11 7.46±0.83
48 HOURS									
Control	1.78±0.05	24.24±2.0	7.34±1.80	6.64±0.21	4.72±1.42	1.243±0.02	26.51±1.86	37.30±0.47	140.67±0.63
0.05	1.52±0.05	13.68±.54	11.11±1.70	6.24±0.17	4.76±2.00	1.368±0.22	31.31±2.60	41.05±0.38	131.09±0.26
0.25	1.48±0.09	12.80±2.40	11.56±3.27	5.84±0.29	3.80±2.11	1.315±0.11	25.67±2.63	39.45±1.31	153.68±0.66
0.5	1.43±0.10	10.40±3.40	13.75±4.30	4.96±0.49	3.60±2.20	1.155±0.07	25.17±2.65	34.68±0.71	137.77±1.20
72 HOURS									
Control	1.77±0.07	22.32±2.20	7.93±2.80	6.60±0.40	4.52±1.20	1.242±0.09	25.53±1.66	37.28±0.78	146.01±0.91
0.05	1.50±0.10	16.56±3.12	9.05±3.86	5.96 ±0.48	4.00±0.94	1.324±0.19	26.66±2.11	39.73±0.73	149.00±0.94
0.25	1.38±0.07	14.08±2.40	9.80±3.08	5. 24±0.48	3. 44±0.96	1.265±0.07	24.92±4.13	37.97±0.92	152.32±0.66
0.5	1.36±0.08	11.60±0.09	11.72±2.44	5. 16±0.28	3. 44±1.20	1.264±0.09	25.29±5.55	37.94±0.69	150.00±1.05
96 HOURS									
Control	1.69±0.09	24.72±0.31	6.83±2.14	5.92±0.88	4.40±1.72	1.167±0.06	26.03±5.93	35.02±0.71	134.54±0.62
0.05	1.46±0.40	18.06±0.21	8.11±2.72	5.441±0.78	3.88±0.99	1.241±0.05	26.57±6.11	37.26±1.10	140.201±0.72
0.25	1.40±030	14.80±0.31	9.45±1.81	4.281±1.22	3.64±0.71	0.907±0.05	24.97±3.95	29.31±1.31	117.58±1.40
0.5	1.37±0.09	12.48±0.41	10.97±1.91	3.921±1.23	3.32±0.91	0.953±0.04	24.23±2.85	38.61±0.89	118.07±1.61

DISCUSSION

Blood analysis forms an useful tool for determining pathological conditions of the fishes. The pathological changes were noticed mostly in the peripheral blood of erythrocytes and leucocytes as observed by Reichenbach-Klinke. (1967). In the present study changes in ervthrocyte count was observed after the fish was treated for longer duration. Wlasow and Dabrowska (1989) have reported inhibition of erythroblastosis in circulating blood due to ammonia intoxication in carp. In the present study similar changes of erythrocyte formation was noted in treated fishes. The haematological parameters like Hb content, RBC and WBC count showed significant decrease in treated fishes. The decrease in Hb content and RBC number may result in hypochromic microcytic anaemia, which could be due to deficiency of iron content (Bhai *et al.*, 1971). Since Hb and RBCs are oxygen carrying devices, the quantitative decrease in their levels might have led to the dearrangement of the oxidative metabolism with a concomitant decrease in the tissues of respiratory potential.

The anaemic condition recorded in the present study could be due to the destruction of mature RBC or inhibition of erythrocyte production. Such a decrease in RBC and anaemic response have been observed by Koundinya and Ramamurthy (1979) in *Sarotherodon mossambicus* and *Tilapia mossambicus* after exposure to lethal concentrations of sumithion; Lai *et al*, (1986) in *Heteropneustes fossilis* exposed to malathion: and Sakthivel and Sampath (1989) in *Cyprinus carpio* treated with sublethal levels of tannery. It is evident that the reduction in number of RBC and Hb content and decrease in MCV values might have caused microcytic anaemia in *G. giuris.*

An increase in the total WBC count and MCHC content in *G. giuris* after exposure to different sublelhal concentrations of malathion coincides with the observations made by Joshi (1987); Radhakrishnan and Prasad (1994). In the present investigation MCV significantly decreased which coincide with the findings of Rai and Quayyam (1981) and. Surendra Singh and Bhati (1994). *G. giuris* exposed to different temperatures (20 °C and 30 °C) and sublethal concentrations (0.05. 0.25 and 0.5 ppm) of malathion the heamatological parameters exhibited variations. Among the various external factors, temperature considered to be as controlling factor and regulating the metabolism (energy utilization) in fishes (Prosser and Brown, 1961 and Kasim, 1979).

The decrease in RBC count Hb content and PCV in *G. giuris* subjected to temperature (20°C) and malathion toxicity are in agreement with the observations made by Slieher and Pickford (1968) in *Fundulus heteroclitus* and Anthony (1961) in *Carassius auratus.* The results in the present study indicate that the fishes exposed to malathion at 20°C the total number of leucocytes was significantly increased and RBC count Hb concentration and PCV gradually declined during 24 and 48 h. On the other hand RBC count and PCV increased gradually when the fishes were exposed to malathion with the increased period of exposure at 20°C. However, the total WBC. MCHC and MCH significantly decreased when compared to those of controls. The significant decrease in MCV with simultaneous increase in MCHC might have led to microcytosis in *G.giuris* as pointed out by Haws and Goodnight (1962) and Joshi *et al,* (1980).

The fishes exposed to malathion at a temperature of (30°C) and malathion-temperature combination, the haematological parameters except MCHC were found to have decreased with the increase in the exposure period. This phenomenon suggests that a high temperature and malathion toxicity may cause anaemic and macrocytic changes in *G.giruis* when compared to low temperature and malathion stress. Similar results have been recorded by Joshi *et al.*, (1980) in *Clarias batrachus.* In *G. giuris* a decrease in WBC count could have led to a fall in the resistance for malathion and temperature (at 30 °C) stress, and this suggests that malathion in combination with high temperature has an adverse effect resulting in anemic and macrocytic conditions.

ACKNOWLEDGEMENTS

The author is grateful to Dept. of Zoology Bangalore University, Bangalore for providing necessary research facilities and sincere thanks to Dr. S.Krishnan, professor, Dept. of Zoology, Bangalore University, Bangalore for their kind help.

REFERENCES

Anthony. E.I I. 1961. The oxygen capacity of gold fish, *Carassius auratus* in relation to the thermal environment, J. Exp, Biol. 38: 93-107

Apha Awwa and Wpcf 1985. Standard methods for the examination of water and waste water. 16th Edition. American Public Health Association, Washington, D.C.

Bhai, L., Nath, N. and Nath. M.C. 1971. Haematological changes in rats injected with acetoacetate. Proc. Soc. Exp. Biol. Med., **138**: 597-599.

Blaxhall. P.C. and Daisley. K.W. 1973 Routine haematological methods for use with fish blood, *J. Fish Biol.* 5: 771-782.

Dalela, R.C.. Bhatnagar. M.C., Tyagi, A.K. and Verma, S.R. 1978. Adenosme triphosphatasc activity in few tissues of a freshwater teleost. *Channa gachua,* following *in vivo* exposure to endosulfan. Toxicology. **11**: 361-368.

Haws, T.G. and Goodnight, C.J. 1962. Some aspects of the haernatology of two species of catfish in relation to their habitats. Physiol. Zool, **35**: 8-17.

Joshi, B.D., Chaturvedi, L.D. and Dabral, R. 1980. Some haematologic values of *Clarias batrachus* following its sudden transfer to varying temperature. Indian J. Exp. Biol., **18**: 76-77.

Joshi. B.D. 1987. Changes in some blood values of *Clarias batrachus* exposed to lead nitrate. Him, J. Env. Zool., 1: 33-36,

Kasim. H.M. 1979. Ecophysiological studies on fry and fingerlings of some freshwater fishes with special reference to temperature tolerance. Ph.D., thesis. Madurai Kamaraj Univ.. India.

Koundinya, P.R and Ramamurthy. R, 1979. Effect of Organophosphate sumithion (Fenthion) on some aspects of carbohydrate metabolism in a freshwater fishes *Sarotherodon, mossambicus* and *Tilapia mossambicus* (peters). J.Expt.. 15:1632-1633.

Kumar, A.V. 1994. Endosulfan induced biochemical and pathophysiological changes in freshwater fish, *Clarias batractus* (Linn.) Ph.D. Thesis. Osmania University, Hydrabad, India.

Lal, A.S.B.. Anitakumari, S and Sinha, R,N, 1986, Biochemical and haemotological changes following malathion treatment in the freshwater catfish *Heteropneustes fossilis* (Bloch). Environ. Pollut.. 42(A): 151-156

Mount D.I and Stephan. C.E. 1967. A method for establishing acceptable toxicant limits for fish malathion and the butoxy-ethanol ester of 2.4-D. Trans. Amer. Fish. Soc. **21**: 185-193.

Nemcosk, J., Orban, L., Asztalos. B. and Vig. E. 1987. Accumulation of pesticides in the organs of carp *Cyprinus carpio* L. at 4 and 20°C. Bull. Envion. Contam. Toxicol.. **39**: 370-378.

Pascoe, D.. Evans, S. and Woodworth. J. 1986.Heavy metal toxicity to fish and the influence of water hardness. *Arch. Environ. Contam. Toxicol.,* 15: 481-485.

Prosser. C.L. and Brown. F.W. 1961.Comparative animal physiology (2nd Edn.) Saunders, Philadelphia.

Radhakrishnan. S. and Prasad, G, 1994, A unique haematological response of *Oreochromis mossambicus* to sublethal concentrations of the insecticide, Ekaluk EC-25, Proceedings of the Indian National Science Academy Part B. Biological Sciences **60(6):** 499-503.

Rai, R. and Quayyam, M.A. 1981. Haematological studies of mercury intoxicated fleets fish *Catla Catla* (Hamilton). Ind. J. ZooL 9: 87-91.

Ramalingam. K. and Ramalingam. K. 1982. Effect of sublethal levels of DDT, malathion and mercury on tissue proteins of *Sarotherodon mossambicus* (Peters). Proc. Indian. Acad. Sci. (Anim Sci) **91:** 501-503.

Reichenbach-klinke, H.H. 1967. Untersuchungen uber die Einwirking des Ammoniakgehalts aufden fish organisums. Arch. Fisch. Wisc. 17: 122-130.

Sakthivel. M. and Sampath. K. 1989. Haematological responses of *Cyprinus carpio* in relation to starvation. Geobios. **16:** 61-65.

Slicher, A.M. and Pickford, G.E., 1968. Temperature controlled stimulation of haemopolesis in a hypophysectomized cyprinodont fish *Fundulus heterocitus.* Physiol. Zool. **41:** 293-297.

Surendra Singh and Bhati D.P.S. 1994. Effect of zinc chloride on certain morphological parameters of the blood in *Channa panctatus* (Bloch.). Poll Res. **13(14):** 381-384.

Wintrobe, M.M. 1977. Clinical haematology. Henr Kimpton. London. 448.

Wlasow. T. and Dabrowska, H. 1989. Cellular changes in the blood and haemopoietic tissues of common carp exposed to sublethal concentration of ammonia. Aquat. Living Resour., **2(3):** 169-174.

CHAPTER 30

Length - Weight Relationship and Condition Factor of *Oreochromis Mossambicus* (Peters) from a Domestic Pond, Nazira, Upper Assam

G. Hatikakoty and S.P. Biswas*

Department of Biology, Nazira H.S. & M.P. School, Nazira - 785685, Assam
**Department of Life Sciences, Dibrugarh University, Dibrugarh - 786004, Assam*

ABSTRACT

The length-weight relationship and condition factor in different life stages and in different seasons were analyzed. The entire length-weight data were expressed as Log W = -1.815 + 3.005 Log L (Y=0.99). The value of exponent 'b' in the equation $W = aL^b$ was 3.005 which followed the cube law. The 'b' value of the female was found to be generally higher than the males. 'K' values were highest during monsoon and lowest during winter. The 'b' values and condition of the fish were attributed to different factors, food availability, gonadal maturity and abiotic factors of the system.

Keywords: *Lengtlh-weight relationship. Condition factor, Oreochromis Mossambicus, Assam*

INTRODUCTION

The study of the length-weight relationship and condition factor help to understand the general well being of a fish in a particular habitat. Length-weight relationship of fish varies depending upon the condition of life in the environment (Basheer et.al.1993). Individual variation from general length-weight relationship had been studied under the general name condition (Le Cren, 1951). Such changes in 'condition' had usually been analysed by means of condition factor or 'k' factor or ponderal index. Hart (1946) stated that the K- factor or the ponderal indices might give a very good idea of the broad outline of the seasonal cycle for the species concerned. The same was found true in a great number of fish species, in view to assess the general well being and maturity cycle of mouth breeding tilapia (*Oreochromis mossambicus*) from a pond in Assam.

MATERIALS AND METHODS

A total of 244 specimens of *O.mossambicus* (size range of 0.5 to 21.8 cm) were used on a seasonal basis to calculate length-weight and condition factor of *O.mossambicus*. The specimen were measured and weighted to the nearest centimeter (cm) and gram (gm) respectively using measuring broad and physical balance. A functional relationship between length and weight has been established based at the following empirical formula :

$W = a L^{b}$ (Huxley, 1924)

Where 'W' stand for weight in gram and 'L' for total length in centimeter,'a' is a multiplying constant and 'b' is the exponent or growth co-efficient. The logarithmic transformation of this relationship was expressed by Le Cren (1951) -

Log W = Log a + b Log L

The values of 'a' and 'b' are determined empirically from the data. The observed average weight was plotted against the average observed length to examine the nature of parabola. Correlation coefficient 'r' was calculated by using the following formula (Biswas, 1993) : -

$$r = \frac{\sum XY - n\overline{xy}}{\sqrt{(\sum X^2 - n\bar{x}^2)(\sum Y^2 - n\bar{y}^2)}}$$

Where, X and Y stand for length and weight respectively, $\bar{x}$ and $\bar{y}$ are the means of X and Y and n is the sample size.

In the present study conditions (K) factor was determined using the following equation :

$K = \frac{W}{L^3} \times 100$ (Beckman, 1945)

Where K= condition factor and W and L are weight (gm) and length (cm) respectively.

RESULTS

Length-Weight Relationship

The entire length-weight data of *O.mossambicius* were pooled into a single equation : Log W = -1.815+3.005 log L (r=0.99)

and derived parabolic equation : $W = 0.015311\ L^{3.005}$

Where W = total weight; and L = total length.

On plotting the observed weight observed length a parabolic curve had been obtained (Fig 30.1a) and when plotted logarithmically, a straight-line (Fig. 30.1b) relationship was evident in each case.

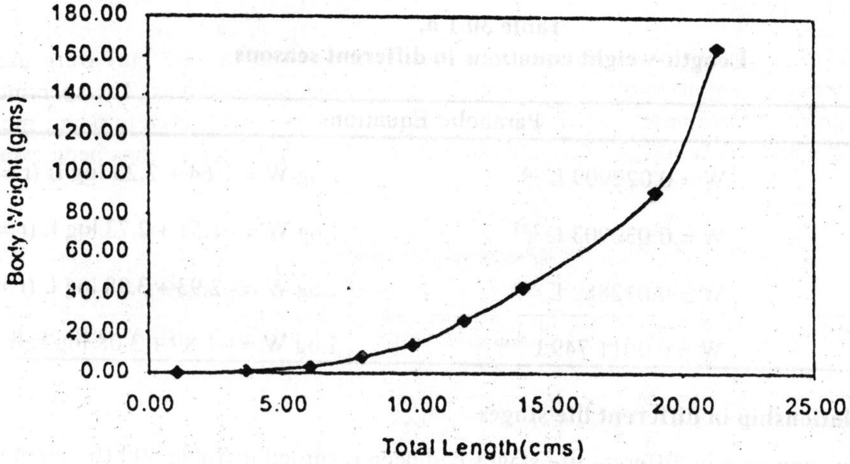

Fig. 30.1a
Length & Weight Relationship (Parabolic Curve)

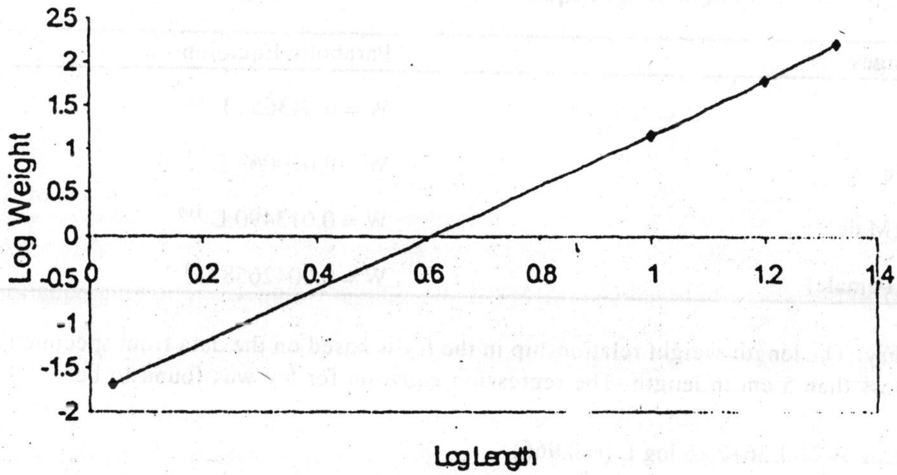

Fig. 30.1b
Length & Weight Relationship (Parabolic Curve)

Length-Weight Relationship during different seasons:

The length weight relationship during different seasons was calculated to see any impact if any, of seasons on this relationship. The four seasons namely viz., Premonsoon, monsoon, autumn and winter had been taken into consideration.

The parabolic equation in different season had been recorded in Table. 30.1 (a). It is evident from the table that the value of regression coefficient (b) was maximum (3.98) during monsoon and minimum (2.23) in winter. While it was 3.08 in autumn. However, the value progressively decreased in winter and pre-monsoon (Table 30.1 a).

Table 30.1 a.
Length-weight equations in different seasons

Seasons	Parabolic Equations	
Winter	$W = 0.022909\ L^{2.23}$	$\text{Log } W = 1.64 + 2.23 \log L\ (r = 0.76)$
Pre-monsoon	$W = 0.030903\ L^{2.73}$	$\text{Log } W = -1.51 + 2.73 \log L\ (r = 0.913)$
Monsoon	$W = 0.012882\ L^{3.98}$	$\text{Log } W = -2.93 + 3.98 \log L\ (r = 0.999)$
Autumn	$W = 0.0011\ 749\ L^{3.008}$	$\text{Log } W = -1.89 + 3.08 \log L\ (r = 0.97)$

Length weight relationship in different life stages

The parabolic equation in different life stages had been recorded in Table.30.1 (b) and described below:-

Table 30.1b.
Length-weight equations in different life stages

Life Stages	Parabolic Equations
Fry	$W = 0.043652\ L^{2.16}$
Juvenile	$W = 0.010069\ L^{3.32}$
Adult (Male)	$W = 0.013490\ L^{3.09}$
Adult (Female)	$W = 0.0042658\ L^{3.45}$

(a) Fry: The length-weight relationship in the fry is based on the data from specimens having less than 5 cm in length. The regression equation for fry was found to be:

$$\text{Log } W = -1.36+2.16 \log L\ (r=0.966)$$

(b) Juveniles: The regression equation in case of juveniles (having a length of 5-10 cm.) irrespective of sex were calculated to be:

$$\text{Log } W = -1.997+3.32 \log L\ (r=0.998)$$

(c) Adults: The length-weight relationship of adult specimens had been worked out separately for males and females. The regression equation derived for males and females of *O.mossambicus* are: -

Male $\text{Log } W = -1.87+3.09 \log L\ (r = 0.998)$
Female $\text{Log } W = -2.37+3.45 \text{ Log } L\ (r = 0.999)$

Condition factor

The value of mean condition factor (K) of the species, (Table 30.2a) showed almost a uniform trend of variations. It may be seen from the table that 'K' value showed maximum during monsoon (2.89) and minimum in winter (1.48).

The condition factor 'K' had been calculated for each 5 cm. length groups on seasonal basis and presented in Table 30.2 (b). It is observed that the highest 'K' value (3.67) was observed in case of fry followed by juvenile (1.76), adult female (1.59) and adult male (1.55).

Table 30.2a.
Seasonal variation in mean k-factor

Seasons	Condition Factor (K)
Winter	1.48
Pre-monsoon	2.01
Monsoon	2.89
Autumn	2.19

Table 30.2b.
Seasonal variation in Condition factors in different life stages

Seasons	Fry	Mean	Juvenile	Mean	Male	Mean	Female	Mean
Winter	2.35		1.25		1.11		1.21	
Premonsoon	369		1.55		1.36		1.45	
		3.67		1.76		1.55		1.59
Monsoon	4.58		2.55		2.28		2.14	
Autumn	4.06		1.69		1.45		1.56	

DISCUSSION

Length Weight Relationship

Ideally, the growth coefficient (b) of a fish species should be close to 3.0 (Allen, 1938). However, Hile (1936) and Martin (1949) reported that, the value of exponent '*b*' usually ranged between 2.5 and 4.0. In the present case, the value of 'b' differs with different life stage and in different seasons, the range being 2.16 to 3.98. The regression co-efficient 'b' showed variation in different seasons and even in the same population was reported by Rao & Rao (1972), Pathak (1975) and Biswas (1982).

In the present study it had been observed that the mean 'b' value of the females was found to be higher than the males in all seasons indicating that the females had better condition and growth than the males. Higher values of 'b' in females was also reported by Kulshrestha, et.al., (1993). The difference in the 'b' value of male and female indicated that the females were heavier than the males of the same length and the growth of the female fish which was slightly affected due to weight of the ripe ovary. Homiara (1992), reported similar result on *Arins jella* and *Arins tenuisfinis* respectively.

A significant difference in length-weight relationship was found between seasons especially in winter and premonsoon months which may be because of onset of maturity during the summer months as viewed by Dhasmana and Lai, (1993). Further, the 'b' value was less than 3.0 in winter season might be due to the fact that the fish would be in recovery phase after spawning in autumn season as observed by Basheer et.al. (1993).

In the present study the variation of 'b' values in different seasons were directly proportional to the water temperature (r= 0.94), pH (r=0.72), dissolved oxygen (r=0.98), plankton density (r=0.97), density of macrophytes (r=0.93) and density of macrofauna (r=0.57) indicated that 'b' value was highly influenced by these environmental factors of the pond. Similar observation were also made by the several investigators (Jhingran,1968; Rao and Rao,1972; Sinha,1973; Pathak, 1975; Quadri and Mir,1980; Roy,1987 and Kulsherestha et.al.,1993) in different fish species from different environmental conditions. The variations in the exponential value (b) of fish was generally supported to be under the influence of numerous factors viz., seasonal fluctuation in environmental parameters, physiological condition of the fish at the time of collection, sex, gonadal development and nutritive conditions of the environment of the fishes (Sinha, 1973; Kaur, 1981 and Dasgupta, 1982). Hence, it was quite likely that the variation in the exponential values might be due to the reasons stated above as the specimens were collected from four different seasons.

Tesch (1971) viewed that the exponent 'b' value of '3' indicates that a fish grows isometrically while Le Cren (1951) expressed that the value of the exponent gives an indication of condition, if the value of the exponent remains same in each season. Vasnetsov (1953) stated that during their development fish typically pass through several stages or stanzas, each of which may had its own length-weight relationship. Total weight of fish may also be altered by the weight of stomach content depending on the food ingested just before weighing (Muth and Smith, 1974). During winter, food scarcity may occur due to cool environmental condition resulting the death of many phyto and zooplankton. This effects the food and feeding intensity of the fish may influences the growth (Hatikakoty, 2002).

The value of 'b' in the *O.mossambicus* was found to be less than 3.0 in case of fry during winter and pre-monsoon, which might also be due to their feeding preference towards algal and animal matter. In the juvenile and adult specimens the higher 'b' value might be due to the presence of detritus and good amount of mud particles in their stomach which supports the view made by Soni and Kathal (1979) and Kaur (1981).

It was clear from the observation that exponential value of the length-weight relationship (b) in *O.mossambicus* as a whole followed the cube law (b=3.005), indicating thereby an isometric pattern of growth of the fish in the studied pond and the computed value of coefficient of correlation (r =0.99) was suggestive of a very close relationship between length and weight of the fish. Incidentally, Saroja et.al.(2002) also observed that *O.mossambicus* followed 'cube law' in pond.

Season-wise averages of 'K' values were highest during monsoon and lowest during winter. During wet months (pre-monsoon, monsoon and autumn) 'K' values showed higher fluctuation. Higher values of condition factor in monsoon was recorded in the fish *Tor putitora* (Nautiyal, 1985 and Dasgupta,1991). The variation in the condition factor may be attributed to different factors such as food availability, gonadal maturity and abiotic factors of the system, and had also been suggested by many workers (Le Cren 1951; Jhingran 1972; Bashirullah 1975; Kaur and Nasar 1984). Further, Le Cren (1951) and Weatherley (1987) pointed out that factors like food availability, gonadal maturity and biotic component of the system are responsible for the fluctuation in 'K' values.

The changes in the 'K' values with the increase in length may yield evidences reflecting the size at first maturity, while the seasonal fluctuation may reflect the spawning cycle of the fish as the 'K' was influenced by gonadal condition (Menon, 1950; Natarajan and Jhingran, 1963). But in the present study, it was observed that the smaller size groups had higher condition vlues indicating fry and juvenile had better condition. Many workers (Menon 1950; Pillay 1954; Sarojini 1957; and Varghese, 1961 and Biswas, 1982) had also observed higher 'K' values in the juveniles. The changes in the values of condition factor with the increase in length may yield evidences about the size at first maturity while the seasonal fluctuation may reflect the fatness and gonads (Jhingran, 1972 and Basirullah, 1975). The result further suggested the increase in the weight of the body due to weight of mature gonads followed by a decrease due to spawning, was also reflected in the 'K' values of the fish. According to Biswas (1982), increase weight of the body of the fish was due to the weight of the maturing gonads followed by decrease due to spawning was also reflected in the 'K' values of the fish. In the present study, the mature male and female fish showed decrease in the 'K' values in April, June, August, and October but increase of this two values in March, May, July, September may indicates that the fish spawns four times in a year.

The condition factor calculated from eviscerated weight eliminated the influence of gonads and food present in the alimentary canal. The pattern of fluctuation in these indices was almost similar that of the indices obtained from the fish having gut and gonads, the difference being only in magnitude. This indicated that condition factor was not only influenced by the maturation of the gonads and food present in the alimentary canal, but also by the environmental factors. Similar observations were made by Jhingran (1972) on *Setipinna phada* and Dasgupta (1982) and Nautiyal (1985) on *T. putitora*. The fish population display considerable changes in average condition, reflecting normal seasonal fluctuations in their metabolic balance and in the pattern of maturation and subsequent release of reproductive products; even the state of fullness of the alimentary canal may influence 'K' factor. In the present study the 'K' value of males and females and mean 'K' followed a similar pattern (Fig. 30.2). Therefore, it may be inferred that the changes in condition factor may be due to either sexual cycle or feeding intensity or it might also due to some other environmental factors. In all probability various factors exert their influences on the well being of the fish species.

The seasonal variation of 'K' value and certain others physico-chemical parameters such as Water temperature, pH, dissolved oxygen, total alkalinity, plankton density, macrophytic density and density of macro fauna are positively correlated which are indicated by the correlation coefficient between the 'K' value and above mentioned parameters (r = 0.96, 0.88, 0.99, 0.92, 0.97, 0.96 and 0.96 respectively).

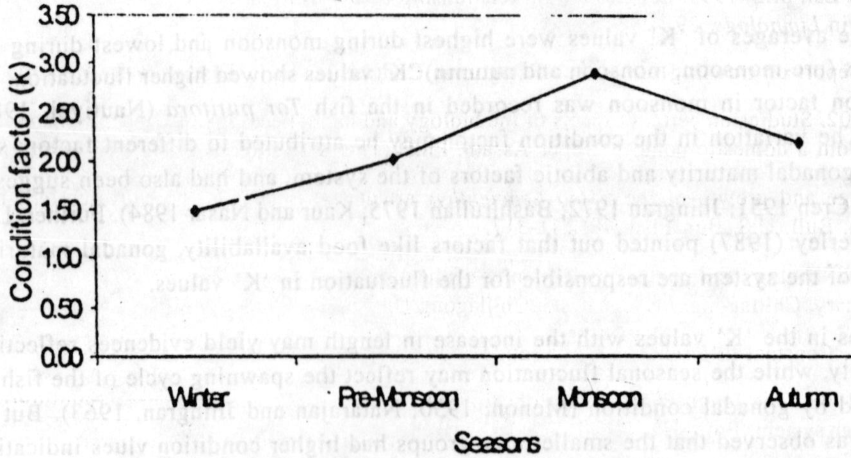

Fig. 30.2
Seasonal Variation of Mean Condition factors (K)

ACKNOWLEDGEMENTS

Authors are thankful to the staff water chemistry laboratory, ONGC Sivsagar, APL & HFC, Namrup for providing the facilities in working water analysis and also to the Head, Dept. of Life Sciences for providing necessary facilities.

REFERENCES

Allen, K.R. 1938. Some observations on the biology of the trout *(Salmo trutta)* in Windermere. J.Anim. Ecol., 7: 333-349.

Basheer, V.S., Khan, A.A. and Siddiqui, I.A., 1993. On the Length-Weight relationship of *Channa punctatus* (Bloch) from the river Yamuna. Advances in Limnology. 241-246.

Bashirullah, A.K.N.1975. Biology of *Lutjanus griseus* (L.) of the Cubagua Island, Ven-zuela. 1. Length-weight, body length - gut length relationships and condition factor. Biological Institute of Oceanography University Oriente. 14(1): 101-7.

Beekman, W.C. 19-15 : The length-weight relationship, factors for conversions between standard and total lengths and coefficients of conditions for seven Mitchigan fishes. Trans. Amer. Fish. Soc.,75 : 237-256.

Biswas, S.P., 1982 : Studies on certain aspects of the biology of *Labeo pangusia* and *L.dero* from North Eastern India. Ph. D. Thesis, North - Eastern Hill university, Shillong.

Biswas, S.P. 1993. Manual of Methods in Fish Biology. South Asian Publishers Pvt. Ltd. Delhi. 157.

Dasgupta, M. 1982. An investigation on the biology of Mahseer from the North Eastern India. Ph.D. Thesis. North-Eastern Hill Univ.

Dasgupta, M. 1991. Length-Weight relationship and condition factor (K-value) of Tor putitora (Hamilton) from Garo Hills, Meghalaya. Indian J. Fish 38(1): 35-38.

Dhasmana, N. and Lal, M.S. 1993. Length-Weight relationship of a Hill-Stream fish *Garra gotyla gotyla* (Gray). Advances in *Limnology*. 291-298.

Hart, J.L 1946. Report on trawling surveys on Patagonian Continental shelf. Discovery Rep; 23: 223-408.

Hatikakoty, G. 2002. Studies on certain aspects of the biology and habitat ecology of *Oreochromis mossnmbicus* (Peters) from a domestic pond in Upper Assam. Ph.D. Thesis Dibrugarh University, Dibrugarh. 139.

Hile, R. 1936. Age and growth of the *Cisco Lencichthys* artedt (Le Sucur) in the North-eastern highlands, Wisconsin. Bull. U.S. Bur. Fish. 48: 211-317.

Homiara, S.J. 1992. Biology and population dynamics of the cat fish, *Arius jella* Day, 1877 from the Karnafully River Estuary, Chittagong. M.Sc. Thesis, Chillagong University, Bangladesh.

Huxley, J.S. 1924. Constant differential growth ratios and their significance. Nature, 114:895-896.

Jhingran, A. G. 1968. The length-weight relationship and K factor of *Gudusia chapra* (Hamilton) from the Ganga river system. Proc. Nat. Acad. Sci. India. B38: 3-4.

Jhingran, A.G. 1972. Fluctuations in the Ponderal index of the Gangetic anchovy *Satipinna phasa* (Ham.) Journal of Inland Fisheries Society of India 4: 1-9.

Kaur, S. 1981 : Studies on some aspects of the ecology and biology of *channa gachua* (Hamilton) and *Channa stewartii* (Playfair). Ph.D. Thesis. North-Eastern Hill University.

Kaur, S. and Nasar, S.A.K. 1984. On the length-weight relationship and condition factor of *Channa gachua* (Ham.) Arquivos do Museu Bocage.

Kulshrestha, S.K., Srivastava, M. and George, M.P. 1993. Length-Weight relation-ship of a major carp, *Catla catla* (Ham.) from two water bodies of Bhopal. Advanced in Limnology. 329-332.

Le Cren, E. D. 1951. The length-weight relationship and seasonal cycle in gonad weight and condition in the perch *(Perca fluviatilis). Jour.* Anim. Ecol. 20(1): 201-219.

Martin, W.R. 1949. The mechanics of environmental control of body form in fishes. Univ. Toronto Stud. Biol; 58 Publ. Out. Fish. Res. Lab., 70:1-91.

Menon, M.D. 1950. Biometrics of the poor Cod *(Gadus minutus* L.) in the plymouth area. J. mar. biol. Ass. U.K., 29:185-239.

Muth, K. and Smith L.L.Jr. 1974; The Burbut Fishery in Lake of the Woods. Technical Bulletin 296 : 33

Natarajan, A.V. and A.G. Jhingram. 1963 : On the biology of *C.catla* from the river Jamuna. Proc. nat. Inst. Sci. India, 29 B (3): 326-355.

Nautiyal, P. 1985. Length-weight relationship and relative condition factor of the Garhwal Himalayan Mahseer with reference to its fishery. Indian. Indian. Sci. 5591 : 65-70.

Pathak, S.C. 1975. Length-weight relationship, condition factor and food study of *Labeo calbasu* (Ham.) from Loni reservoir (M.P). J.Inland. Fish. Soe, Indian, 8: 58-64.

Pillay, T.V.R. 1954. The Biology of grey mullet, *Mugil tade* Forskal with notes on its fishery in Bengal. Proc. natn. Inst. Sci. India, 17:187-217.

Quadri, M. Y. and Mir, S. 1980. Length-weight relationship of *Oreinus plagiostotmus* (McCl). Geobios. 7: 158-159.

Rao, G.R.M. and Rao, L.N.M. 1972: On the biology of *Labeo Calbasu* (Ham.) from the river Godvari, J.Inland Fish Soc. India 4: 74-86.

Roy, P.K. 1987 : Length-weight relationship of a major carp., *Cirrlius mrigala* (ham.). Matsya 12-13 : 39-44

Saroja, S.P.; Sivakumar, A.A and Chandran, R. 2002. Studies on length-weight relationship and Sex-ratio in *Oreochromis mossambicus*. Indian J.Environ and Eco- plan.6(2): 251-254.

Sarojini, K.K. 1957. Biology and fisheries of the grey mullets of Bengal. I. Biology of *Mugil parsia* (Ham.) with notes on its fishery in Bengal Indian j.Fish. 4(2): 254- 283.

Sinha, A.L.1973: Length-Weight relationship of a freshwater cat fish *Clarias batrachus* (Linn) Indian J.Zoology, 14:97-102.

Soni, D.D. and M. Kathal. 1979. Length-weight relationship in *Cirrhinus mrigala* (Val) and *cyprinus carpio* (Hamilton) Matsya, 5 : 69 - 72.

Tesch. F.W. 1971: Age and growth, in Methods for Assessment of Fish Production in Freshwater (2nd ed.) Ed. W.E. Ricker. IBP Handk book No.3. Blackwell Scientific Publications, Oxford and Edinburgh. 93-130.

Varghese, P.J. 1961. Some observation on the Biology of *Coilia borneensis*. Indian J.Fish., 8(2): 312-325.

Vasnetsov, V.V.1953. Patterns of fish growth In.Ocherki poobshchim vaprosm *biologini* Academiya Nauk press. Moscow.

Weatherley, A.M. 1987. Growth and ecology of fish population. Academic Press, London. 293.

CHAPTER 31

LENGTH-WEIGHT RELATIONSHIP AND RELATIVE CONDITION FACTOR IN *HILSA ILISHA* (HAMILTON) OF BARAK DRAINAGE IN ASSAM

Devashish Kar and Monjurul Hoque Barbhuiya*

Department of Life Science, Assam University, Silchar.

**Department of Biology, Sonai, N.G.H.S. School, Sonaimukh, Cachar, Assam*

ABSTRACT

The Length-Weight (L-W) relationship and relative condition factor (Kn) in Hilsa ilisha (Hamilton) were worked out on samples collected from Barak draingage in Assam during the period from June, 2001 to October, 2002. The value of exponent 'n' in the equation $W=cL^n$ was 2.29 for the species Hilsa ilisha. The computed 't' value indicated that 'n' value of the species is significantly different from the expected value '3' and hence the Cube Law $W = cL^n$ did not hold good in the case of Hilsa ilisha found in Barak drainage of Assam. The computed value of correlation coefficient (r) in Hilsa ilisha was 0.9833324 which is closer to 1 indicating that there is high positive correlation between length and weight in the species. The relative condition factor (Kn) remained greater than 1 for this species indicating their general well-being to be good in the Barak drainage of Assam.

Key Words: *Length-Weight relationship, Relative Condition Factor, Hilsa ilisha (Hamilton), Barak drainage, Assam.*

INTRODUCTION

The Length-Weight (L-W) relationship of different fish species have been reported from different waterbodies in India and abroad (1-20). However, the Length-Weight relationship and the condition factor of Hilsa ilisha (Hamilton) have not yet been studied so far from the Barak drainage of Assam. As such this investigation was undertaken to study the length-weight relationship and the condition factor (Kn) of the species Hilsa ilisha (Hamilton) collected from Barak drainage situated in the Cachar district (24° 10' N to 93°15' E) of Assam.

MATERIALS AND METHODS

Fish samples were collected during June, 2001 to October, 2002 and the length and weight of the fishes were recorded with the help of meter scale and pan balance. The length-weight relationship was established by fitting equation of the form

$$W = cL^n \quad \text{(i)}$$

where 'W' is the weight of the fish, 'L' its length and 'c' and 'n' are constants. The equation (i) was expressed in the linear form by using logarithms as given below:

Log W = ... (ii)

The L-W relationship was calculated out by the method of Rousenfell (1953). The relative conditon factor (Kn) was then computed as the ratio of observed weight to estimated weight of the species. When the condition factor is greater than 1, the well-being of the fish is considered as good.

RESULT AND DISCUSSION

The formula correlating L-W of *Hilsa ilisha* (Hamilton) is given below:

$$\text{Log } W = -3.11175 + 2.29 \text{ Log } L$$

The exponent 'n' in L-W relationship for Hilsa ilisha was 2.29. The computed 't' values indicated that 'n' value of the species was significantly different from the expected value '3'. Hence, the Cube Law $W=cL^n$ did not hold good in the case of the species Hilsa ilisha.

The computed correlation coefficient (r) value in Hilsa ilisha was 0.9833324 which is closer to 1 indicating that there is high positive correlation between length and weight in the species.

The condition factor (Kn) is an indication of general well-being of fishes. The computed value of Kn (LeCren's condition factor) in case of Hilsa ilisha revealed greater than 1, thus indicating their general well-being to be good. Therefore, all these observations appear to indicate that single value of 'c' and 'n' may not be responsible for the entire size range of the fish.

REFERENCES

Hile, R. 1936. Age and growth of the ciscp Leucichthys artedi (Le Sucer) in the lakes of the North Eastern highlands. Winconsin Bull. U. S. Burg. Fish 48:311-317.

Allen, K.R. 1938. Some observations on the biology of the trout (Salmo trutta in windermere. J. Anim. Ecol. 7:333-349.

Le Cren E. D. 1951. The length-weight relationship and seasonal cycle in gonad weight and condition of perch (*Perca fluviatilis*. J. Anim. Ecol. 21 : 210-219.

Rillay T.V.R. 1958. Biology of the hilsa *Hilsa ilisha* (Ham.) of the river Hooghly Indian J. Fish. 5:201-257.

Swarup K. 1958. The hilsa fishery of Allahabad Proc. Nat. Acad. Sci. India. 28:579-391.

Jhingran V.G. 1952. General length-weight relationship of three major carps of India. Proc. Nat. Inst. India. 17 : 449-460.

Swarup Krishna, 1965. Biometric studies of Hilsa ilisha (Hamilton) of Allahabad waters (1). Length-weight relationship. Japanese J. Ichthyology. 11 : 82-88.

Antony Raja P. T. 1967. Length-weight relationship in the oil sardine *Sardinella longiceps* Val. Indian J. Fish. 14 : 159-170.

Narasimhan K. A. 1970. On the length-weight relationship and relative condition in *Trichiurus liporus* Linn. Indian J. Fish. 17 : 90-91.

Rangarajan K. 1973. Length-weight relationship in the snapper Lutianus kashmira (Forskal). Indian J. Fish. 20 : 205-208.

Rangaswamy, C.D. 1976. Length-Weigth relationship in the grey mullet, *Mugil cephalus* Linn. Matsya 2: 19-22.

Rita Kumari S. D. and Balakrishna Nair. 1978. Length-Weight relationship of the loaches *Nemachilus triangularis* Day and Lepidocephalus thermalis (Cuv Val.) Matsya 4.: 69-72.

Sone D.D. and K.M. Kathal, 1979. Length-Weight relationship in *Cirrhina mrigala* (Val.) and *Cyprinus caprio* (Ham.) Matsya 5 : 69-72.

Srivastava S and A. K. Pandey. 1981. Length-weight relationship and condition factor of three Indian major carps in composite fish farming. Matsya 7 : 70-74.

Ramkrishniah N. 1986. Studies on the fishery and biology of *Pangasius pangasius* (Hamilton) of the Nagarjunsagar reservoir in Andhra Pradesh. Indian J. Fish. 33 : 320-350.

Mukhopadhyay P. K., B. Venkatesh and P. Das. 1986. Growth and some biochemical changes in Clarias batrachus due to methyl testosterone. Indian J. Fish. 33 : 262-284.

Gowda G, S. L. Shanbihogue and K. S. Udupa. 1987. Length-weight relationship and relative condition of grey mullet, *Valmugil seheli* (Forskal) from mangalore waters. Indian J. Fish. 34 : 340-347.

Kar, D. and M. H. Barbhuiya 2000 Length – weight relationship and conditon factor in *Gudusia chapra* (Ham. – Buch.) and Botia Dario (Ham. – Buch.) from Chalta haor (Flood plain wotland) in Cachar district of Assam. Environment & Ecology 18(1): 227 – 229.

Patgiri, A.M.H. Goswami, D. Kar and M. H. Barbhuiya. 2001 Comparative study on length-weight relationship and relative condition factors in major and exotic carps in ponds of Guwahati, Indian J. Environ. & Ecoplan. 5 (1) : 179 – 180.

Rousenfell. G. A. and Everhari, W. H. 1953. 'Fishary Science' the method and application. John Wiley & Sons, New York.

CHAPTER 32

EFFECT OF WATER pH ON TOXICITY AND ACCUMULATION OF INORGANIC AND METHYL MERCURY IN A FISH *GAMBUSIA AFFINIS*

P.V. Paulose

Aquatic and Fish Biology Laboratory, Department of Zoology, University of Rajasthan, Jaipur-30204.

ABSTRACT

Bioconcentration and toxicity of inorganic and methyl mercury in mosquito fish *Gambusia affinis* are studied at different pH levels. The toxicity and whole body mercury accumulation are found greater in lower pH, but the effect is more prominent for methyl mercury as compared to inorganic mercury. The increase in accumulation is 40% for methyl mercury as against 26% for inorganic mercury on 8th day as the pH is lowered to 6.0 from 7.5. But in higher pH of 9.0 the accumulation is found decreased by 14% and 7% for methyl mercury and inorganic mercury respectively. The increased uptake of mercury at lower pH may be due to gill permeability changes.

Keywords: *Inorganic mercury, Methyl mercury, Fish, pH, Toxicity, Accumulation, Gambusia affinis.*

INTRODUCTION

Although literature is abundant on the toxicity of mercury in the aquatic environment, these reports often differ widely. The wide range of concentration at which mercury is toxic to aquatic organisms indicates that the chemical and physical factors as well as the normal environmental variables affect the degree of toxicity of a particular mercurial. (DItri, 1973). Mercury toxicity has been shown to a function of several water quality parameters like temperature, hardness, alkalinity, and dissolved oxygen (Spry et al., 1981). Water quality can change the toxicity of a metal either by modifying the chemical form of the metal (Taylor, 1779) or the physiological response of the animal (Winfrey and Rudd, 1990).

In the aquatic ecosystem pH is an important environmental variable which can adversely affects the fish and fisheries (Brown, 1982). Acid and mercury stress may occur simultaneously due to direct discharge of acidic effluents into water bodies. But very little is known about the physiology of mercury toxicity, or of the possible interactions between mercury and acid stress in fish. The present paper reports the effect of varying water pH on toxicity and accumulation of both inorganic and methyl mercury in mosquito fish *Gambusia affinis* (Baird & Girard).

MATERIALS AND METHODS

Gambusia affinis procured from a local pond were acclimated to the laboratory condition for 15 days and fed with pellets prepared from prawn powder, fish powder, and minced liver in the ratio 2:2:1. Bioassays for mercuric chloride and methyl mercury chloride were conducted in three different water pH (6.0, 7.5, 9.0) according to Sprague (1973). Observed mortalities were plotted on log-probit paper and LC_{50} values and their 95% confidence limit was calculated as per Litchfield and Wilcoxan (1949).

For mercury accumulation studies 50 fish were exposed to 0.02 ppm Hg as mercuric chloride or methyl mercury chloride in the same pH as mentioned above. The experiments were carried out in 20 litres glass aquaria and the water was changed daily to maintain the mercury level. The lower pH was achieved by the addition of an aqueous solution of HCl and higher pH by the addition of an aqueous solution of NaOH. The pH was continuously monitored by a pH meter and the variation was within the limit of ± 0.3 throughout the experimental period. The water was analyzed according to APHA and the values (mg/l) were: total alkalinity, 141±9; hardness, 220+:24; chloride, 78±6; dissolved oxygen, 7.6±0.4. The water temperature varied between 19°C- 23°C throughout the experimental period. The water was analyzed for background level of mercury and the values were found <0.02µg/l.

Whole body mercury accumulation was determined on first, second, fourth and eighth days. Nine fishes from each group were removed, washed and divided into 3 samples of 3 fish each for mercury determination. The fish were processed wet according to Uthe et al. (1970) and the mercury content was determined by cold vapour atomic absorption spectrophotometry using a Mercury Analyzer (EC 5800). Details of tissue processing are same as reported earlier (Paulose, 1987).

RESULTS AND DISCUSSION

The LC_{50} values and their 95% confidence limits are presented in Table 32.1. Both mercuric chloride and methyl mercury chloride was found more toxic in lower pH as compared to higher pH. If the LC_{50} values for 96 hr are taken into consideration the toxicity of methyl mercury was found almost doubled when the pH was lowered to 6 from-7.5. But the increase was only 39% in inorganic mercury. The effect was not so prominent when the pH was raised to 9.0.

Table 32.1

LC_{50} values in ppb and their 95% confidence limit (CL) for $HgCl_2$ (mc) or CH_3HgCl (mmc) at different pH levels

pH	Form of Hg	24h		48 h		96h	
		LC_{50}	95% CL	LC_{50}	95% CL	LC_{50}	95% CL
6.0	mc	385	332-411	220	192-245	140	127-162
	mmc	77	52-98	51	37-70	45	31-56
7.5	mc	560	479-655	340	302-387	230	103-150
	mmc	138	116-170	120	98-151	90	48-118
9.0	mc	588	492-670	382	301-440	285	200-366
	mmc	141	123-172	138	115-174	112	96-132

It is obvious from the data of whole body mercury accumulation presented in Table 32.2 that the lowering of the pH also resulted in an increased accumulation of both inorganic and methyl mercury. When the pH was reduced to 6.0 from 7.5 inorganic mercury accumulation was increased by 26% on 8th day whereas it decreased by 7% when the pH was raised to 9.0. This increased rate of accumulation of inorganic mercury was found at all intervals of time from 1st to 8th day. As far as methyl mercury is concerned the effect was more pronounced. Thus 40% increase in accumulation was observed on 8th day at pH 6.0. Higher pH (9.0) on the other hand suppressed the accumulation by 14%. Thus methyl mercury accumulation was found more influenced by pH variations as compared to inorganic mercury.

Table 32.2

Whole body mercury concentration in *Gambusia affinis* after exposure to $HgCl_2$ (mc) or CH_3HgCl (mmc) at different pH levels ± SE of the mean

pH	Form of Hg	Residue levels (μg/g fish) Days 1	2	4	8
6.0	mc	2.40±0.30	3.79±0.45	6.11±0.50	8.20±0.70
	mmc	4.95±0.63	9.02±1.01	16.88±1.45	26.25±2.01
7.5	mc	1.62±0.28	2.81±0.50	4.60±0.85	6.50±0.89
	mmc	3.84±0.45	7.28±1.10	12.52±1.26	18.80±1.88
9.0	mc	1.28±0.20	2.55±0.35	4.12±0.55	6.02±0.78
	mmc	2.80±0.43	5.51±0.99	9.05±0.98	16.25±1.22

Drammond, et al., (1974) also reported higher accumulation of methyl mercury by the fish *Salvenilus fontinalis* in lower pH but no significant difference was observed for mercuric chloride. Their studies were not based on whole body mercury accumulation but only on gill and blood mercury content. Many earlier workers have reported increased gill permeability in fish at lower water pH. (Beamish, 1976, Brown, 1982). This increased permeability can enhance mercury bioaccumulation because gills are the main organs of mercury uptake in fish from water (Olson et al.,1973, Paulose, 1989). The gill surface has a substantial net negative charge, therefore having a high affinity for cationic metals like mercury.

Decreased swimming performance, growth and increased cost of maintaining blood ions have been attributed to mercury exposure. Similar effects have been seen with environmental acid exposure also (Packer and Dunson 1972, Spry et al., 1981), and so the effects may be addictive. Acid base disturbance accompanied by dramatic sodium losses from fish rainbow trout has been reported by McWilliams (1980). Low pH environments in both laboratory and field studies have caused a number of fish species to lose electrolytes, thereby proving plasma electrolytes to be a fairly reliable indicator of sublethal acid stress. In lower pH there is a net Na^{+}efflux which consists of inhibited active influx and stimulated passive efflux. There is also a concurrent Cl- loss from fish. These ion losses are concentration dependent upon H+ present in water. Another reason for the increased toxicity of mercury in lower pH may be the higher body burden of mercury as observed in the present study. Mercury can inflict cellular damage wherever it accumulates in sufficient concentrations. (Miura, et al., 1999).

In case studies reported by Hakanson (1980) on the effects of mercury contamination in lakes show a significant increase of damage to fish populations with the lowering of the pH of the aquatic ecosystem concerned. Higner levels of mercury has been reported in fishes caught from acidic lakes as compared to circumneutral lakes (Mason and Reinfelder, 1995). Low pH systems generally promote higher speciation and mobility of mercury in water (Richman et al., 1988).

It is evident from the present study that laboratory experiments can be useful in corroborating and explaining the dynamics of mercury in natural environment. Many of the details of the aquatic mercury cycle are still unknown (Morel, et. al, 1998) and need further research. It is essential to understand the pharmacokinetics of mercury in fish and then incorporate the effects that low pH have on the kinetics of uptake, release and assimilation efficiency. In the natural environment the animals are always under stress because of constant changes in environmental conditions. The present results also clearly show that simple monitoring of mercury concentration in aquatic environment is meaningless unless and otherwise the numerous physical, chemical and biological parameters are taken into consideration.

REFERENCES

APHA, 1985. Standard methods for the analysis of water and wastewater. 16th edition. American Public Health Association, Washington, D.C.

Beamish, R.J. 1976. Acidification of lakes in Canada. Water Air Soil.Pollut. 6:501-514

Brown, D.J.A. 1982. The effect of pH and calciun on fish and fisheries. Water Air Soil Pollut. 18:343-351.

Dltri, F.M. 1973. Mercury in the aquatic ecosystem. In Bioassay Technique and Environmental Chemistry. Ann Arbor Publishers Inc. pp. 1-67.

Drummond, R.A.,Olson,G.F. and Batterman.A.R. 1974. Cough response and uptake of mercury by brook trout *Salvelinus fontinalis*, exposed to mercuric compounds at different hydrogen ion concentrations. Trans. Amer. Fish. Soc.2: 244-249.

Hakanson, I. 1980. The quantitative impact of pH bioproduction and Hg concentration on the mercury content of fish.(Pike). Environ. Pollut. (B). I: 285-304.

Litchfield, J.T. and Wilcoxon,F. 1949. A simplified method for evaluating doze effects experiments. J. Pharmacol. Exp.Ther. 96: 399-408.

Mason, R.P. and ReinfelderJ.R. 1995. Bioaccumulation of mercury and methyl mercury. Water Air Soil Pollut: 80: 915-921.

McWilliams, P.G. 1980. Effect of pH on sodium uptake in Norwegian brown trout *(Salmo trutta)* from an acid river. J. Expl. Biol. 88: 259-267.

Miura, K. Koida,N. and Himeno,S. 1999. The involvement of microtubular disruption in methyl mercury induced apoptosis in neuronal and non-neuronal cell lines. Toxicol. Appl. Pharmacol. 160: 279-288.

Morel, F.M. Kraepiel, A.M.L. and Amyot, M.1998. The chemical cycle and bioaccumulatio of mercury. Annual. Rev. Ecol. Syst. 29: 543-566.

Olson, K.R. and Fromm,P.O. 1973. Mercury uptake and ion distribution in gills of rainbow trout *(Salmo gairdneri):* Tissue scan with an electron microprobe. J. Fish Res. Bd. Can. 30: 1575-1578.

Packer, R.K. and Dunson,W.A. 1972. Anoxia and sodium loss associated with the death of brook trout at lower pH. Comp.Biochem.Physiol. 41: 17-26.

Paulose, P.V. 1987. Bioaccumulation of organic and inorganic mercury in a fresh water mollsc, *Lymnaea acuminata.* J. Environ. Biol. 8: 185-189.

Paulose, P.V. 1989. Histological changes in relation to accumulation and elimination of inorganic and organic mercury in gills of *Llabeo rohita.* Hamilton. Ind. J. Expl. Biol. 27: 146-150.

Richman, L.A.,Wren,C.D. and Strokes,P.M. 1988. Facts and fallacies concerning mercury uptake by fish in acid stressed lakes. Water Air Soil Pollut.37: 465-473.

Sprague, J.B. 1969. Measurement of pollutant toxicity to fish. Part. 1. Bioassay methods for acute toxicity. Water Res. 3: 793-821.

Spry, D.J., Wood,C.M. and Hodson, P.V. 1981. The effects of environmental acid on fresh water fish with particular reference to the soft water lakes in Ontario and the modifying effects of heavy metals. A literature review. Canadian Technical Report of Fisheries and Aqutic Sciences. No.999.

Taylor, D, 1979. A review of the lethal and sublethal effects of mercury on aquatic life. Residue Rev. 72: 33-70.

Uthe, J.F., Armstrong,F.A.J. and Stainton.M.P.1970. Mercury determination in fish samples by wet digestion and flameless atomic absorption spectrophotometry. J.Fish.Res. Bd. Can.27: 805-811.

Winfrey, M.R. and Rudd, J.W.M. 1990. Environmental factors affecting the formation of methyl mercury in low pH lakes. Environ. Toxicol. Chem. 9: 853-869.

CHAPTER 33

Impact of Dairy Effluents on The Behavioural Responses of Fish, *Glossogobius Giuris* (HAM)

Krishna Ram, H. and M. Ramachandra Mohan

Department of Zoology, Jnana Bharathi, Bangalore University, Bangalore-56, (Karnataka) India.

ABSTRACT

The present experimental studies, revealed the effect of dairy effluents on the behavioural responses of fresh water edible fish, *Glossogibus giuris* (HAM). Today's world is more dependent on environment resources and development man's activities are continuously polluting all such resources. Lake water pollution is one such problem, which is gaining concern, especially in fast developing countries. Balaganahalli lake of Kolar (Karnataka) was highly polluted with dairy effluents, it's water are not suitable for either aquatic life nor agriculture. In this contest the behavioral responses of gobid fish has been taken and fish exhibited various behavioural responses like, locomotary, physiological, neurological, morbidity under the influence of dairy effluents. The present study also indicated the possible utility of the behavioural responses as rapid biomonitoring assessment technique for qualitative evaluation of various parameters of dairy effluents.

Keywords: *Dairy effluents, fish behaviour, Glossogibus giuris*

INTRODUCTION

The lake pollution is a serious water pollution problem, which needs concern. Fish behaviour has been an interesting area of study for aquatic biologists. The physical, chemical, biological analysis alone donot provide adequate information in respect of environment contamination. The altered environmental influence can be detected and evaluated by studying the behaviour of fish under appropriate condition. Study on behaviour of fishes under environmental stress was carried out by Furlinger and Beamish (1977), Wildish *et al*, (1977), Trivedy and Dubey (1978), Srinivas *et al.* (1984), Waterpaugh and Beitinger (1985), Nakamura (1986), Vijayram *et al.*, (1991), Sen *et al* (1991), Amudha *et al*, (1997) and B.K.Baruah & M.Das (2002). Review of literature indicated very limited information regarding the effect of dairy effluent on the behaviour of fish, therefore, the present experiment was carried out to find out behavioural responses of a freshwater fish exposed to dairy effluent.

MATERIALS AND METHODS

Mature healthy live fish *G.giuris* of size 11-13cm were procured locally and acclimatized in glass aquarium (size 30 x 45 x 30cm) for one month and fed with earthworms. The dairy effluent of Kolar (Karnataka) was collected from the point of discharge in plastic container and brough to laboratory. Eight fishes from control aquarium were transferred in to test aquarium containing dairy effluents the fishes were exposed to the effluent for a period of 2 hrs, under normal temperature and natural light and different data regarding behavioural responses were recorded.

RESULTS AND DISCUSSION

The result of the observations are given in Table 33.1. The experiment clearly revealed that the behaviour of fishes remarkably changed due to the influence of dairy effluent. The various locomotary responses exhibited by the fishes at initial stage of exposure include restlessness, erratic and faster swimming, abrupt change in position and direction, jumping and overall hyperactivity. The fishes showed surfacing tendency throughout the experimental period. The overall hyperactivity suggested their tendency to get out of the test aquarium containing dairy effluent. The dairy effluent perhaps caused irritation, inflammation and injuries to the skin, gills and buccal cavity tissues. The overall change in activities may be related to the alteration of metabolic pathway to yield energy under oxygen depleted environment. This observation were in conformity with earlier experiments in fish

Table-33.1

Behavioural responses of fish *G.giuris*

Responses	1-10min	10-20min	20-40min	40-80min	80-120min
(1) Locomotary					
I. Restlessness	Observed	Observed	Observed	Not Seen	Not Seen
II. Erratic Movement	Observed	Not Seen	Not Seen	Not Seen	Not Seen
III. Fast Movement	Observed	Not Seen	Not Seen	Not Seen	Not Seen
IV. Position change	Frequent	Observed	Observed	Observed	Not Seen
V. Direction Change	Observed	Observed	Observed	Not Seen	Not Seen
VI. Sudden Swimming	Occasionally	Occasionally	Not Seen	Not Seen	Not Seen
VI. Surfacing	Observed	Observed	Observed	Observed	Observed
VII. Jumping	Not Seen	Not Seen	Observed	Observed	Observed
VIII. Sustained Swimming	Not Seen	Not Seen	Not Seen	Not Seen	Not Seen
IX. Hyperactivities	Observed	Occasionally	Occasionally	Not Seen	Not Seen
(2) Neurological					
I. Jerks	Not Seen	Not Seen	Not Seen	Not Seen	Not Seen
II. Frightened	Not Seen	Not Seen	Not Seen	Observed	Observed
III. Balance Loss	Not Seen	Not Seen	Not Seen	Observed	Observed
(3) Physiological					
I. Gulping	Observed	Not Seen	Observed	Observed	Observed
II. Operacular Movement	Rapid	Rapid	Rapid	Rapid	Rapid
(4) Morbidity					
I. Reduced Swimming	Occasionally	Observed	Observed	Observed	Observed
II. Slow Movement	Observed	Observed	Observed	Not Seen	Observed
III. Stationary	Observed	Not Seen	Not Seen	Not Seen	Not Seen

under different pollutant stress. Wildish *et al.* (1977), Ferlinger and Beamish (1977), Trivedy and Dubey (1978), Sen *et al.* (1991) and Baruah *et al.* (1998).

The physiological responses of fish as evidenced by gulping and rapid opercular movement were indicative of respiratory inconveniences triggered by the absence of dissolved oxygen in the effluent. Similar findings in relation to respiration were also recorded by Panigrahi and Mishra (1978), Malyarevkya (1978), Watenpaugh and Beitinger (1985) and Sen *et al.*, (1991). The neurobehavioural changes relfected by loss of balance during the exposure period due to the influence on sensory-motor interaction of the nervous system. The effluent might act as neurotoxic leading to disorganization of nervous co-ordination. This findings were in the line of earlier report of Panigrahi and Mishra (1978) and Ram and Gopal (1991) in fish intoxicated by pollutants. In the later stages of experiment, the fishes showed morbidity responses like reduced activities may be explained by the infiltration of toxic component of the effluent into the body systems resulting inhibitory influences to a variety of bio-chemical mechanisms. Reduced activity of fishes under environmental stress were also recorded. Our results are concides with Panigrahi and Mishra (1978) and Malyarevskya (1978).

The present experiment clearly demonstrated behavioural changes in the *G.giuris* under dairy effluent stress. The different types of responses may be useful in obtaining qualitative data regarding the biotoxic nature of different industrial effluent. The behavioural responses in fish may be utilized as rapid effluent quality assessment technique in biomonitoring study by the dairy environmental workers to acquire immediate preliminary information in relation to quality of effluent or polluted water.

REFERENCES

Baruah, B.K; Das.M. 2002. Study on Behavioural Responses of Fish Heteropneustes fossilis exposed to paper mill effluent. Indian J. Environ. 85 Ecoplan. 6(2): 263-266.

Baruah, B.K; Baruah, D. and Das, M. 1996. Sources and characteristics of paper mill effluent. Environment and Ecology. 14:686-689.

Baruah, B.K.; Sengupta, S. and Das, M. 1998. Toxocological impact of paper mill effluent on muscle glycogen profile of *H. fossilis* (Bloch.) Gau. Univ. J. Sci. Golden Jubilee Volume. 1-5.

Farlinger, S. and Beamish, F.W.H. 1977. Effect of time and velocity increments of the critical speed of large mouth brass *(Micropterous salmoid).* Trans. Ass.Fish. Soc. 106:436-439.

Malyarevskya. A.K. 1978. Specific and non specific changes induced in fish by various toxic agents. Hydrobiological Journal. 14:52-59.

Nakamura, F.:1986. Avoidance behaviour and swimming activity of fish to detect pH changes. Bull. Environ. Contam. Toxicol. 37 L 808- 815.

Panigrahi, A.K. and Mishra, B.N. 1978. Toxicological effect of mercury on a freshwater fish *Anabas scandens* (Cuv and Val) and their ecological implications. Environ. Pollut. 16:31-39.

Ram, M.D. and Gopal, K. 1991. Neurobehavioural changes in freshwater fish *Channa punctatus* exposed to fenitrothion. Bull. Environ. Contam. Toxicol. 47: 455-458.

Sen, G; Behera, M.K. and Patel, P.N. 1991. Toxicity of Zinc to the fish *Channa punctatus* (Bloch) with behavioural, morphological and skeletal abnormalities. Environment and Ecology. 9: 1023-1027.

Srinivas, P.; Benerjee, G. and Rao, P. 1984. Preliminary observations on the effect of industrial effluents to certain animals. Indian J. Environ. Hlth. 26: 57-60.

under different pollutant stress. Wildish *et al.* (1977), Ferlinger and Beamish (1977), Trivedy and Dubey (1978), Sen *et al.* (1991) and Baruah *et al.* (1998).

The physiological responses of fish as evidenced by gulping and rapid opercular movement were indicative of respiratory inconveniences triggered by the absence of dissolved oxygen in the effluent. Similar findings in relation to respiration were also recorded by Panigrahi and Mishra (1978), Malyarevkya (1978), Waiwappaugh and Bettinger (1985) and Sen *et al.*, (1991). The neurobehavioural changes reflected by loss of balance during the exposure period due to the influence on sensory-motor interaction of the nervous system. The effluent might act as neurotoxic leading to disorganization of nervous co-ordination. This findings were in the line of earlier report of Panigrahi and Mishra (1978) and Ram and Gopal (1991) in fish intoxicated by pollutants. In the later stages of experiment, the fishes showed morbidity responses like reduced activities may be explained by the infiltration of toxic component of the effluent into the body systems resulting inhibitory influences to a variety of bio-chemical mechanisms. Reduced activity of fishes under environmental stress were also recorded. Our results are coincides with Panigrahi and Mishra (1978) and Malyarevskya (1978).

The present experiment clearly demonstrated behavioural changes in the *G. giuris* under dairy effluent stress. The different types of responses may be useful in obtaining qualitative data regarding the biotoxic nature of different industrial effluent. The behavioural responses in fish may be utilized as rapid effluent quality assessment technique in biomonitoring study by the dairy environmental workers to acquire immediate preliminary information in relation to quality of effluent or polluted water.

REFERENCES

Baruah, B.K., Das, M. 2002. Study on Behavioural Responses of Fish *Heteropneustes fossilis* exposed to paper mill effluent. Indian J. Environ. & Ecoplan. 6(2): 263-266.

Baruah, B.K. Baruah, D. and Das, M. 1995. Sources and characteristics of paper mill effluent. Environment and Ecology. 14:686-689.

Baruah, B.K., Sengupta, S. and Das, M. 1998. Toxecological impact of paper mill effluent on muscle glycogen profile of *H. fossilis* (Bloch). Gau. Univ. J. Sci. Golden Jubilee Volume. 1-5.

Bettinger, S. and Beamish, F.W.H. 1977. Effect of time and velocity increments of the critical speed of large mouth bass (*Micropterous salmoid*). Trans. Ass. Fish. Soc. 106:136-139.

Malyarevskya, A.K. 1978. Specific and non specific changes induced in fish by various toxic agents. Hydrobiological Journal. 14:52-59.

Nakamura, F. 1986. Avoidance behaviour and swimming activity of fish to detect pH changes. Bull. Environ. Contam. Toxicol. 37 L 808-815.

Panigrahi, A.K. and Mishra, B.N. 1978. Toxicological effect of mercury on a freshwater fish *Anabas scandens* (Cuv and Val) and their ecological implications. Environ. Pollut. 16:31-39.

Ram, M.D. and Gopal, K. 1991. Neurobehavioural changes in freshwater fish *Channa punctatus* exposed to fenitrothion. Bull. Environ. Contam. Toxicol. 47:455-458.

Sen, G. Behera, M.K. and Patel, P.K. 1991. Toxicity of Zinc to the fish *Channa punctatus* (Bloch) with behavioural, morphological and skeletal abnormalities. Environment and Ecology. 9. 1023-1027.

Srinivas, P., Banerjee, G. and Rao, P. 1984. Preliminary observations on the effect of industrial effluents to certain animals. Indian J. Environ. Hlth. 26. 57-60.

CHAPTER 34

ACUTE TOXICITY AND BEHAVIOURAL RESPONSES OF NICKEL SULPHATE, TO THE FISH *HETEROPNEUSTES FOSSILIS*

M. Choudhary & M.M. Jha

Department of Zoology, M.L.S.M. College, Darbhanga

ABSTRACT

Release of heavy metals into atmosphere contaminate and the aquatic system. Such contamination deteriorates the water quality and results into large scale fish mortality. Further more the toxic potential of a particular chemical reduces the fitness of a population. It was thus, felt necessary to have an accurate method for finding out toxic calculations of a given chemical to a given animal in a specific time. Acute toxicity test in form of LC_{50} are the most widely practiced and accepted methods for the determination of such toxic levels. The toxicity of Nickel sulphate to *H. fossilis* was evaluated by (Tab-34.2) and (Tab-34.3). The LC_{50} values obtained by former method were 425.3 mg/l, 382.9 mg/l, 351.1 mg/l and 333.6 mg/l for 24, 48, 72 and 96 hours where as the letter method give LC_{50} values to be 450 mg/l, 400 mg/l, 380 mg/l and 334.8 mg/l for 24, 48, 72 and 96 hours respectively. Surfacing, rapid uppercular movements, onvulsions, profuse mucus secretions and its deposition over the gills gyratery, swimming movements by moribund fish and loss of equilibrium before death were the prominent behavioral changes noticed during present study (Table-34.1).

INTRODUCTION

The nickel toxicity has assumed great interest because of the widespread environmental occurrence of nickel and the incidence of accidental parsoning in nickel workers (Sunderam et al 1988a). Though quite a good number of toxicity tests on nickel have been performed on various animal's including fishes (Gill and Pant 1981, Saxena & Parshari, 1983, Khangarot & Ray, 1990, Jha & Jha 1995, Roy 2003) . No study has yet been undertaken to determine the toxicity of nickel to *H. fossilis*, one of the most important for ever water fishes.

Contact Address: Dr. M.M. Jha, Mohalla - Mirzapur Dist.- Darbhanga- 846004

The present study has, therefore, been undertaken to establish the acute toxicity in term of LC_{50}, of Nickel sulphate to this fish in relation to the behavioural responses various methods like logit, regression equation, graphic etc. but these methods have been shown to have a number of statistical deficiencies (Roy, 1988), therefore an attempt has been made in the present study to evaluate the LC_{50} of Nickel sulphate by yet another but more accurate logistic response model of Byron & Brown (1967). The accuracy of the values so obtained have been confirmed and compared with the LC_{50} values, calculated by establishing regression lines following the method of Downie & Heath, 1970.

MATERIALS AND METHODS

The adult living species of *H. fossilis* were collected from a local fish pond and were brought to the laboratory in wide-mouthed large earthen pots filled half with natural water. The fish were washed with 0.1% known solution to remove dernal infections if any. Healthy fish of average length (9-11 cm) and weight (21-24 gm) were selected and acclimated in 40 litre glass aquaria to the laboratory conditions for a fortnight. Running tap water was used in all the experiments.

Bioassays were conducted for the determination of LC_{50} value of nickel sulphate ($NiSO_4$, $6H_2O$, BDH analytical reagent grade) for 24, 48, 72 and 96 hours following the methods of APHA, AWWA & WPCF (1985).

A stock solution of Nickel sulphate was prepared by similar dilution technique as described by Doudoroff et al. (1957). The fish were exposed to different concentrations of nickel sulphate & their percent survival at different time intervals (24, 48, 72 & 96 hrs), were recorded as a measure of acute toxicity.

RESULTS

Table 34.1

Physio-chemical characteristics of the test water.

Characteristics	Unit	No. of observation	Mean	Range
pH		10	7.7	7.2-7.9
Temperature	O°C	10	23.6	24-28
Total solids	mg/l	10	37.5	30.6-38.5
Dissolved solids	mg/l	10	12.0	10.0-12.5
Dissolved oxygen	(Do) mg/l	10	6.0	5.0-7.3
Free CO_2	mg/l	5	2.0	1.2-3.0
Total hardness	mg/l as $CaCO_3$	5	175.0	160.0-200.0
Total alkalinity	mg/l as $CaCO_3$	5	150.0	136.0-165.0

Table 34.2

LC_{50} Values of Nickel sulphate to *H. fossilis* calculated by Logistic Response Model.

S.N.	Time Intervals (Hours)	Antilog	LC_{50} (mg/l)
1.	24	2 .6287	425
2.	48	2.5831	382.9
3 .	72	2.5456	351.2
4.	96	2.5233	333.6

Table 34.3

LC_{50} values of Nickel sulphate to *H. fossilis* calculated by Regression Equation.

S.N.	Time Intervals (Hours)	Regression Equation	LC_{50} values
1.	24	- 189 + .53x	450 (mg/l)
2 .	48	-155 + .5x	400 (mg/l)
3 .	72	- 140 + .5x	380 (mg/l)
4.	96	- 2930 + 8.9x	334 (mg/l)

DISCUSSION

The behavioural observations noticed during the present study are in agreement with those of Gill & Pant (1981), Saxena & Parashari (1983), Jha & Jha (1995), Roy 2003. The rapid opercular movements and culvelsions observed in the present case may be due to the hypotoxic condition created by nickel toxicity. Similar was also observed Lal *et al* (1984). Secretion of mucus is a normal feature under stressful conditions. Jha & Jha (1995) also observed similar with Nickel chloride to *A. testudineus.* The gyratory movements as observed in the present study may be attributed to the pathogenic effect of nickel chloride such as fatigue and comatose behaviour (Roy,. 1988) . The apparent loss of equilibirium in the nickel exposed fish is indicative of the fact that nickel also disturbs and influence the central nervous system. It is, therefore, calculated that the toxicity of nickel depends upon a number of physical, chemical and biological factors. Each of which may be used as a tool for the prediction of metal toxicity to fishes.

REFERENCES

Sunderman, F.W. Jr., Dingle, B., Hopfer, S.M. and Swift, T. (1988a), Acute nickel toxicity in electroplating workers who accidentally ingested a solution of nickel sulphate and nickel chloride. Amer. J. Ind. Med. 14; 257-266.

Gill, T.S. and Pant J.C. (1981), Toxicity of nickel to the fish Puntius conchonius and its effects on blood glucose and liverglycogen. Com. Physiol. Ecol. 6(2), 99-102 .

Saxena, O.P. and Parashari, A. (1983), Comparative study of the toxicity of six heavy metals to Channa punctatus. J. Environ. Biol. 4(2): 91-94.

Khangarot, R and Ray, P.K. (1990), Acute toxicity and toxic interaction of chromium and nickel to common gurpy Poecilia reticulata (Peters). Bull. Environ. Contam. Toxicol. 44 : 832-839.

Jha, B.S. and Jha, M.M. (1995), Acute toxicity of nickel chloride to freshwater climbing perch, Anabas testudineus (Bloch). Coluban Journal of Life Science, pp 11, fig.-l, Table-2.

Roy, D.N. (2003). Studies on the effect of chromium to the fish, *H. fossilis*. The Ph.D. Thesis, L.N. Mithila University, Darbhanga.

Roy, D. (1988), Toxicity of an anionic detergent, dodecylbenzene sodium sulfonate, to a freshwater fish, Rita rita. Determination of LC_{50} values by different methods. Ecotoxicol. Environ. Saf. 15, 186-194.

Byron, W. and Brown, J.R. (1967), Quantal response assay In: statistics in endocrinology (eds. J.U. Mc Arthur and T. Colton). MIT Press, Cambridge, MA/London.

Downie, N.M. and Heath, R.W. (1970). Basic statistical methods, Harper international Edition, New York, Evanston and London, 129-131.

APHA, AWWA and WPCF (1985). Standard methods for the examination of water and waste water. American Public Health Association, American Water Works Association and Water Pollution Control- Federation, 16th ed, American Public Health Association Washington, D.C.

Doudoroff, P. Anderson, E.G. Burdick, G.E. Galtsoft, P.S., Hart, W.B., Partick, R., Strong, E.K., Surber, E.W. and Vonhorn, W.M. (1957). Bioassay methods for the evalaution of acute toxicity of industrial wastes to fresh sewage Ind. Wastes 23(11): 1380-1397.

Lal, H., Mishra, V. Vishwanathan, P.W. and Krishnamurty, C.R. (1984). Effect of synthetic detergents on some of the behavioural patterns of fish fingerlings (Cirrhina Mrigala) and its relation to ecotoxicology. Bull. Environ. Contam. Toxicol. 32: 109-115.

CHAPTER 35

BIOLOGY OF LENGTH-WEIGHT AND CONDITION IN A BRACKISH WATER FISH *LIZA TADE* (FORSKAL)

S. K. Das

Departmnent of Fishery Biology and Resources Management, Faculty of Fishery Sciences (W. B. University of Animal & Fishery Sciences) Mohanpur, Nadia, India, Pin - 741252

ABSTRACT

The length weight relationship and condition factor of a brackish water fish *Liza tade* (Forskal) was studied for a period of one year. It was found that the growth pattern of fish was allometric with exponent 'b' value significantly less than 3. The condition factor also fluctuated widely which might be due to maturity and feeding.

Key words : *Length weight relationship, Condition factor, Liza tade.*

INTRODUCTION

Growth fluctuations are more frequent in fishes of tropical and subtropical water bodies due to several factors such as environmental variations, multiple spawning, dynamics of food compositions, physiological changes etc. Growth of fish is a mathematical function of length and weight and of varies due to biological changes and seasonal dynamics. The study of length weight relationship in fishes is of primary importance in setting up yield equations, in estimating the number of fish landed as well as in comparison of population in time and space (Chanchal *et al.*, 1978).

The general expectation is that weight of fish would vary as the cube of its length (Pandey and Sharma, 1998 and Pathak, 1975). But actual relationship may deviate significantly from it (Le Cren, 1951) as fishes normally do not retain the same shape through out their life span.

Studies on length weight relationship of mullets are relatively scarce. Sarojini (1958) found that there was no marked difference in length weight relationship between males and females of *Mugil parsia* and *M cunnesis.* Studies on length-weight relationship of *Mugil cephalus* indicated that the relationship does not vary significantly between years and environment (Rangaswamy, 1976). This fish exhibited non linear relationship between length and weight and it does not change as the fish grows (Das, 1977). Here is an attempt of studying the length weight relation and condition factor in a brackish water fish *Liza tade* (Forskal).

MATERIALS AND METHODS

The length weight relationship *of Liza tade* was calculated from the data of 165 specimens having a length range of 72 to 320 mm. The fishes were collected from Ganjam and Berhampur fish markets of Orissa during the period of October 1996 to September, 1997 and it were mainly procured from Rushikulya estuary and southern part of Chilka lake. Length of fish was recorded up to nearest one mm and weight of fish was measured to nearby one gram for the purpose. The allometric equation fitted into these data was $W = aL^b$ where W = weight of fish in gram, L = length of fish in cm, a = initial growth index of fish, b = equilibrium constant and exponent value. The linear equation after logarithmic conversion becomes log W = log a + b log L. For convenience of discussion, the fishes were categorised into three groups arbitrarily such as Group I (less than 120 mm of total length or TL), Group II (120 to 240 mm TL) and Group 111 (more than 240 mm TL).

The condition factor (K) was calculated using the equation $K = W/L^3 \times 100$, where W = observed body weight (g) and L = body length (cm).

In order to test isometric growth, Fisher's 't' test was applied to test whether 'b' value is significantly different from '3' or not.

RESULTS AND DISCUSSION

The details of biological features of *Liza tade* are given in Table-35.1. The correlation between log values of length and weight was significant ($P < 0.01$). It was also found that 'b' value significantly ($P > 0.05$) deviate from isometric value of '3'. It indicated negative allometric growth pattern in the fish. It resembles with the earlier studies of Rangaswamy (1976) for *Mugil cephalus* from the Ennorre and Adyar estuaries of India the 'b' value was 2.7788. The value of exponent 'b' in the equation usually lies between 2.5 and 4.0 (Hile, 1936). Relatively higher value of 'b' in group I might be due to faster growth rate of the fish because they were in early stage. The decline in 'b' value in groups II and III might be due to maturation stage of fish when somatic growth because slower.

Table 35.1
Biological characteristics of *Liza tade*

	Group - I	Group - II	Group - III
Size range (mm)	less than 120mm	120 -240mm	more than 240mm
Sample size (n)	62	55	48
Initial Growth Index (a)	0.03106	0.02952	0.02914
Exponent (b)	2.7102	2.7015	2.6895
Allometric Equation $W = aL^b$	$W = 0.03106L^{2.7102}$	$W = 0.02952L^{2.7015}$	$W = 0.029\,14L^{2.6895}$
Linear Equation logW = loga + blogL	logW=-1.5078+ 2.7102logL	logW=-1.5299+ 2.7015logL	logW=-1.5355+ 2.6895logL
Condition Factor (K)	1.56	1.23	1.03

The condition factor (K) of fish exhibited variations at different stages. Le Cren (1951) opined that K was affected by length, nutritional and biological factors, degree of parasitism etc. Here, K value was relatively higher in smaller size fish which is in agreement with earlier studies in other species (Gairola *et al.*, 1990). In the present study, significant variations ($P < 0.05$) in condition factor of *Liza tade* might be due to different biological factors especially feeding and maturation.

ACKNOWLEDGEMENTS

The author thanks to Dr. P.C. Thomas, Ex Director, College of Fisheries, Rangailunda, Berhampur – 7 (Ganjam) for facilities.

REFERENCE

Chanchal, A. K., Paandey, B. N. and Singh, S. B. 1978. Studies on some aspects of biology *of Anabas testudineus* (Telcostei: Anabantidae). *Matsya, 4* 15 - 19.

Das, H. P. 1977 Length weight relationship and relative condition of grey mullet *Mugil cephalus. Mahasagar,* 10(3 - 4) : 145 - 149.

Gairola, D. Singh, O. V., Malhotra, S. K., Nanda, S. and Ghildiyal, R. P. 1990. Parasites, Ponderal Index and length weight relationship of *Barilius bendelisis* (Ham.) and *Puntius ticto* (Ham.). *Indian J. Fish.*, 37(4) : 361 - 365.

Hile, R. 1936. Age and growth of the Cisco, *Leucichthys artedi* (Le Suur) in the lakes of three northern highland, Wisconsin. *Bull. U. S. Bur. Fish.*, 48 : 209 - 317.

Le Cren, E. D. 1951. The length weight relationship and seasonal cycle in gonad weight and condition in perch *(Perca fluviatilis). J. Animal Ecol,* 20 : 201 - 219.

Pandey, A. C. and Sharma, M. K. 1998. Applicability of cube law in the length weight relationship of the four species of exotic carps. *Proc Natl. Acad Sci. Indian Biol.Sci,* 68(1) :67-71.

Pathak, S. C. 1975. Length weight relationship, condition factor and food study of *Labeo calbasu* (Ham.) from Lone reservoir (M. P). *J. Inland Fish. Soc. India,* 8:58-64

Rangaswamy, C. P. 1976. Length weight relationship in grey mullet, *Mugil cephalus* Linneaeus. *Matsya,* 2:19-22.

Sarojini, K. K. 1958. Biology and fisheries of grey mullet of Bengal, II Biology of *Mugil cunnesius* Valenciennes *Indian J. Fish,* 5 : 56 - 76.

CHAPTER 36

Bioconcentration of Chromium, Copper and Iron in an Indian Major Carp, *Labeo Rohita* (Hamilton)

Arun Kumar Singh, K. S. Rana and Sunita Solanki

Department of Zoology, Agra College, Agra-282002, (U. P.), India

ABSTRACT

Both fingerling and adult of *Labeo rohita* (Hamilton) was exposed to sublethal concentration of chromium, copper and iron in a static bioassay after the LC_{50} value was determined. Equal amount of tissue samples were taken from the important organs viz. liver, muscle and heart after exposing the fishes for 28 days at sublethal concentrations. The high toxicity and bio-concentration of chromium, copper in the heart was remarkable, and high concentrations lead to the regulation of the metal in an attempt to minimize the toxic effect. The convulsion and other behavioural changes are also explained in the text.

Keywords: *Metal pollution, Bioconcentration, Labeo rohita.*

INTRODUCTION

Water is a very essential and vitally important substance. Environmental poisoning by heavy metals has increased in recent years due to extensive use of heavy metals in agriculture, chemical and industrial processes, becoming a threat to living organisms. Among the heavy metals copper, chromium and iron being the important pollutants from industrial effluents and agricultural wastes in aquatic environment cause effects on non-target aquatic organism resulting inbalance of ecosystem.

The effect of heavy metals on aquatic organisms is currently attracting widespread attention, particularly in studies related to industrial pollution. Heavy metals are being introduced into aqueous environment through industrial and urban effluents, soil leaching and rainfall. Fishes occupy high tropic level in accumulating various xenobiotics in the aquatic environment. Several investigations have already shown that metals accumulate at higher concentration in freshwater and marine food chain. Sandwar (1991) reported harmful effects of solid wastes on human health. The wastewater from the textile, dyeing, printing, tannery etc., are highly coloured and can cause an imbalance in the ecological system (Dutta, 1994, Birunda and Nagarajan, 1997).

Therefore, in the present study the level of heavy metals in the some important organs was examined after exposing the fish for 28 days at sublethal concentration.

MATERIALS AND METHODS

The living specimens of *Labeo rohita* (Ham.) were collected from local resources around Agra. They were treated with 0.1% $KMnO_4$ solution to remove dermal infections if any. The fishes were grouped by their size during acclimatization and feeding was done with readymade fish food and boiled eggs. Fish were exposed to 0.25 ppm and 0.35 ppm of chromium, copper and iron respectively for a period of 28 days. A known amount of tissue was dissected out after the experimental period. The tissues examined are liver, muscles and heart. The tissues digested in a mixture of nitric and perchloric acid. Chromium, copper and iron were determined by flame atomic absorption spectrophotometer. The values obtained were compared with that of control animals. The data were analysed using students 't' test.

RESULTS AND DISCUSSION

96hr LC_{50} obtained using chromium, copper and iron and different sublethal concentrations (SLC) for the *Labeo rohita* (Ham.) are given in Table 36.1. The highest level of chromium and copper were found in the heart, while the highest concentration of iron was present in the liver only (Table. 36.1). The muscle, however, contained highest concentration of chromium, copper and iron than the heart. Distinctive behavioural patterns such as convulsion, fin thickening and rubbing against the wall of the aquarium were evident during the exposure to the metals. Metallic ions present in the river water enter the fish body through permeable surfaces and they get accumulated in various organs like liver and muscle during the study conducted by Al-Mohanna (1994). These metallic ions are the main probable cause of the bioconcentration in fish.

Table 36.1

Levels of Chromium, copper and iron in the liver, muscle and heart of the control and experimental in an Indian major carp, *Labeo rohita* (Hamilton) (Mg/Kg-1 Wet. Weight).

Heavy Metals	Liver	Muscle	Heart
Control Chromium Exposed	2.24±0.18	1.97±0.25	3.24±0.22
	5.39±0.30	4.11±0.46	7.71±0.33
Control Copper Exposed	2.98±0.21	2.51±0.28	3.78±0.25
	6.48±0.33	5.20±0.49	8.80±0.36
Control Iron Exposed	3.77±0.24	4.37±0.33	380±0.31
	8.80±0.59	7.24±0.52	7.17±0.63

The high toxicity and bio-concentration of chromium and copper in the heart was remarkable in the present investigation. This high concentrations lead to the regulation of the metals in an attempt to minimize the toxic effect. Metallic compounds in these effluents accumulate in various parts of animal body (Barman and Lal, 1994) causing huge mortification of inhabiting biomass including palatable fish. These metallic ions are the main probable cause of the physiological anomalies in fish. On the other hand, the increased bioconcentration of iron in the liver proved that fish are more tolerant to this metal.

As homeostatic control organ of the body, the liver eliminate toxic substances and metabolites to reduce the detrimental effect of toxicant on normal physiological process. Chromium and copper may possibly bind to metallothioneine, a metal binding protein, to form a non-toxic compound. Iron can effectively be used for synthesis of haemoglobin, since it is less toxic, it may be a profitable explanation for bio-concentration. Compared with most other investigations, the concentrations may be quite high enough to impair several physiological activities in fish.

REFERENCES

Al-Mohanna, M.M. 1994. Residues of some heavy metals in fishes collected from Red sea coast Jizan, Saudi Arabia. *J. Environ. Biol.* 15 7: 149-257.

Barman, S.C. and LaL, M. M. 1994. Accumulation of heavy metal Zn, Cu, Cd and Pb in soil and cultivated vegetables and weeds grown in industrially polluted fields. *J. Environ. Biol.* 152: 107-115.

Birunda, S. and Nagarajan, S. 1997. Decolourizing napthal green B with sodium perborate ferrous ion system. *Ind. J. Environ. Protec.* 18: 174-176.

Dutta, P.K. 1994. An overview of textile pollution and its remedy. *Ind. J. Environ. Pollut.* **14:** 443-446.

Sandwar, B.B. 1991. Hazardous waste management: A global problem. *J. Ind. Assoc. Environ. Manag.* **18:** 186-189.

CHAPTER 37

SUBLETHAL EFFECTS OF BHC ON OXYGEN CONSUMPTION BY FISH *CYPRINUS CARPIO* AND ROLE OF AN AQUAPHYTE (*CHARA*) AS DETOXICANT

R. Punitha, N. Krishnan and G.N. Subbiah*

Department of Zoology, Madura College, Madurai - 625 0011, India.

** Department of Zoology, Thiagarajar College, Madurai - 625 009, India.*

ABSTRACT

Exposure of carp, *Cyprinus carpio* to a sublethal concentration of an organochlorine insecticide, BHC for 30 days under laboratory conditions decreased the rate of oxygen consumption by fish. However addition of an aquatic plant, *Chara* (Family : Charophyceae) to intoxicated test medium (100g /10 litres) enhanced the rate of oxygen consumption by fish. It is interpreted that *Chara* could be tested further as phytodetoxicant in reducing the insecticidal toxicity to cultivable fishes.

Key Words: *Cyprinus carpio*, BHC, oxygen consumption, *Chara*, *detoxicant*.

INTRODUCTION

Environmental pollution in freshwater ecosystems is caused by a variety of pollutants. In an agricultural country like India, insecticides constitute the major components of aquatic chemical pollutants. Laboratory investigations in fish models have documented a variety of physiological, biochemical, haematological and behavioural responses of fishes under toxic stress of insecticides (review : Pritchard, 1993). Oxygen consumption by fishes under toxic stress of various aquatic pollutants including insecticides was estimated quantitatively by several workers in the past (Jeyachandran and Chockalingam, 1986; Roy and Datta Munshi, 1988; Baskaran et al., 1989; Khillare, 1990; Jayasuriya et al., 1991). Use of aquatic plants as phytodetoxicants or biofilters their by reducing the toxic stress on fishes is one of the known method in the control of insecticidal pollution in aquatic ecosystems (Subbiah and Suryaprabha, 1986; Jampani, 1987; Chandrasekaran et al., 1995). In this background, the present study ascertains the valve of aquatic algae, *Chara* as a detoxicant in sharing the toxic stress of BHC by *Cyprinus carpio* in relation to its oxygen consumption.

MATERIALS AND METHODS

Healthy fingerlings of *C. carpiò* (4 + 0.5 g / live weight) were obtained from local fish farm and acclimatized to the laboratory conditions at 29C° for 7 days before experimentation.

They were fed daily with chopped goat liver but food was withheld 24 hrs before bioassay test. The water was changed daily with an intermittent aeration, *chara* (Family : Charophyceae), commonly known as stonewart was selected as phytodetoxicant. It was collected from the college pond, washed with tap water and stored in aquarium. For the present study BHC, an organochlorine insecticide was selected as experimental toxicant. The commercial grade BHC (50% wp; trade name Hexidole) is obtained from Rallis India Ltd. Its stock solution in distrilled water (1g in 500ml) is prepared and then diluted to required test concentrations (8-15 ppm). The acclimatized fishes, ten in each group were exposed to test concentrations along with suitable control. Three replicates were maintained for each treatment. Water and concentration of BHC in aquaria were renewed every 24hr. Mortality rate was observed for 96 hr in each test concentration. LC_{50} for 96 hr was computed by plotting the percentage mortality against the concentration of BHC (Abbott, 1925). The 96 hr LC_{50} was found to be 10ppm (Punitha, 1990). Oxygen consumption by fish was studied under sublethal concentration (5ppm) by closed - chamber method. Fishes were exposed to 10 litres of insecticide treated water with adequate replicates for 30 days. Simultaneously another set was maintained along with *Chara* (100g /10 litres). Oxygen consumption was determined at weekly interval in both groups by Winkler's method. The results are expressed in mg / g wet weight / hr.

RESULTS AND DISCUSSION

In fishes treated with BHC but without *Chara,* the rate of oxygen consumption decreased from an initial value of 7.457 in first week to 3.442mg/ g wet weight / hr. The percentage reduction data in every week of exposure supports this conclusion (Table 37.1). Similarly earlier works have estimated a significant reduction in oxygen consumption by many fish species under the toxic stress of aquatic pollutants such as insecticides (Roy and Datta Munshi, 1988; Khillare, 1990), detergents (Jayasuriya et al., 1991), textile dye effluent (Baskaran et al., 1989) and tannery effluent (Jeyachandran and Chokalingam, 1986). The gills of fishes are in direct contact with the toxic test medium. Histopathological studies have revealed marked alternations in gill structure of the fishes exposed to organochlorine insecticides (Sahai, 1991). For instance, BHC induces an acute inflammation, loss of epithelial layer and fusion of gill lammellae in *Rosbora daniconius* (Singh and Sahai, 1984). It is presumed that BHC may induce similar alternations in gills of *C. carpio* and this could be the reason for the drop in oxygen consumption.

Table 37.1

Oxygen Consumption by *C. Carpio* (in mg/g wet weight/hr) exposed to sublethal concentration of BHC (5 PPM)

Fish Group	Exposure duration (in weeks)				
	I	II	III	IV	Grand Means
Without Chara	7.457 (1.088)	6.310 (1.060)	4.302 (0.630)	3.442 (0.103)	5.4 (1.581)
% Reduction	34.82	29.33	19.99	16.01	
With Chara	7.681 (2.002)	7.465 (0.620)	7.101 (0.098)	7.001 (0.860)	7.3 (0.082)
% Reduction	26.62	25.52	24.27	23.93	

Values are mean (SD); n = 3 difference between grand means is significant of $p < 0.05$ by t test.

In constant, fishes exposed to BHC with *Chara*, it was dropped from 7.681 to 7.001mg/ g wet weight / hr. Eventhough there were no significant differences between the weeks in mean values when compared to groups treated without *Chara*, the difference between the overall mean values is significant at $P<0.05$. It showed that addition of *Chara to* the test medium enhances the rate of oxygen consumption by fish. This alage may act as detoxicant or biofilter possibly accumulating a considerable quantity of BHC. This presumption needs further study. For instance, Chandrasekaran et al. (1995) have shown that *Chara* accumulated endosulfan, an organochlorine insecticide from the test medium significantly their by reducing its accumulation in kidney, liver and gills of *Macrones keletius*. Similarly aquaphytes like *Hydrilla verticillata* and *Pistia* stratiotes have been shown to be detoxicant of endosulfan (Subbiah and Suryaprabha, 1986). Detoxification of pesticides such as phosphomidon and monocrotophos by the blue green algae, *Nostac colci* was reported by Jampani (1987). Further investigations are needed to confirm such BHC accumulation in *Chara* and degree of protection by this algae to other species of fishes.

ACKNOWLEDGEMENTS

This work was carried out by RP at Thiagarajar College under the guidance of Prof. GNS. We are grateful to Board of Madura College, HOD, Principal and all faculty members for encouragement. NK gratefully acknowledge Prof. Dr. Arvind Kumar, Editor for his constant support and encouragement.

REFERENCES

Abbott, W.S. 1925. A method of computing the effectiveness of an insecticide. *J.Enta* 18: 263-267.

Baskaran, B., Palanichamy, S. and Arunachalam, S. 1989. Effects of textile dye effluent on feeding energitics, body composition and oxygen consumption of the fish *Oreochromis mossambicus*. *J. Ecobiol.* 1 : 203-207.

Chandrasekaran, N., Murugavel, P. and Krishnan, N. 1995. Control of endosulfan residue accumulation in *Macrones keletius* with *Chara* (Charophyceae). *J. Appl. Zool. Res.* 6 : 193-194.

Jampani, C.S.R. 1987. Detoxification of insecticides phosphomidon and monocrotophos by the blue green algae *Nostoc cold*. *Natl. Acad. Sci. Lett.* 10:381-382.

Jayasuriya, S., Subramanian, M.A. and Varadaraj, G. 1991. Effect of detergents on the oxygen consumption of the cat fish *Mystus vittatus*. *J. Ecobiol.*, 3:217-220.

Jeyachandran, K.P.S. and Chokalingam, S. 1986. Effects of tannery effiuent on respiratory parameters and biochemical constituents of fish *Channa punctatus*. *Comp. Physiol. Ecol.* 12 : 197-200.

Khillare, Y.K. 1990. Effects of pesticide on fresh water fish, *Puntius stigma*. In *Trends in Ecotoxilogy*. Academy of Environmental Biology, Karad. pp. 185-188.

Pritchard, J.B. 1993. Aquatic toxicology : past, precent and prospects. *Environ. Health Persp.* 100 : 249 - 257

Punitha, R. 1990. The role of the aquaphyte *Chara as a* detoxicant. M.Phil. Thesis, Madurai Kamaraj University, India.

Roy, P.K. and Datta Munshi, J.S. 1988. Oxygen consumption and ventilation rate of major carp *Cirrhinus mrigala* in malathion treated water. *J. Environ. Biol.* 9:5- 13.

Subbiah, G.N. and Suryaprabha, R. 1986. A study on the control of insecticide residue accumulation in a freshwater fish *Macrones keletius* with aquaphyte. *Proc. Natl. Symp. Fish. Environ.* Madras, pp. 19-23.

Sahai, S. 1990. Pesticide pollution and its impact on some fish tissues - a review. In : *Trends in Ecotoxicology*. Academy of Environmental Biology, Karad, pp 63-72.

Singh, S. and Sahai, S. 1984. Histopathological changes in the gills of *Rasbora daniconius* induced by BHC. *J. Environ. Biol.* 6 : 65-69.

CHAPTER 38

FISH DIVERSITY AND FISHERIES OF GAMBHIR DAM, UJJAIN M.P. INDIA

Anil Pandey, Arvind N. Shukla and Sharad Shrivastava

Limnology and Environmental Biology Unit, School of Studies in Zoology
Vikram University, Ujjain 456 010 (M.P.) India

ABSTRACT

During the making first limnological studies on Gambhir dam (latitude 20°11'N and longitude 75°37'E) Ujjain, M.P. India. 30 species of fish biodiversity were recorded which belong 6 orders, 10 families and 17 genera. The commercially important fishes were grouped according to their economic importance into 3 major categories viz. catfish *(Mystus seenghala, Mystus tengara, Wallago attu* and *Ompok bimaculatus)*, major carp *(Catla catla, Cirrhinus mrigala* and *Labeo rohita)* and miscellaneous *(Channa* sps., *Mastacembelus* sps. and *Notopterus notopterus)*. Three zones were classified according to their morphometry, water flow pattern and hydrobiological characteristics viz. zone I (upstream zone) contributed 29.33%, zone II (reservoir zone) contributed 62.18% and zone III (downstream zone) contributed 8.56% of the total fish catch. The total annual fish yield in this water body 24448.17 kg yr^{1} in 1995-96 and 24490.52 kg yr^{1} in 1996-97 were recorded. The monthly fish catch ranged from 59.98 to 502.51 kg month-^{1} in 1995-96 and 57.75 to 476.33 kg month-^{1} in 1996-97. The maximum catch was observed in May while minimum in July. The percentage composition of catfishes shows increasing trend and major carp show declined trend during two years study period. Primitive type gears viz. cast net, gill net, hooks, line and traps were used by fishermen. Cast net was the most suitable gear which contributed highest catch. Catch per unit effort (CPUE) ranged between 0.080 to 0.688 kg/hrs. The average annual fish landings per fisherman ranged between 99.67 to 330.39 kg/ fisherman/year. An inverse relationship has been pointed out between water level fluctuations and fish production.

Key Words: *Topography, Fish diversity, Cast net, Fishermen, Gambhir dam.*

INTRODUCTION

India ranks next to China and Russia in inland fish production in the world. It is estimated that contribution from inland and marine sector amounts to 0.90 and 1.40 million tonnes respectively of the 2.3 million ztonnes of total fish production (Shrivastava and Vathsala, 1984). Eventhough the inland sector contributed nearly one third of the total fish production, its share to the total domestic supply was 57% in contrast to marine sector with 43%. Thus fish production from inland sector is of great significance as it contributes the major share in the protein rich food for domestic consumption. Madhya Pradesh ranks second in the reservoir area of 2.5 lakhs hectares for 56 reservoirs and was giving a total production of 2164 tonnes year^{-1}. The development of the reservoir fisheries was taken up in the state by the State Fishery Department since 1962. The average stocking intensity of major reservoirs of Madhya Pradesh is only 40% as suggested by State Fishery Study Group. Reservoir fisheries management in India is comparatively poor when compared with some of the developed reservoirs in the other countries. Lack of fish seed introduction of commercial fishes, fishing gears and reservoir management measures seems to be greater impending factors in the development of a viable reservoir fishery in the country. A number of foreign workers have contributed to our knowledge in the fisheries aspects (Johnson and Walsh, 1989; Herpman, 1990; Fisher *et al.*, 1991 and Herpman *et al.*, 1993). Prominent studies in this regard in India are those of Jayram (1981), Jhingran (1982), Natrajan (1985), Indra (1992), Singh and Yazdani (1992), Rawat and Nautiyal (1995), Saxena (1997), Devi and Raghunathan (1999) and Nanda and Tiwari (2001). State of Madhya Pradesh is neglected in this regard eventhough gross work of this aspects were made by Dubey and Mehra (1959), Choudhary (1977), Gupta and Rao (1978), Kartha (1987), Rao *et al.* (1988, 1991, 1993) and Saxena (1997). No attempts has been made on Gambhir dam in this regards so far first limnological studies in this dam has been conducted and details of fish biodiversity and potential structure were reported which will helpful in the management and development of fishery in the reservoir.

MATERIALS AND METHODS

The present study was carried out for a period of two years from March 1995 to February 1997. 11 fish collection centres were selected viz. Aslana, Bhundwas, Chikali, Nalwa, Chankheri, Kharotiya, Amirpur, Bamnapati, Barwai, Kantharkhedi (Dam site) and Down stream site. Besides this fish collection and marketing surveys were made for the period of 7 days in every month. The main fishing gear operated in the dam were gill nets of mesh ranging from 23-180 mm bar. Further the fish caught by small meshed gill nets, hooks, lines and trapes. The fishes soon after collection were cleaned and preserved in 5% formalin and glycerine. The fishes were identified to genus/species using keys provided by Day (1958), Shrivastava (1968) and Jhingran (1992). Catch Per Unit Effort (CPUE) computed based on the mean monthly fluctuation of fish landings.

RESULTS AND DISCUSSION

The Gambhir dam managed by the Madhya Pradesh Public Health Engineering Department, which is a major source of drinking water and irrigation, for inhabitants of Ujjain area. Fish culture in this dam is totally neglected till now by M.P. Government and private sector. However, first limnological studies has been conducted in this water body for focussing the fish biodiversity and potentials which will useful to fishery department, Govt. of M.P. for future planning of fisheries development in Gambhir Dam.

(i) Topography

Gambhir dam is located near Kharotiya village, 20 kms. away from Ujjain city (MP). This dam is located at latitude of 21° 11' North and longitude of 75°37' East and altitude of 483.35 m MSL in an orientation from North to South. It is constructed in 1990 on river Gambhir which is the biggest tributary of river Kshipra. As regards the physical features of the dam (Fig. 38.1) it can be seen that the dam site is broad and tail end of the dam is narrow. The central basin muddy having region of deep

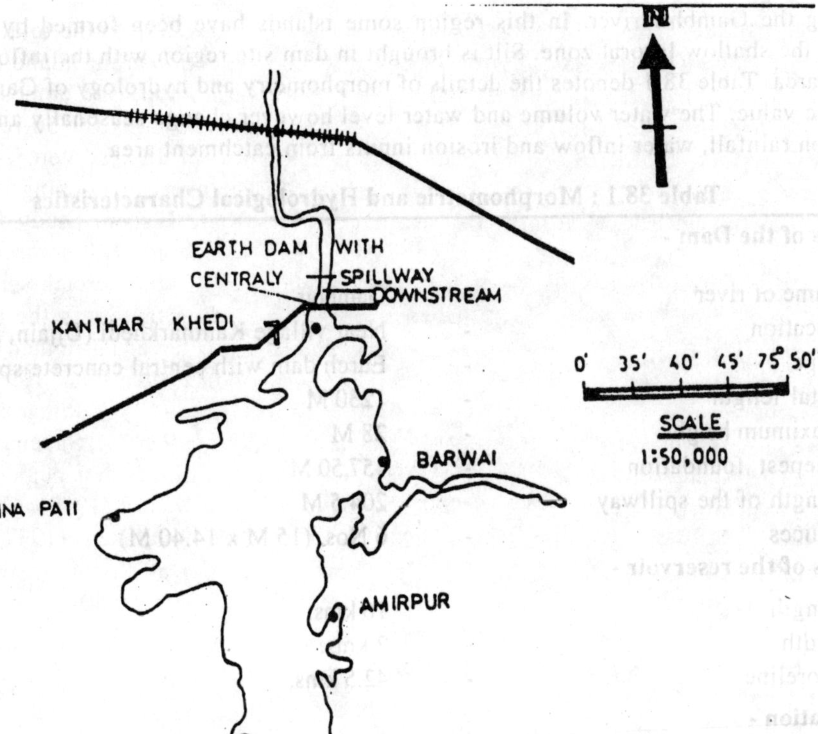

Fig 38.1

Map showing the catchment area and fish collection centres of Gambhir Dam.

water along the Gambhir river. In this region some islands have been formed by submergence of knoles and the shallow littoral zone. Silt is brought in dam site region with the inflowing water from catchment area. Table 38.1 denotes the details of morphometry and hydrology of Gambhir dam which shows static value. The water volume and water level however change seasonally and to some extent depend upon rainfall, water inflow and irosion inputs from catchment area.

Table 38.1 : Morphometric and Hydrological Characteristics

Particulars of the Dam -		
Name of river	-	Gambhir
Location	-	Near village Kantharkhedi (Ujjain, M.P.)
Type	-	Earch dam with central concrete spillway
Total length	-	1230 M
Maximum height	-	28 M
Deepest foundation	-	457.50 M
Length of the spillway	-	204.5 M
Sluices	-	6 Nos. (15 M x 14.40 M)
Particulars of the reservoir -		
Length	-	16 kms.
Width	-	2 kms.
Shoreline	-	42.5 kms.
Water elevation -		
Maximum water level	-	483.35 M
Full reservoir level	-	483.35 M
Crest of overflow level	-	469.50 M
River bed level	-	466.50 M
Foundation gallery elevation	-	461.50 M
Top bundh level	-	486.00 M
Area-		
Gross storage level area	-	1674 hectares
Full reservoir level	-	1674 hectares
Average level	-	83.7 hectares
Dead storage level	-	13.9 hectares
Capacity -		
Gross storage	-	61.50 M Cu M (2175 Mc Ft.)
Live Storage	-	48.50 M Cu M (1855 Mc Ft.)
Dead Storage	-	13.00 M Cu M (320 Mc Ft.)
Water supply dependibility	-	95%
Depth -		
Maximum at FSL	-	15.85 meters
Maximum at DSL Average	-	2 meters
Average	-	6.92 meters

(ii) Fish Diversity

The diversity of fishes mainly depend upon the biotic and abiotic factors and types of the ecosystem (lentic and lotic habitat). Age of the water body, mean depth, water level fluctuations, dam design and morphometric features have great ecological implications. During making limnological studies on Gambhir dam 30 species of fish biodiversity were recorded which belong to 6 orders, 10 families and 17 genra (Table 38.2). Dubey and Mehra (1959) reported 54 species from the Chambal river during preimpoundment of Gandhisagar reservoir. Dubey and Chatterjee (1977) and Choudhary (1977)

Table : 38.2
Qualitative list of fishes recorded from Gambhir Dam During March 1995 to February 1997

ORDER - CYPRINIFORMES
DIVISION - CYPRINI
FAMILY - CYPRINIDAE
SUB-FAMILY - CYPRININAE
SCIENTIFIC NAME

1. *Catla catla*
2. *Cirrhinus mrigala*
3. *Cirrhinus reba*
4. *labeo rohita*
5. *Labeo calbasu*
6. *Labeo gonius*
7. *Labeo bata*
8. *Puntius sarana*
9. *Puntius ticto*
10. *Puntius sophore*
11. *Oxygastar bacalia*

SUB-FAMILY - RASBORINAE

12. *Rasbora doniconius*

SUBF-AMILY - COBITIDAE

13. *Nemachalins - botia*

DIVISION - SILURI
FAMILY - SILURIDAE

14. *Ompok bimaculatus*
15. *Wallgo attu*

FAMILY - BAGRIDAE

16. *Mystus seenghala*
17. *Mystus tengara*
18. *Mystus aor*

FAMILY - SCHIELBEIDAE

19. *Eutropiichthys vacha*

FAMILY - SACCOBRANCHIDAE

20. *Heteropneustes fossilis*

ORDER - CLUPEIFORMES
SUB ORDER - NOTOPTEROIDEI
FAMILY - NOTOPTEROIDAE

21. *Notopterus notopterus*

ORDER - BELONIFORMES
FAMILY - BELONTDAE

22. *Xenentoden cancila*

ORDER - OPHIOCEPHALIFORMES
FAMILY - OPHIOCEPHALIDAE

23. *Channa marullus*
24. *Channa gachua*
25. *Channa striatus*
26. *Channa punctatus*

ORDER - PERCIFORMES
FAMILY - CENTROPOMIDAE

27. *Ambasis nama*
28. *Ambasis ranga*

ORDER - MASTACEMELEFORMES
FAMILY - MASTACEMELIDAE

29. *Mastacembelus arrnatus*
30. *Mastacembelus puncalus*

observed 39 species after impoundment of Gandhisagar reservoir. Kartha (1987) reported 41 species of fishes from Gandhisagar reservoir. Saxena (1997) reported 42 fish species from upstream region and 35 species from down stream region in river Satlaj. Indra (1992) observed 20 species under 6 families at Kamarajar and Pasumpon Ramalingam districts of Tamil Nadu. Singh and Yazdani (1992) recorded 24 species of ichthyofauna of Nasik district, Maharashtra. Devi and Raghunathan (1999) reported 49 species from Dharampuri district of Tamil Nadu. Nanda and Tiwari (2001) observed 18 fish species from Sambalpur-Hirakund— Burla regoin of Orissa. When compared to above water bodies with Gambhir dam, it showed less fish diversity except the observations of Indra (1992) and Nanda and Tiwari (2001). In the absence of a perimpoundment study in Gambhir dam it was not possible to determine whether the piscifauna is depleted or not. But in Gandhisagar reservoir Kartha (1987) pointed out that during the change of the lotic to lacustrine environment the fish diversity decreases.

(iii) Commercial fisheries

The common fishery of Gambhir dam is unorganised due to lack of fisheries management techniques. The fishery of Gambhir dam represents the natural stock. The dominant fisheries includes *Mystus seenghala, Wallago attu, Catla catla, Labeo rohita, Cirrhinus mrigala, Channa* sps., *Mastacembelus* sps., *Notopterus notopterus* and *Ompok bimaculatus.* The commercially important fishes of this dam were grouped according to their economic importance into 3 major categories viz. major carp *(Catla catla, Cirrhinus mrigala* and *Labeo rohia),* catfish *(Mystus seenghala, Mystus tengara, Wallago attu* and *Ompok bimaculatus)* and miscellaneous *(Channa* sps., *Mastacembelus* sps., *Notopterus notopterus).* Rao *et al.* (1988a) reported 9 commercially important fishes from Gandhisagar reservoir. In the present study percentage of commercially important fishes like *Catla catla* (7.11%), *Labeo rohita* (4.52%) and *Cirrhinus mrigala* (3.71%) were very low of the total catch. This may be due to the fact that there is no seed stocking operations made by outside sources in the reservoir. For the development of commercial fisheries in Gambhir dam it is therefore recommended that the species of *Catla catla, Labeo rohita* and *Cirrhinus mrigala* should be introduced from exogenous sources which would certainly dominant the endogenous species and accommodate themselves successfully in this reservoir as recorded by Kartha (1987) in Gandhisagar Reservoir,

(iv) Zonewise Annual Fish Yield

The hydrobiological features of the collection centres play an effective role in fisheries output to a greater extent. At Gambhir dam three zones were classified according to their morphometry, water flow pattern and hydrobiological characteristics. Zone I (upstream zone) exhibits riverine conditions and contributed 29.23% of the total fish catch. Zone II (reservoir zone) which exhibits lentic habitat and contributed to 62.18% of the total fish catch, due to rich growth of phytoplankton, zooplankton and macrozoobenthos. Zone III (down stream zone) represent rapid water currents and contributed 8.56% of the total fish catch (Table 38.3). The centrewise data of annual catch showed minimum 1196.09 kg yr^{-1} at Aslana collection centre and maximum 4089.89 kg yr^{-1} at Dam collection centre. In the present study 14.60 and 14.62 kg/ha yr^{-1} landings were estimated in 1995-96 and 1996-97 respectively. The total fish yield at Gambhir dam 24448.17 kg yr^{-1} in 1995-96 and 24490.42 kg yr^{-1} in 1996-97 were reported. Kumar (1990) reported 78.6 kg/ha yr^{-1} annual landings from Gobind sagar reservoir during 1988-89. Sinha (1990) reported 71.55 kg/ha yr^{-1} annual landing from Gumti reservoir Tripura during 1987-88. Estimates regarding actual yield per hectare of Indian reservoir have varied from 5 to 8 kg/ha by National Commission of Agriculture in 1976 and 16 kg/ha by Jhingran (1983). A decline trend in the per hectare yield of Gambhir dam was noticed. Various problems responsible for such as unsatisfactory state of affairs can be broadly classified as biological, technological and economic administrative and legislative (Saxena, 1990).

Table : 38.3
Zone and Collection Centrewise details of Annual Fish Yield in Gambhir Dam During March 1995 to February 1997

S.No.	Name of Catchment Centres	1995-96		1996-97	
		Catch	%	Catch	%
I	**Upstream Zone**	**7240.44**	**29.61**	**7068.2**	**28.86**
1	Aslana centre	1218.03	4.98	1196.09	4.88
2	Bhundwas centre	1568.29	6.41	1579.97	6.45
3	Chikali centre	1439.06	5.88	1395.06	5.69
4	Nalwa centre	1370.01	5.6	1329.04	5.42
5	Channkheri centre	1645.02	6.72	1568.02	6.4
II.	**Reservoir Zone**	**15125.48**	**61.86**	**15311.02**	**62.51**
06.	Kharotiya centre	2287.01	9.35	2312.77	9.44
07.	Amirpur centre	2326.05	9.51	2345.06	9.57
08.	Bamanapati centre	3363.71	13.75	3378.23	13.79
09.	Barwai centre	3065.01	12.53	3185.01	13.00
10	Kantharkhedi centre	4085.52	16.71	4089.93	16.70
III.	**Down Stream Zone**	**2082.25**	**8.51**	**2111.28**	**8.62**
11	Down stream site	2082.25	8.51	2111.28	8.62
	Total	24448.17	100	24490.52	100

(v) Monthwise catch structure

The monthly fish catch ranged between 59.98 to 502.51 kg month^{-1} in 1995-96 and 57.75 to 476.33 kg month^{-1} in 1996-97. The maximum catch was observed in May at Dam while minimum in July at Bhundwas. At downstream the maximum catch 235.12 kg month^{-1} was observed in August 1997 due to migration of fishes from lower sites but due to presence of dam wall fishes were unable to cross the dam and caught by fishermen at downstream site (Fig. 38.2a and 38.2b). The minimum catch (108.64 kg) at this study site was observed in June 1995. Kartha (1987) reported highest fish landings of 192 tonnes month^{-7} and 272 tonnes month during March 1984-85 and 1985-86 respectively at Gandhisagar reservoir. The maximum catch in summer months at Gambhir dam may be due to rich nutrients levels, maximum growth of planktonic and macrozoobenthic organisms which were consumed by fishes as food.

(vi) Species Composition

In the present study percentage composition of *Mystus seenghala* was 49.35% in 1995-96, this was increased upto 49.64% in 1996-97. *Mystus tengara* contributed by 19.45% and 20.12% in 1995-96 and 1996-97 respectively. *Wallago attu* contributed by 8.80% in 1995-96, this was increased upto 10.35% in 1996-97, *Ompok bimaculatus* contributed by 1.58% and 1.66% during 1995-96 and 1996-97 respectively. Among the major carps *Catla catla* contributed by 7.45 in 1995-96 which was decreased upto 6.78% in 1996-97. *Labeo rohita* contributed 4.72% and 4.33% of the total landings during 1995-96

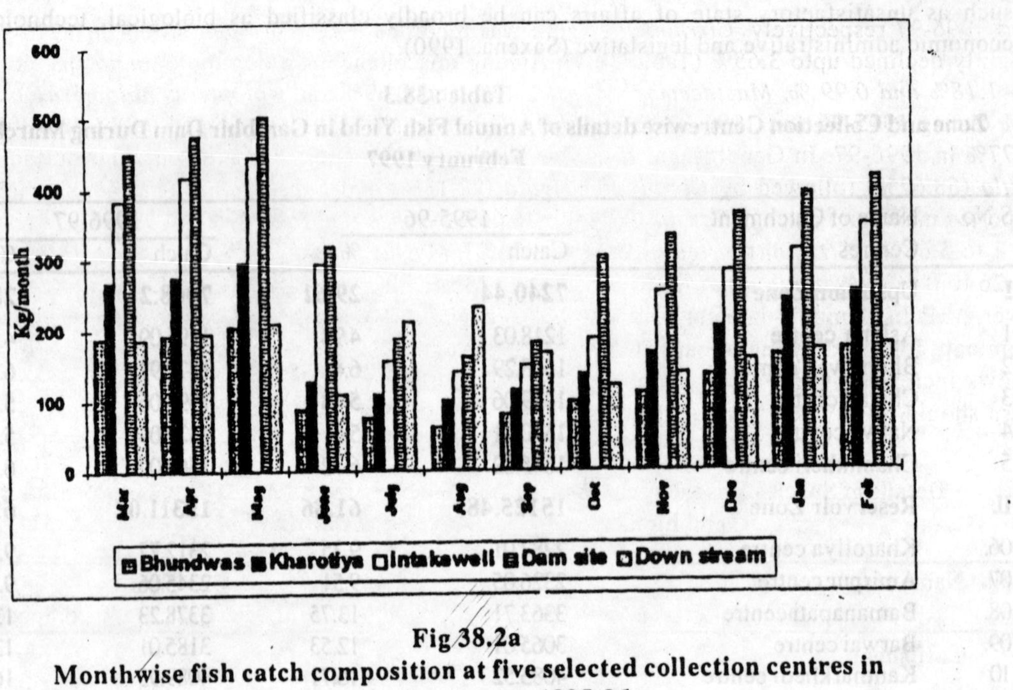

Fig 38.2a

Monthwise fish catch composition at five selected collection centres in Gambir dam during 1995-96.

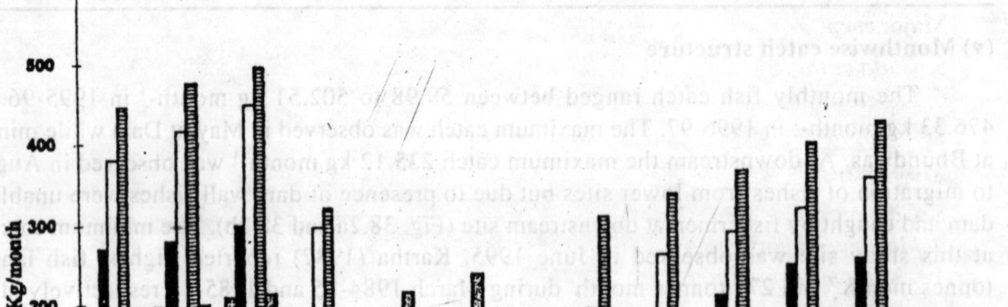

Fig 38.2b

Monthwise fish catch composition at five selected collection centres in Gambhir dam during 1996-97.

and 1996-97 respectively. *Cirrhinus mrigala* was recorded 3.77% in 1995-96 but in 1996-97 it was slightly declined upto 3.65% (Table 38.4). Among miscellaneous group the *Channa sps.* contributed by *1.18% and 0.99 %, Mastacembelus sps.* 1.46% and 0.69% and *Notopterus notopterus* 1.21% and 1% during 1995-96 and 1996-97 respectively. Other small fishes contributed 0.98% in 1995-96 and 0.77% in 1996-97. In Gandhisagar reservoir Kartha (1987) reported maximum contribution of *Catla catla* (58.67%) followed by *Cirrhinus mrigala* (22.18%) and *Labeo rohita* (17.14%). Singh (1993) reported maximum catch of *Tor tor* (23.84 to 26.66%), *Labeo fimbratus* (18.1 to 18.8%), *Labeo calbasu* (5.2 to 5.66%), *Cirrhinus mrigala* (2.26 to 3.06%), *Mystus seenghala* (7.7 to 8.6%), *Mystus tengara* (0.26 to 0.4%), *Wallago attu* (7.3 to 8.14%) and Miscellaneous (5.1%) in Sardar sarovar of Narmada river, M.P. In Gambhir dam cat fishes *Mystus seenghala, Mystus tengara* and *Wallago attu* were most dominant followed by major carp and miscellaneous groups. The percentage composition of catfishes shows increasing trend while major carps show declined trend. It is therefore suggested that major carp should be stocked from exogenous sources to fulfil the stocking requirements.

Table 38.4

Details of Species Wise Percentage Composition of Fishes at Five selected Study Sites in Gambhir dam During March 1995 to February 1997

S.No.	Name of fish	1995-96		1996-97	
		Catch (Kg)	%composition	Catch (Kg)	% composition
1	**Catfishes**				
	1. *Mystus seenghala*	1331.48	49.35	1335.69	49.64
	2. *Mysus tengara*	520.87	19.45	541.43	20.12
	3. *Wallago attu*	235.70	8.80	278.54	10.35
	4. *Ompok bimaculatus*	42.42	1.58	44.67	1.66
2	**Major carp**				
	5. *Catla catla*	199.61	7.45	182.45	6.78
	6. *Labeo rohita*	126.41	4.72	116.54	4.33
	7. *Cirrhinus mrigala*	101.09	3.77	89.22	3.65
3	**Miscellaneous**				
	8. *Channa sps.*	31.66	1.18	26.84	0.99
	9. *Masta cembelus sps.*	39.24	1.46	18.57	0.69
	10. *Notopterus notopterus*	32.54	0.98	20.84	0.77
4	**Other small fishes**	**26.28**	**0.98**	**20.84**	**0.77**
	Total	2677.3	100	2690.72	100

(vii) Fishery of Important Species

In the present study dominant fish species were selected on the basis of their constant monthly occurrence for evaluation of monthly catch fluctuations. The monthly fluctuation of fishes were reported at all selected study sites but for the sake of briefty only Dam site is considered for discussion. Its monthly landings ranged between 83.56 to 255.76 kg month-^{1}. The minimum catch was observed in August and maximum in May. *Mystus seenghala* was the first dominant species in Gambhir dam which contributed by 49.48% of total landings. *Mystus tengara* represent as second dominant fishery which contributed by 19.45 to 20.12%, its monthly landings varied between 32.05 to 64.03 kg month-^{1}. *Wallago attu* was the third dominant fishery of Gambhir dam which contributed by 8.73% of total catch. Its monthly landings ranged between 11.26 to 50.80 kg month-1. This species was abounded in upper stretch of river Yamuna and Ganga at Agra, U.P. (24.01 tonnes) and Kankur U.P. (7.99 tonnes) reported by Jhingran (1982). *Catla catla* was first dominant species among the major carp and contributed by 7.42% of total landings of Gambhir dam. Its monthly landings ranged from 4.83 to 40.01 kg month-1. *Catla catla* were observed as first dominant fishery at Ranapratap Sagar Rajasthan (Jhingran, 1982), Hirakund reservoir, Orissa (Varghese *et al.*, 1983) and Gandhisagar reservoir M.P. (Rao *et al.*, 1988a). *Labeo rohita* was second dominant. species among major carp and contributed by 4.71% of total fisheries of Gambhir dam. Monthly landings ranged from 1.93 to 21.85 kg month -1 with maximum in May and minimum in March. *Cirrhinus mrigala* was contributed by 3.76%, monthly catch fluctuation ranged from 5.71 to 18.81 kg month-1 (Table 38.5). The percentage contribution of *C. mrigala* was very high ranging upto 21.1% at Ranapratap Sagar Rajasthan (Jhingran, 1982).

(viii) Gearwise composition of Fish Catch

The recent advances in fishery technology is call for the acute need to modernize the present activities in the country. This would enhance to a great extent the meager reservoir fishery output from 3 million hectares area in our country. David *et al.* (1969), Natrajan (1976) and Kartha (1987) have stressed the need for a rational approach towards the gill netting practices. The most important aspects considered by GIFT, Cochin have been to asscss the fishery potential and to evolve more efficient fishing gear and methods for its exploitation. At Gambhir dam the fishermen were not so expert to operate the good quality of gear. Their efforts were limited only in primitive type of gears viz. cast net, gill net, hooks, line and trapes. For this purpose they used small fishing boats which were plank built canoes locally known as Dongas, about 6-13 ft. in length 1.5 to 2.5 ft wide and 1.5 to 3.0 ft. deep designed to carry two persons. In the present study it was observed that cast net as the most suitable gear which contributed 57.78% the total catch in 1995-96 and 57.65% in 1996-97. The gill net contributed 36.86% and 37.40% in 1995-96 and 1996-97 respectively. The gear like hooks, lines and trapes were contributed 5.29 to 5.34% of the total fish catch (Fig. 38.3). The standardization of gear and their proper use in Gambhir dam is recommended to increase the fish catch.

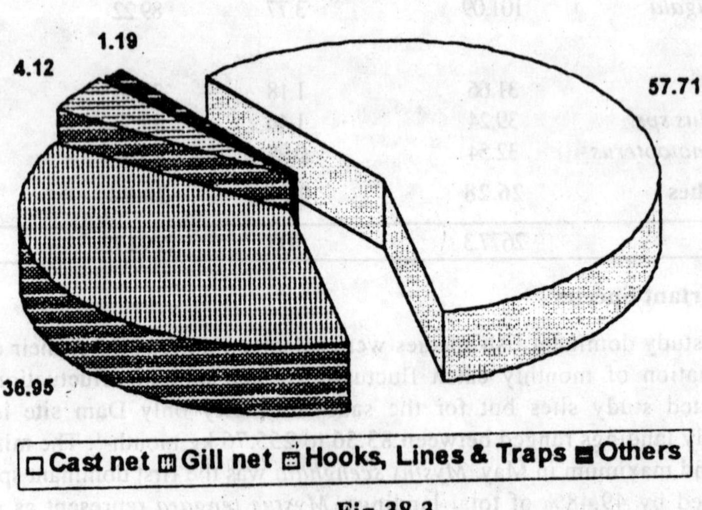

Fig 38.3

Gearwise percentage composition of fish catch at Gambhir Dam

Table 38.5

Average Monthly Catches (kg month-[1]) of Various Fish Species From Gambhir Dam at Kantharkghedi (Dam) Collection Centre During March 1995 to February 1997.

Name of Fish	Mar. 95	Apr.	May	June	July	Aug	Sep.	Oct.	Nov.	Dec.	Jan.'96	Feb.	Mar.	Apr.	May	June	July	Aug.	Sep.	Oct.	Nov.	Dec.	Jan.'97	Feb.
Catfishes																								
Mystus seenghala	226.15	232.83	250.36	162.92	92.01	83.56	86.52	152.06	162.03	177	3 194.25	206.6	230.55	234.11	255.76	164.21	93.34	83.89	86.21	154.31	164.81	175.01	197.91	211.01
Mystus tengara	88.64	94.06	98.02	64.03	36.04	32.05	34.56	59.02	62.01	70.2	76.02	82.13	89.03	93.75	97.72	63.85	36.38	33.36	34.28	58.61	62.21	69.82	75.53	81.81
Wallugo attu	46.03	49.53	50.63	24.01	14.62	11.26	17.03	27.01	32.04	34.5	35.88	39.02	45.42	49.41	50.88	24.31	14.27	11.47	16.81	26.53	31.59	34.92	35.42	37.86
Ompok bimaculaters	4.72	5.51	6.21	5.18	4.23	3.87	4.21	4.62	5.09	5.22	5.61	5.73	4.85	5.32	6.06	4.97	4.58	3.98	4.01	4.48	4.91	5.41	5.58	5.86
Carp fishes																								
Catla catla	34.84	38.09	39.78	24.01	15.03	9.83	15.21	23.04	28.05	30.5	32.24	33.35	34.63	38.31	40.01	23.71	15.28	10.01	15.08	23.35	27.72	30.91	32.48	33.89
Labeo rohita	1.93	21.06	21.85	12.93	8.27	7.05	9.07	15.02	16.25	18	18.32	18.69	19.23	21.37	21.65	13.37	8.06	7.26	9.17	15.23	16.31	17.58	18.09	19.08
Cirrhinus mrigala	16.13	17.94	18.81	11.27	6.89	6.01	6.84	11.06	13.02	14.1	14.92	15.44	16.45	17.56	17.91	11.49	6.57	5.71	6.51	11.21	13.39	13.72	15.22	15.59
Miscellaneous																								
Channa sps.	4.19	4.38	4.59	3.25	2.81	2.34	2.64	2.94	3.28	3.44	3.69	3.81	4.2	4.59	4.78	3.36	3.02	2.11	2.48	3.08	3.55	3.68	3.75	4.06
Mastacembelus. sps	4.42	4.62	4.91	3.36	2.94	2.59	2.81	3.15	3.39	3.75	3.91	4.14	4.28	4.78	5.22	3.51	2.88	2.28	2.68	3.05	3.28	3.61	4.02	4.37
Notopterus sps.	3.53	3.61	3.93	3.27	2.59	1.84	2.08	2.41	2.81	3.26	3.13	3.36	3.73	3.85	4.27	3.36	2.31	2.01	2.28	2.31	2.65	3.01	3.23	3.47
Others	3.26	3.01	3.41	3.09	2.38	2.14	2.45	2.41	3.23	3.51	4.12	3.71	3.36	3.25	2.56	3.65	2.48	1.98	2.31	2.61	3.04	3.42	4.05	3.86
Total	451	474.98	502.51	317.36	187.87	162.59	183.48	303.21	331.25	363.11	1 392.14	415.96	446.77	476.33	495.85	319.87	189.22	164.09	181.87	304.83	333.54	361.38	395.31	420.91

(ix) Catch Per Unit Effort (CPUE)

Catch per unit effort (CPUE) per hour has been estimated on the basis of monthly mean fluctuations of fish catch. CPUE also depends upon many physico-chemical factors, current velocity and volume of water. The factors like availability of food organisms, nutrients and temperature variation etc. are also responsible for movement of fish stocks. In the present study CPUE was computed in five selected study sites of Gambhir dam. CPUE varied between 0.080 to 0.688 kg/hrs with highest in May at Dam while lowest in August at Bhundwas (Table 38.6). Kartha (1987) reported CPUE range between 21.57 to 31.0 kg/hrs at Gandhisagar reservoir. Singh (1993) recorded CPUE values ranged from 3.3 to 4.43 kg/hrs at Sardar sarovar area of Narmada river. When compared with the above cited observations Gambhir dam showed less values it may be due to lack of fish seed stocking in this dam.

Table 38.6
Average Catch Per Unit Effort (CPUE) in kg, at Five Collection Centres in Gambhir Dam During March 1996 to February 1997.

Months	Upstream	Kharotiya	Intakewell	Dam	Down stream
March	0.241	0.375	0.528	0.620	0.271
April	0.273	0.387	0.578	0.661	0.275
May	0.291	0.416	0.622	0.688	0.296
June	0.125	0,180	0.408	0.444	0.153
July	0.105	0.153	0.21	0.262	0.296
August	0.080	0.140	0.195	0.227	0.326
September	0.111	0.158	0.211	0.252	0.235
October	0.155	0.192	0.263	0,423	0.170
November	0.151	0.235	0.358	0.463	0.195
December	0.184	0.286	0.395	0.501	0.223
January	0.232	0.336	0.433	0.549	0.240
February	0.247	0.349	0.486	0.584	0.251

(x) Impact of water level fluctuations on fish landings

The present study made observations with regards to fish landings and its relation to water level fluctuations on Gambhir dam. The maximum and minimum water level for the year 1995-96 were computed as 482.20 m and 478.97 m and the average level 480.79m respectively. For the year 1996-97 the maximum and minimum water level fluctuations were computed as 481.86 m, 477.78 m and average level 480.06 m respectively. The total fish catch in 1995-96 and 1996-97 were 24448.7 kg and 24490.52 kg respectively. After comparing the average water level and fish production the present study has pointed out an inverse relationship (Fig. 38.4).

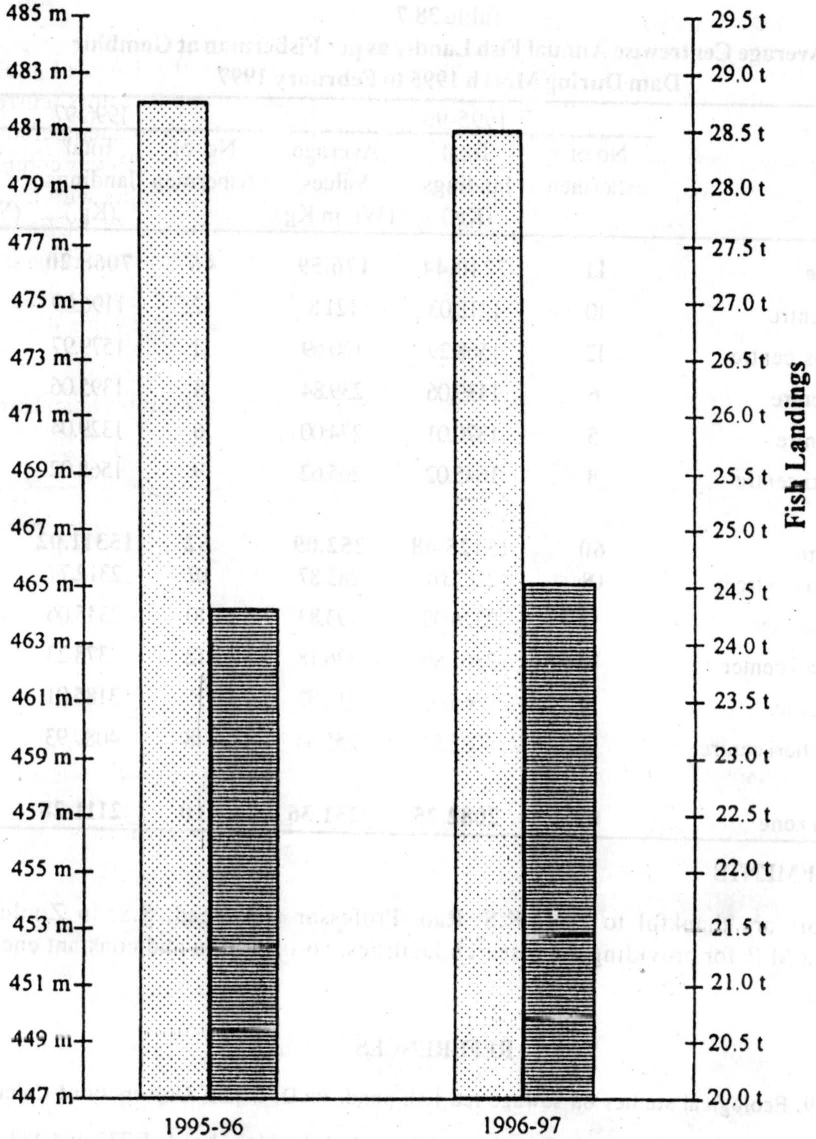

Fig 38.4
Average water and fish landings in Gambhir dam for 1995-96 and 1996-97

(xi) Centrewise annual fish landings per fishermen

The details of average catch per fishermen per year are presented in Table. 38.7. The average centrewise annual fish landings per fisherman ranged from 99.67 to 330.39 kg/fisherman/yr (Table 38.7). The maximum catch was observed at Kharotiya centre while minimum catch was observed at Aslana centre. Similar observation was observed by Kartha (1987) in Gandhisagar reservoir, M.P.

Table 38.7
Average Centrewise Annual Fish Landiy as per Fisherman at Gambhir Dam During March 1995 to February 1997

Zone	Centres	1995-96			1996-97		
		No.of fishermen	Total landings (Kg.)	Average Values (Wt. in Kg)	No. of fishermen	Total landings (Kg.)	Average Value (Wt. in Kg.)
I.	**Upstream zone**	**41**	**7240.44**	**176.59**	**46**	**7068.20**	**153.65**
1	Aslana centre	10	1218.03	121.8	12	1196.09	99.67
2	Bhundwas centre	12	1568.29	130.69	11	1579.97	143.63
3	Chikali centre	6	1439.06	239.84	8	1395.06	174.38
4	Nalwa centre	5	1370.01	274.00	6	1329.04	221.50
5	Chankheri centre	8	1645.02	205.62	9	1568.02	174.22
II.	**Reservoir zone**	**60**	**15125.48**	**252.09**	**62**	**15311.02**	**246.95**
6.	Kharotiya centre	08	2287.01	285.87	07	2312.77	330.39
7.	Amirpur center	12	2326.05	193.83	10	2345.06	234.50
8.	Bamanipati center	10	3361.86	336.18	12	3378.23	218.50
9.	Barwai center	14	3065.01	218.92	15	3185.01	212.33
10.	Kantharkheri centre	16	4085.52	255.34	18	4089.93	227.21
III.	**Down stream zone**	**09**	**2082.25**	**231.36**	**10**	**2111.28**	**211.12**

ACKNOWLEDGEMENTS

The authors are thankful to Prof. K.S. Rao, Professor and Head, S.S. in Zoology, Vikram University, Ujjain M.P. for providing the research facilities, co-operation and constant encouragement during this study.

REFERENCES

Choudhary,S. 1989. Ecological studies on sewage fed fish pond. Ph.D. thesis, University of Calcutta, India.

Day,F. 1958. The Fishes of India. William Dawson and Sons. Ltd. London. Vol.1, P.777 and Vol.11, p. 198.

Devi,K.R. and Raghunathan,M.B. 1999. The ichthyofauna of Dharampuri district, Tamil Nadu. Records of the Zoological Survey of India, 97, Part D : 145-162.

David,A.; Ray, P; Govind, B.V; Rajgopal, K.V and Banerjee, R.K. 1969. Limnology and Fisheries of Tungbhadra reservoir. Bull. Cont. Inland Fish Res. Inst, 13 : 1-118.

Dubey,G.P. and Mehra,R.K. 1959. Fish and fisheries of Chambal river. Proc. All India Zool. Congr. 1(2) : 647-665.

Dubey,G.P. and Chatterjee,S.N. 1977. Case studies on Gandhisagar reservoir, M.P. India. In : Symposium on the development and utilization of inland fishery resources. Fisheries of natural lakes and man-made reservoirs. Proc. IPFC 17(3): 232-245.

Fisher ,A.; Hanneman,W.N. and Keeler,A.G. 1991. Integrating fishery and water resource management: A biological model of California Salmon fishery. J. Environ. Econ. Manage. 20(3) : 234-260.

Gupta,D.R. and Rao,K.S. 1978. Experimental studies on tagging as an effective technique in ageing and fish growth on Gandhisagar reservoir. Bioresearch. 2(182): 15-18.

Herpmann, D.A. 1990. The value of in stream flow use to produce a recreational fishery. Ph.D. thesis, Deptt. Agric. and Reser. Econ. Colorado State Univ. Fort Collins.

Herpman, D.A.; Sparling, E.W. and Waddle, T.J. 1993. A methodology for quantifying and valuing the impacts of flow changes on a fishery. Water Resour. Res. 29(3) : 575-582.

Indra, T.J. 1992. Report on the Ichthyofauna from the mopping surveys of Kamarajar and Pasumpon Ramalingam districts (Tamil Nadu). Rec. Zool. Surv. India. 90(1-4) : 61-67.

Jayaram, K .C. 1981. The freshwater fishes of India. Zoological Survey of India, Calcutta, quoted by Gopinath and Jayakrishnan, 1984.

Jhingran, V.G. 1982. Fish and Fisheries of India. 2nd Edn. Hindustan Publ. Corp. India Ltd., New Delhi.

Jhingran, V.G. 1992. Fish and Fisheries of India. 3rd Edn. Hindustan Publ. Corp. India Ltd., New Delhi.

Johnson, D.M. and Walsh,R.G. 1989. Effects on participant skill on the value of alternative fishery management practices, In: Transactions of the 54th North American Wildlife and Natural Resources Conference, p. 247-254., WMI, Washington D.C.

Kartha, K.N. 1987. Studies on experimental trawl fishing in Gandhisagar reservoir Mandsaur, MP India. Ph.D. thesis, Vikram University, Ujjain India.

Kumar, A. 1990. Management and development of Gobindsagar reservoir : A case study. Reservoir Fisheries of India. Ed. A.G.Jhingran and B.K.Unnithan, p. 13-20.

Nanda, S.N. and Tiwari,T.N. 2001. A survey of fish fauna in the Sambalpur-Hirakund— Burla region, Orissa. Environment and Pollution, 8(1): 43-44.

Natrajan,A. V. 1976. Ecology and the state of fishery development in some of the manmade reservoir of India. In : Symposium on the development and utilization of inland fisheries reserouces, Colombo, Sri Lanka. Proc. Indo. Pacific Fish Coun. 17(3) : 1-15.

Natrajan,A.V. 1985. Potential and prospects of inland fisheries in India. In : Harvest and Post Harvest Technology of Fish (Ravindran,K. *et al.)* Society of Fisheries Technologists (India), Cochin, p. 16-18.

Rao,K.S.; Kartha,K.N.; Gupta,D.R.; Pandya,S.S. and Iyer, H.K. 1988a. Studies on morphometry and hydrology of Gandhisagar reservoir with special reference to fisheries. Fisheries Technol. 25(1) : 21-28.

Rao,K.S.; Chatterjee,S.N. and Singh, A.K. 1991. Studies on preimpoundment fishery on Narmada basin (Western region) in the context of Indira Sagar, Maheshwar, Omkareshwar and Sardar Sarovar reservoirs. J. Inland Fish Soc. India. 23(1) : 34-41.

Rao,K.S.; Shrivastava,S.; Shukla Arvind N. and Thilak J. 1993. A report on severe parasitic incidence in the fish population on Gandhisagar reservoir, MP, India. Hydrobiol. 9(1): 49-53.

Rawat, VS. and Nautiyal,P. 1995. Natural history of hillstream cyprinid *Barilius bendelisis :* Food and feeding habits in relation to growth rate. Inland Fish Soc. India. 27(2) : 6-13.

Saxena, A. 1999. Fishes and human population. Ecodevelopment and Environment (eds. S.P.Singh, R.Singh and R.Prakash) Bindra Publ. Jalgaon India.

Srivastava,Gopal Ji. 1968. Fishes of Eastern Uttar Pradesh. Vishwavidyalaya Prakashan, Varanasi, U.P. p. 163.

Srivastava,U.K. and Vathsala, S, 1984 : Strategy for development of inland fishery resources in India. IIM, Ahmedabad, p.630.

Singh, A.K. 1993. Preimpoundment studies on Sardar sarovar area of Narmada river (western zone) with special reference to fisheries. Ph.D. thesis, Vikram University, Ujjain, India.

Singh, D.F. and Yazdani,G.M. 1992. Studies on the ichthyofauna of Nasik district Maharashtra, India. Rec. Zool. Surv. India, 90(1-4) : 195-201.

Varghese, M.D.; George,V.C.; Khan,A.A. and Iyer, H.K. 1983. Shore Science for reservoirs, Part II. Studies on optimum mesh size. Fish Technol., 20 : 91-100.

CHAPTER 39

Alterations Induced by Lead Toxicity in Blood Glucose and Tissue Glycogen Level of Heteropneustes fossilis (Bloch)

A.N.Jha

Deptt. of Zoology C.M.Sc. College, Darbhanga, 846004
Lalit Narayan Mithila University, Darbhanga

ABSTRACT

Biochemical response of lead acetate $(CH_3COO)_2$ Pbx $3H_2O$)in fresh water Indian teleost *Heteropneustes fossilis* was studied for sub chronic (15 days) & Chronic (30 days) period. The blood glucose and tissue glycogen estimation were done in stress of sub lethal concentration (227.70 mg/L) of lead acetate and compared to the control group of fish. The blood glucose level for sub chronic period showed significant increase by 11% where as Chronic exposure showed glucose depletion by 40%. The hepatic and gonadal glycogen showed depletion in sub chronic as well as chronic period of exposure but the significant decline of hepatic and testis glycogen was noticed under chronic exposure as compared to ovary glycogen. Hyperglycemia is supposed to be due to glycogenolysis and gluconeogenesis where as hypoglycemia can be related to the utilization of reserve glycogen and impaired glycogenesis. The tissue glycogen reduction in present study is reflection of tissue acidosis and breakdown of glycogen in stress.

INTRODUCTION

The harmony of the nature has been continuously exploited by menkinds for their own comforts and conveniences. More and more nutritional requirements and infrastructure are required to fullfill the demands of growing population: These basic needs have led to the expansion of Agriculture, Industries and unpaused process of Urbnigation. Indeed, these activities have enhanced good food availibilities and better life amenities but only at the cost of degradation of the Environment and depletion of Natural rasources. This degradation of environment has threatened the biological world in the form of Pollution.

Since time immemorial the quality of natural water bodies is being polluted by different types pollutants viz. pesticides, heavy metals & others. In recent years heavy metal pollution has become subject of great interest in the field of aquatic toxicology. Toxic heavy metals are dispersed in the biosphere are carried one place to other by winds and finally washed out by rain on the land or the surface of water ways.

Number of works in heavy metal toxicology have been done (Forstner & Wittman 1979; et al 1984; Goyer 1986; Roy Choudhury et al 1986; Roy & Dubey 2001;). The alltration in biochemical profiles of fish on lead exposure have also been worked out by Jha & Pandey (1989) and Jha (1991), and Katti & Sathyanesan(1985). In recent years lead pollution attracted attention in view of its large scale industrial use and toxic potential.

A critical review of previous works reveals that most of the works are confined to bioassay and TLM determination (Vallee & Ulmer 1 972; Sexena & Parashari 1983, Jha & Pandey 1990). In the light of above facts the fish *Hecteropneustes fossilis* (Bloch), commonly called "Singhi" was selected for Biochemical study under sublethal concentration of lead acetate (227.70 mg/l) for the subchronic (15 days) period and Chronic (30 days)period.

MATERIAL & METHODS

The fish *Heteropneustis fossilis* (Bloch) were procured from local pond with the help of fisherman. They were fed and maintained as per suggestion of APHA, AWWA & WPCF (1985). They were fed daily *ad libitum* with goat liver. The feeding was stopped during the period of toxicity assesments but during the tenure of other expriments feeding schedule was strictly maintained in order to avoid starvational effect. The test Chemical used in this experiment was lead-acetate $[CH_3COO)_2$ Pb X $3H_2O$ BDH]. A static bio-assay test was done to determine LC_{50} of lead acetate to H - fossils following the methods of APHA, AWWA and WPCF (1385). Sublethal concentration was calculated by adopting the formula of Hart *et al* (1945).

For each experiment the fish (average length 18-20 cm. & wt. 32 -35 gm.) were exposed to a sublethal concentration of lead acetate (227.70 mg/L) for a period of 15 and 30 days. Side by side a control group was also run in equal volume of water (p_H - 7.8, Temp. 32.6° C hardness in mg./L as $CaCo_3$ - 184.32). The exposure medium was renewed every 24 hours. At the end of exposure period that is on day 15 and 30 the fish were anaesthetized with 1:4000 MS 222 (tricane Methane Sulphonate, Sandoz) and then processed for quantitative estimation. Blood samples were callected in heparanized vials with the help of 1 ml. disposable syringe equipped with 22 gauge hypodermic needle by puncturing the ventral aorta (Ruparelia *et al* 1980). Blood obtained from indiridual fish was centrifuged (3000 rpm for 5 minute) and plasma was callected in different vials and stored at - 20° C until analysed (Ruparelia *et al* 1989). The anaesthtized fish were dissected to take out liver and gonads. The hepatic tissue and gonads were quickly weight and processed for the quantitative estimation.

Blood glucose was estimated by o - toluidine reagent (Dubowski, 1962) at 630 nm in spcetrophotometer (spectronic 20). The tissue glycogen was estimated by the method of Carrol *et al* (1956) at 620 nm in spectrophotometer. 1974). In present study statical analysis and t test (Campbell - 1974) were done.

OBSERVATION

Blood Glucose

Both 15 and 30 days exposed fish exhibits interesting results in the blood glucose profiles. These are being mentioned below under separate subheadings.

15 Days Treatment Period

The fish of this exposure period reflects hyperglycaemic response. The blood glucose level elevates to 67.6 ± 1.01 mg/100 ml of blood against the control value of 60.90 ± 0.64 mg/100 ml of blood. This amounts to an increase by 11% and is significant at P<0.01 (Table. 39.1).

30 Days Treatment Period

Contrary to the observations of 15 days exposure period, the fish of this group become hypoglycaemic denoting exhaustion of glucose to meet high energy demand under the toxic stress of lead. The extent of glucose depletion is highly-significant (P < 0.001) and has been calculated to be nearly 40%. The blood glucose level in control fish is 62.71 ± 1.56 mg/100 ml and it lowers down to 37.80 ± 1.82 mg/100 ml of blood (Table. 39.1).

Glycogen

The alteration in the glycogen content of liver, testis and ovary has been shown in Table -39.2. In all these tissues there has been a significant depletion in the glycogen content in both the exposure period (15 and 30 days). The depletion has been recorded to be more pronounced in 30 days exposed fish. The liver reveals maximum depletion in both the exposure period amounting to 29.3,2 and 47.08% respectively.

15 Days Treatment Period

The glycogen content of liver, testis and ovary of the control fish group has been estimated as 34. 10 ± 1.4, 19.0 ± 1.02 and 15. 7 ± 0.35 mg/g wet tissue respectively. Whereas in the experimental group it depletes to 24.1 ± 1.32,14.08 ± 0. 88 and 12. 20 ± 0. 23 mg/g wet tissue in liver, testis and ovary respectivly. All these values significant depletion at P < 0.001.

30 Days Treatment Period

Glycogen profiles of the treated fish group have been estimated as 19.05 ± 0.82; 12.58 ± 0.46 and 10.45 ± 0.40 mg/g wet tissue in liver, testis and ovary as against their control values of 36.0 ± 0.41; 17.2 ± 0.77 and 15.6 ± 0.47 mg/g wet tissue respectively. The glycogen depletion has been found to be highly significant (P < 0. 001) in liver and testis as compared to that of ovary (P < 0. 01).

Table 39.1

Profiles of Blood glucose, Plasma protein and Serum cholestorel in normal and lead acetate treated fish, Heteropneustes fossilis (values are ± SE of five fish in each group).

Parameter	15 Days Control	15 Days treated	Student 't' Test P value	% Increase (+) or Decrease(-)	30 Days Control	30 Days Treated	Student 't' test P Value	%Increase (+) or Decrease(-)
Blood								
Glucose	60.90	67.6	<0.01	+11%	62.71	<0.001	-39.72%	-39.72%
Mg/100ml	±0.64	±1.01			±1.56	±1.82		

Table 39.2

Alteration in the glycogen content (mg/g wet tissue ± SE of five fish) in liver, testis and ovary of *Heteropneustes fossilis* following sublethal treatment (227.70 mg/l) of lead acetate.

Sl. No.	Tissue	Days	Control	Experimental	Student 't' test P Value	% Increase(+) or Decrese(-)
(1)	Liver	15	34.10±1.4	24.1+1.32	< 0.001	-29.32 %
		30	36.0±0.41	19.05±0.82	< 0.001	-47.08 %
(2)	Testis	15	19.0±1.02	14.08±0.88	< 0.001	-25.89 %
		30	17.7±0.77	12.58±0.45	< 0.001	-26.86 %
(3)	Ovary	15	15.7±0.35	12.20±0.23	< 0.001	-22.29 %
		30	15.6±0.47	10.45±0.40	< 0.01	-33.01 %

DISCUSSION

The present study reflects a very interesting result in blood glucose and tissue glycogen profiles. Tables - 39.1 and 39.2 reveal that there has been a significant increase in the blood glucose level accompanied by a highly significant decrease in hepatic and gonadal glycogen content in the lead treated fish sacrificed on day 15. However, 30 day exposed fish unlike their 15 day counterparts snowed hypoglycaemia and highly significant glycogen depletion amounting to 47.08%, 26.86% and 33.01% in liver, testis and ovary respectively. The highest depletion of glycogen has been recorded in the hepatic tissue in both the exposure periods. Liver is the principal site of glycogen storage and is referred to as the seat of metabolism, its maximal loss from this organ seems to be convincing under adverse situations. The significant increase in blood glucose level in 15 days trated fish points towards a hyper matabolic state of the fish in response to lead intoxication, thus glucose hunger of the fish is increased to meet the excessive demand of energy under metallic stress. At the same time hypoglycaemic condition met within 30 days treated fish lead us to think of the two possibilities, one being the increased tolerance of the fish to lead stress and the another being exhaustion of the reserve energy.

The present findings on the tissue glycogen and blood glucose resemble with those of Gill and Pant (1981); Larsson and Haux (1982); Jha and Pandey (1989b); Rajan (1990) and Jha (1992).

Various workers such as Black *et al.*, (1960); Heath and Pritchard (1965) follwing physical excersise and hypoxia and Silbergald (1974); McLeay and Brown (1979); Srivastava and Singh (1980), under the influence of various pollutants, have reported hepatic glycogenolysis with concomitant hyperglycemia in fish. Larsson and Haux (1982) noted cadmium induced significant increase in blood glucose level in rainbow trout and flounder. They corelated this with reduced insulin secretion and stimulated gluconeogenesis. This fact can further be supported by the contention that metals accumulate in fish pancreas and cause necrotic lesion in the insulin producing B-cell in the islet of Langerhans (Havu, 1969).

Shaffi (1978) noted decrease glycogen content in liver, muscle and brain of a number of fish and associated it with lowered capacity of the blood to carry oxygen to internal organs due to toxic stress.

Choudhary and Nath (1985) observed elevated blood glucose level in *Colisa fasciatus* treated with nickel. They described this to be due to decreased hepatic glycogen mediated by increased rate of catecholamines and due to depressed insulin secretion and hyperglucogonemia. Similarly, decreased glycogen value has been recorded in *Sarotherodon mossambicus* under chromium intoxication by Ghosh and Chatterjee (1985). Jha and Pandey (1989) noticed a significant decline in the total carbohydrates of intestine, liver gonads of lead intoxicated Channa punctatus. They have suggested utilization of the total carbohydrates to meet enhanced energy requirements of the cell due to toxic stress of lead nitrate. They further explain that this is acheived by rapid glycogenolysis and possible inhibition of glycogenesis. Similarly, Choudhary (1991) and Jha (1992) have recorded decreased glycogen level in liver, testis and ovary along with a significant elevation of blood glucose level in the chromium treated *Heteropneustes fossilis* and nickel exposed *Anabas testudineus* respectively. They correlate such changes with either enhanced glycogen breakdown or involvement of pancreatic and adrenal hormone. All these findings more or less corroborate the present findings.

Garg et al., (1989) observed hyperglycaemia in manganese treated Heteropneustes fossilis. Decreased glycogen content in liver and kidney of the manganese exposed *Ophiocephalus punctatus* has also been recorded (Garg and Garg, 1990), Rajan (1990) noted decreased total carbohydrates in muscle, liver and intestine of *Cyprinus corpio* exposed to textile mill effluent. The european eel, *Anguilla anguilla*, hyperglycaemia under penta chlorophenol exposure as low as 0.1 elevated blood glucose level persisted even after 55 days of recovery in (Holmberg,1972).

A few more explanations have been given by a number of workers to account for glycogen depletion and blood glucose elevation. Under the toxic stress of various pollutants in general and metals in particular which would make the prasent discussion more clear and therefore, needs to be discussed.

Sastry and Gupta (1978b and 1979c) in mercury and lead Ophiocephalus and lead and cadmium treated *Heteropneustes fossilis* noted a decreased activity of glucose -6- phosphatase in their digestive system. Similar results were Obtained by Chandra and Imam (1973) in gastrointestinal mucosa of guinea pig treated with manganese. Glycogen depletion as noted in present observation may also be correlated with inhibition of glucose-6- phosphatase activity by lead acetate. Particularly in case of liver it may be associated with the lost capacity of hepatocytes to mobilize glycogen normally as suggested by Orr *et al.*, (1948).

Jafri and Shaffi (1975) and Shaffi (1978) suggest tissue hypoxia and changes in the buffering system resulting in tissue acidosis and breakdown of glycogen the possible reasons for glycogen depletion.

It is assumed that lead acetate might be enhancing the activity of key gluconeogenic enzymes which may also be contributing to the hyperglycaemia in the present case. Another possibility may be that there might have occurred a reduction in calcium permeability at the level of B-cell membrane which would inhibit insulin release since calcium is a prerequisite for insulin secretion (Malaisse *et al.*,1971).

Higher blood glucose level recorded in the present study suggests that lead influences the glucose homeostasis. In addition, it also seems to affect ion balance as suggested by Larsson *et al.*, (1967).

Stressful stimuli elicit rapid secretion of both glucocorticoids (Wedemeyer, 1969; Grant and Mehrle 1973) and catecholamines (Nakano and Tomblinson, 1967) from the adrenal tissue of the fish. Both the hormones induce hyperglycaemia (Robertson, et al., Oguri and Nace, 1966 and Larsson, 1973).

Mazeaud *et al.*, (1977) explains that all types of stress in fish results in increase the circulating catecholamines and plasmatic corticosteroid referred to as the primary effects.

These endocrine changes induce secondary biochemical responses such as pronounced increase in blood sugar level.

Similar to the present observation Bhattacharya *et al.*, (1987) recorded a depletion in blood glucose level on day 30 of the murcury exposure in Channa punctatus. They have opined that hepatic glycogen is the source of hyperglycaemia as maximal glycogen depletion corresponded to a dramatic increase in blood glucose level. They further suggest that some of the hepatic glycogen was converted to glucose which enters the circulation. Again they put forth that glycogen depletion may be a manifestation of an initial regulatory step which increases intermediary metabolism resulting in protection of hepatocyte under xenobiotic insult.

The depletion of blood glucose as recorded on day 30 of exposure in the present study supports the contention that hexose is given to other tissue and glycogemolysis can follow biochemical pathway differing from those of glucose synthesis (Bhattacharya *et al.*, 1987).

The gluconeogenic activity might be mediated by cortisol (Bentley, 1980; Lidman *et al.*, 1970) and altered cortisol may sufficiently enhance gluconeogenesis. Besides, an increased influence of other hormone such as glucagon, in the situation of insulin defeciency can not be excluded as the reason for obviously higher glycogenolytic activity in lead acetate exposed *Heteropneustes fossilis*.

Thus. we see that diverse interpretations can be given for glycogen depletion and blood glucose elevation. No single reason can efficiently describe the changes noted during the present study. At last we can say that there exists a high catabolic potency during lead acetate exposure. This metal might have caused severe anaerobic stress resulting in breakdown of tissue glycogen.

Biochemical estimation denoted a hyperglycemic response in 15 days exposed fish whereas the 30 days treated fish became hypoglycemic. Hyperglycemia has been attributed to the glycogenolysis and gluconeogenesis whereas hypoglycemia has been associated with utilization of reserve glycogen and inhibition of glycogen synthesis. The glycogen profile of liver, testes and ovary showed highly significant decrease in both the exposure periods and has been associated with tissue acidosis and breakdown of glycogen.

REFERENCES

APHA, AWWA, AND WPCF (1985) : Standard method for the examination of water and waste water American Public Health Association and water Pollution control Federation. 16th Ed. American Public Health Association Washington, D. C.

Bentley, P. J. (1980) : Comparative Vertebrate Endocrinology, Cambridge, University press Cambridge. London, New York and Melbourne, P. 145.

Bhattacharya, T.; Roy, A. K. and Bhattacharya, S. (1987): Blood glucose and hepatic glycogen Interrelationship in *Channa Punctatus* Bloch): A parameter of non lethal toxicity Bioassay with industrial pollutants. Ind. J. Exp. Biol. 25: 539-541.

Black, E.C.; Robertson, A.C.; Hanslip, A. and Chiu, W.G. (1960): Alteration in glycogen, glucose and lactate in rainbow and karalcops trout (*salmo gairdneri*) following muscular activity. J. Fish. Res. Board. Can. 17: 487-500.

Campbell, R.C. (1974): Statistics for Biologists pp. 385. Cambridge University press, London.

Carrol, N. V. ; Longley, R. W. and Roe, J. H. (1956): The determination of glycogen in liver and muscle by use of anthrone reagent. J. Biol. Chem. 220 : 583-593.

Chandra, S. V. and Imam, Z. (1973) Manganese induced histochemical alteration in gastrointestinal mucosa of guinea pig. Acta Pharmacol. Toxicol. 33: 449-458.

Choudhary, H. S. and Nath, K. (1985) : Nickel induced hypergl glycaemia in the fresh water fish, Colisa fasciatus. Water, Air and Soil Pollution. 24: 173-176.

Dubowski, K M. (1962) : An O-toluidine method for body fluid glucose determination. Clin. Chem. 8: 215-235.

Forstner, U. and Wittmann G. T. W (1979) : Metal pollution in the Aquatic Environment, Springer- Verlag. Berlin. Heidelberg, New York, pp-486.

Garg, S. K. and Garg. V. K. (1990) : Metabolic changes in the snake head fish, *ophiocephalus punctatus* (Bloch) to experimental manganese poisoning. Environmental and Pollution and Health Hazard. Environmental Series. I I I 145-149.

Garg, V. K.; S. K. and Tyagi, S. K. (1989) : Manganese induced hematological and biochemical anomalies in Indian cat fish, Heteropneustes fossilis. J. Environ. Biol. (4) : 349.

Ghosh, T. K. and Chatterjee, S. K. (1985): Effect of chromium on tissue energy reverse in a freshwater fish, *arotherodon mossambicus*. Environ. Ecol. 3 (2) 178-179.

Gill, T. S. and Pant, J. C. (1981) : Toxicity of Nickel to fish *Puntius conchonius* (Ham) and its effect on blood glucose and liver glycogen. Comp. physiol. Ecol. 6(2) : 99-102.

Goyer, R. A. (1986) : Toxic effects of metals. In casarett and Doull's Toxicology : The Basic Science of poison, 3ed (rd Doull, J. ; Klassen, GD and Amdur mo) Macmillan, New York p-599.

Grant, B. F. and Mehrle, P.M. (1973) : Endrintoxicosis in rainbow trout (salmo gairdneri). J. Fish. Res. Board. Can, 30: 31-40

Hart, W. B. ; Doudoroff, P. and Greenbank, J. (1945) : The evalution of toxicity of industrial wastes, chemicals other substances of fresh water fishes. Allantic Refining Co. Phil. Part (I) : 317-326.

Havu, N. (1969) : Sulphydryl inhibitors and Pancreatic islet tissue. Acta. Endocrinol Suppl. 139 : 1-231.

Heath, A. G. and Pritchard, A. W. (1965) : Effects of severe hypoxia of carbohydrate energy stores and metabolism in two species of freshwater fish. Physiol. Zool. 38: 3254-334.

Holmberg, B. ; Jensen, S. larsson, A.; Lewander, K. and Olsson, M. (1972) : metabolic effects of technical Pentachlorophenol (PCP) on the eel, *Anguilla anguilla*. Comp. Biochem. Physiol. B 43 : 171-183

Jafri, A. K. and Shaffi, S.A. (1975): Effects of asphyxiation on the serum and kidney' alkaline phosphatase acitivity in *Clarias batrachus* (linn) Broteria XLIV, 59-62.

Jha, B. S. (1991) : Alteration in the protein and lipid contents of intestine, liver and gonads in the lead exposed fresh water murrel, *Channa punctatus* (Bloch) J. Ecobiol. 3 (1) : 29-34.

Jha, B. S. and pandey, S (1989) : Alteration in the total carbohydrate level of intestine, liver and gonadsinduced be lead nitrate in the fish, Channa punctatus. In : Environmental risk assessement (eds. Y. N. Sahai, P.B. Deshmukh, T. A. Mathai, K.S. Pillai.) The Academy of Environmental Biology, India. 207-211.

Jha, B.S. and Pandey, S. (1990) : Toxicity of lead nitrate to the fish *Channa punctatus* (Bloch). Biojournal, 2 (1): 151-154.

Jhn, M. M. (1992) : Studies on the effects of nickel chloride to some organs of the climbing perch, *Anabas testudineus* (Bloch) : Ph. D. Thesis. L. N. Mithila University, Darbhanga.

Katti, S. R. and sathyanesan, A. G. (1985) : Chronic effects of lead and cadmium on the testis of the catfish, *Clarias batrachus*. Environ. & Ecol. 3 (4) : 596-598.

Larsson, A. (1973) : Metabolic effects of epinephrine and norepinephrine in the eel, *Anquilla anquilla*. L. Gen. Comp. Endocrinol, 20: 155-167.

Larsson, A.; Bengtsson, B. E. and Haux, C. (1981) : Disturbed ion balance in flounder, Platichthys flesus L. exposed to Sublethal levels of cadmium. Aquatic.. Toxicol. 1 : 19-35.

Larsson, Ake and Haux, C. (1982): Altered carbohydrate metabolism in fish exposed to sublethal levels of cadmium. J. Environ. Biol. 3 (2), 71-78.

Lidman, U. ; Dave, Johansson- Sjobech, M. L. ; Larsson, A. and lewander, K. (1979) : Metabolic effects of cortisol in the European eel, *Anquilla anquilla*. L. Comp. Biochem. Physiol. 63A: 339-344.

Mahajan, C. L. and Juneja, C. J. and malaisse -legal, F. (1971) : Effects of sulfonylurease on calcium uptake and insulin secretion of islet of Langerhans. Arch. Int. Pharmacodyn. Therap. 192: 205-207.

Mazeaud, M.M. Mazeaud, F. and Donaldson, E. M. (1977) : Primary and Secondary effects of street in fish, some new data with a general review. Trans. Am. Fish. Soc. 106: 201-212.

Mcleay, D. J. and Brown, D.A. (1979) :L Stress and chronic effects of untreated and treated bleached kraft pulpmill effluent on the biochemistry and Stamina of Juvenile coho salmon (*Oncorhvnchus-kisutch*. J. Fish. Res. Board. Canad 36, 1049-1059.

Nakano T. and Torablinson N. (1967) : Catecholamines and carbohydrate concentrations in rainbow trout (*salmo gairdneri*) in relation to physical disturbance. J Fish Res. Board. can. 24: 1701-1715.

Oguri, M. and Nace, P. F. (1966) : Blood sugar and adrenal histology of the gold fish after treatment with mammalian adrenocorticotrophic hormone. Chesapeake Sci 7 : 198-202.

Orr. J. W. ; Price D. E. and Srickland L. W. (1948) : Glycogen Content in rat liver after poisoning with large doses of dimethylaminobenzene. J. Pathol. Bacteriol. 60: 573-581.

Ranjan, M.R. (1990) : Sublethal effects of textile mill effluent on protein, carbohydrate and lipid content o different tissues of the fish. *Cyprinus carpio*. Environ. Ecol. 8 (1) : 54-58.

Robertson, O. H. ; Wexler, B. C. and Rinfret, A. P. (1963) : The effects of hydrocortisone on immature rainbow trout (*Salmo qairdneri*). Gen. Comp. endocrimol. 3 : 422-436.

Roychowdhury, A, ; gautam, A. K. ; Rao, R. V. Sathwara, N. G. ; Parikh, D, J, Chatterjee B. B. (1986) : Changes in the adrenals in lead treated Rats. Bull. Environ. Contain. Toxicol. 37 : 62-69.

Roy, D. N. & Dubey, N. K. (2001) : Haematological response of trivalent chromium poisoning in Indian teleost catfish *Heteropneustes fossilis* (Bl) Indian. J. Environ & Ecoplan, 5 (3) : 607-616.

Ruparelia, S. G. ; Verma, y. ; Kashyap, S. K. and Chatterjee, B. B. (1986) : A new approach for the use of standard fish in toxicological study In : Environmental Biology : Coastal ecosystem, (eds. Dalela, R. E; Madhyastha, M. N. Joseph M. M.). The academy of Environmental Biology, Muzaffar Nagar, India, 89-92.

Ruparelia, S. G. Verma, y. ; Mehta, N. S. and Salyed, S. R. (1989): lead induced biochemical changes in fresh water fish, *Oreochromis mossambicus*. Bull. Environ. Contan. Toxicol. 43. 310-314.

Sastry, K. V. and Gupta, P. K. (1979b) : Mercuric chloride intoxication in digestive system of *Channa punctatus*. J. Toxicol. Environ Health 4 : 777-783.

Sastry, K. V. and Gupta, P. K. (1979c) : Enzymological alteration in the digestive system of Heteropneustes fossilis induced by lead nitrate. Toxicol letter 3: 145-150.

Saxena, O P and parashari, A (1983) : Comparative study of the toxicity of six heavy metals of *Channa punctatus* J. Environ. Biol. 4 (2) : 91-94.

Shaffi, S. A. (1978 a) : Changes in tissue glycogen content due to cadmium intoxication on three freshwater teleosts. Curr. Sci. 47 : 668-670.

Shaffi, S. A, (1978b) : Variations in tissue glycogen, serum lactate and glucose levels due to copper intoxication in three freshwater teleosts. Curr. Sci. 47 (24) : 954-956.

Srivastava, A. K. and Singh N. N. (1980) : Observation on hyperglycaemia in the murrel *(Channa punctatus)* after acute exposure to methylparathion. Comp. Physiol. Ecol. 5 : 100-101.

Vallee, B. L. and Ulmer, D. D. (1972) : Biochemical effects of mercury, cadmium and lead. Ann. Rev. Biochem. 41: 91-128.

Wedemeyer, G. (1969) : Stress-induced ascorbate depletion and cortisol production in to salmonoid fish. Comp. Bio. Chem. Physiol. 29, 1247-1251.

CHAPTER 40

EFFECT OF ACUTE EXPOSURE OF SODIUM ARSENITE (Na_3AsO_3) ON BLOOD GLUCOSE OF *CLARIAS BATRACHUS* (COMMON INDIAN CAT FISH) *IN VIVO*

Divya Bala Sahu, Rohit Kumar, Seema Tripathi, Anil Kumar *

** Asstt. Professor, Department of Zoology, Govt Science and Arts College, Durg (C.G)*

ABSTRACT

The effect of acute arsenic exposure in the form of Sodium arsenite (Na_3AsO_3) was studied on the blood glucose level of *Clarias batrachus*. Three subleathal concentrations of sodium arsenite (5 ppm, 10 pprn, 15ppm,) were exposed to experimental set of 10 fishes in a glass aquarium filled with 40L water. The estimated blood glucose after 120 hrs exposure for each experimental set were subjected to student "t" test. A mild decrease in blood glucose level was reported between cont. set to 5 ppm exposure set (t> 0.05 p) and cont. set to 10 ppm exposure set (t>0.10 p) only. This experiment elicited distrubed glucose metabolic pools probably due to breakdown of (GSH. increased GSSG and reduced ALAD) after arsenic intoxication.

Key Words: *Sodium arsenite (Na_3AsO_3) Blood glucose, Glutathione (GSH), Oxidised glulathione (GSSG), malondialdehyde (MDA), Delta amino levulinic acid hydratase (ALAD).*

INTRODUCTION

The adverse health effects of arsenic are well documented (Chappel et.al.,1994). symptoms of chronic arsenic toxicity (Characteristics of areas with water containing >100µg l-¹As) include peripheral nerve damage, tissue wastage (black-foot disease), dermal hyperpigmentation, kurtosis, skin carcinomas, and gastrointestinal cancers (Abernathy et.al. 1993). Enhanced incidences of these

Correspondence Address
Dr. Anil Kumar B-73, street -12 SMRITINAGAR, BHILAI DIST - Durg (C.G)

disorders, directly attributable to arsenic exposure, have been reported in many gold and base- metal producing countries including Chile (Borgono et.al;1980), Argentina (Astolfi et.al. 1981) and Mexico (Cebrian. et. al; 1983). Additional instances of chronic arsenism unrelated to mining have been reported from Taiwan (Tseng et.al, 1968., Chem et.al., 1988) and India (Das et.al,1994; Chatterjee et.al., 1995; Bhattacharya et.al. 1996, 1997; Chakraborty et.al,1987). The presence of arsenic in ground water has been reported in recent years from serveral parts of the world, like USA, China, Bangladesh and India (Robertson et.al, 1986., Forst : et. al.,1993., Das et.al.,1994., Chatterjee et.al.1995;)

In India arsenic contamination of water is a growing threat to the public health and dvelopment programme of west Bengal, Tripura and recently in Chhattisgarh. The presence of arsenic in ground water is generally associated with the geochemical environments such as basinfill deposits of alluvial lacustrine origin, Volcanic deposits, inputs from geothermal sources, and mining wastes. Uncontrolled anthro pogenic activities such as smelting of metal ores, use of arsenic pesticides and wood preservative agents may so release arsenic directly to the environments. An expanding exploitation of ground water and the change in landuse pattern is believed to have triggered the metabolisation of arsenic into the alluvial aquifers. Occurence of arsenic in ground water depends on the local geology, hydrology and geochemical characteristics of the acquifer materials. The geochemical characteristics of the aquifer play a crucial role in understanding the mobilisation of arsenic (Mattis Von Bromessen, 1999). The concentration of arsenic in ground water in West Bengal, Tripura and Chhattisgarh is in the average 22 times the safe limit (0.01mg / litre recommended by WHO) and at place as high as 370 times of that. The average of high arsenic concentration is 4.5 times of the limits of tolerance by human body 0.5μg / litre recommended by WHO) and at places goes up 75 times of the latter. Most experts agree that the source of such high arsenic anomaly in ground water is geological rather than pesticide (Roy A. Kumar, 1999).

Arsenic from water, sediments, soil and rocks can concentrate in organisms. In some cases it is harmless but in other cases it is harmful to health. Recently. Hering and Elimelech in 1995 have reviewed the international prospective and treatment strategies on the problem of arsenic contamination in ground water while Jain and Ali in 2000 have reported the occurrence, toxicity and speciation techniques for arsenic.

The available literature clearly indicates thai arsenic contamination may cause serious toxicity in human and animal.The present project has been undertaken to evalute various aspects of bio chemical alterations caused by arsenic intoxication. The work of present paper deals with alteration in blood glucose level in *Clarias batrrachus* alter arsenic.exposure.

MATERIAL AND METHOD

Experiment was carried out on a adult fresh water healthy common indian catfish, *clariasbatrachus (*length 12-16 cm, weight 80-120 gram, mixed sex and age group) procurred from local fresh water pound. The fishes were allowed to adjust laboratory condition in tape water for 7 days under a natural photoperiod and ambient temperture (28-32°C) in 40L glass aquarium filled with water (Pandey et. al., 2000). They were fed daily with pellets of carbohydrate and protein. The physico-chemical characteristics of water was maintained.

In order to study the impact of arsenic on blood sugar of *clarias batrachus* arsenic trioxide (A_2O_3) obtained from S.d fine - Chem. Ltd. Mumbai was used. A solution of A_2O_3 with NaOH was prepared before use and finally three subleathal concentration (5 ppm, 10 ppm, 15 ppm) of sodium arsenite (Na_3AsO_3) were exposed to experimental sets of 10 fishes for 5 days (120 hrs). Exposure medium was renewed every 24hrs to maintain the effective cencentration. Parallel groups of control fish were kept in 40L glass aquarium filled with tap water. At the termination of exposure period fish were picked out individually from the aquaria and sampled for blood gluose. Corresponding groups of control fish were also sampled. The blood samples were collected in glass vials with the help of 5ml disposable syreinge equipped with 22 gauge hypodermic needle by caudal peduncle cut and subjected to biochemical analysis using standard methods (Reitman and Frankel,1957). The result were subjected to student 't' test (Snedecor and Cochram, 1967) for significance.

RESULT AND DISCUSSION

The effect of sublethal concentrations (5 ppm,10 ppm, 15 ppm) of sodium arsenite (Na_2AsO_3) on blood glucose of *clarias batrachus* following acute exposure of 5 days (120hrs) are summarised in Table - 40.1 with "t" value Test of significance reveals that actue exposure of three subleathal concentrations of sodium arsenite has resulted mild decrease in blood glucose level. Decrease in blood glucose level between Control to 5 ppm exposure and control to l0ppm exposure was found significant only at 0.50 p level and 0.10 p level.

Some related works has been performed in different parts of the world.

Sodium arsenite exposure of 100 ppm was orally given to male rat for 12 weeks and was found to deplete glutathione (GSH) levels, increase oxidised glutathione (GSSG) and promote malondiaklehyde (MDA) production in both liver and brain samples. In addition to a significant reduction in RBC delta -aminolevulinic acid dehydratase (ALAD) activity and GSH levels, a marked elevation in MDA production may also contribute to arsenic - induced oxidative stress. (Flora-S-J, 1999).

Multiple dose exposure of gallium arsenide(GaAs) produced a significant inhibition of blood delta - aminolevulinic acid dehydratase (ALAD) activity, an increase in urinary ALA excretion and blood zinc protoporphyrin level. Blood glutathione (GSH) contents also decreased on GaAs exposure. (Flore, S-J, et.al., 1996)

Breakdown of GSH and increased GSSG level might be one of the reasion for increased level of reduced insuline. In the specific case of insuline which has three -s-s- linkage, an enzyme called glutathion insuline transhydrogenase catalyse formation of -s-s- bonds of insuline with oxidised glutathione (GSSG) serving as acceptor

$$6GSH + \text{Oxidised insuline} \rightleftharpoons 3\,GSSG + \text{reduced insuline.}$$

Above reaction might be an important cause for fall in blood glucose level under sodium arsenite intoxication of *Clarias batrachus.*

Significant reduction in ALAD activity may adversely affect Shemin rection i.e formation of porphobilinogen a precursors of protoporphyrin from Succinyl Co A. Inhibition of Shemine reaction may leads to accumulation of succinyl CoA and its accumulation may initiate several catabolic pool of sugar oxidation, viz.—

(i) activation of enzymatic pool of kreb cycle for fast oxidation of sugar,

(ii) activation of specific enzymatic mechanism of anaplerotic reactions by which tricarboxylic acid cycle intermediates can be replenished (Kornberg, H.L., 1966).

(iii) activation of several important auxiliary enzymatic reactions may remove specific intermediate of Krebcycle such as - Ketoglutarate, succinate and oxaloacetate from the cycle to serve as precursors of amino acids.

Above all catabolic pools of sugar migh be causative factors for fall in blood sugar level after intoxication of sodium arsenite in *Clarias batrachus.*

TABLE - 40.1

Showing effect of Sodium Arsenite (Na_3 AsO_3) on blood glucose of *Clarias batrachus*.

S.No.	Dose of Exposure	Duration of Exposure	Estimated Blood Glucose (Mean Value)	Sigficance level ("t" Test)
1.	Control		1222.7	t>0.50p level t<0.50p level
2.	5 ppm.	120 hrs.	108.6	
3.	10 ppm	120 hrs	102.8	t>0.10p level t>0.05p level
4.	15 ppm	120 hrs.	116.5	t<0.50p level

REFERENCES

Abernathy, C.O., Draft Drinking Water Criteria Document on Arsenic, U.S. EPA Science Advisory Board Report, Contract 68 - C8 - 0033, 1993.

Borgono, J.M., Venturino, H. and Vincent, P., Clinical and epidemiological study of arsenism in norther Chile, Rev. Med. Chile, 08, 1039 -1048, 1980.

Astolfi, E.A.N., Maccagno, A., Fernandez, J.C., Vaccara, R. and Stimola, R., Relation between arsenic in drinking water and skin cancer, Biolog. Trace Element Res., 3, 133, - 143, 1981.

Cebrain. M.E., Albores, A., Aquilar, M. and Blakely, E. Chronic arsenic poisoning in the north of Mexico, Hum. Toxicol., 2, 121 - 133. 1983.

Tseng,W.P., M.H.. Fong. J.M.. Lin. C.S. and Yeh. S.. Prevalence of skin cancer in an endemic area of chronic arsenicism in Taiwan, J. Natl. Cancer Inst.. 40 (3). 453- 463, 1968.

Chem. C.J.. Kuo, T.L.and Wu, MM. Arsenic and cancers. Lancet, 1(8575/6), 414 - 415; 1988.

Das, D., Chatterjee, A., Samanta, G., Mandal, B.K.. Chowdhary, T.R., Chowdhary, P.P., Chanda, C., Basu, G., Lodh, D., Nandi, S., Chakroborty,T., Mandal, S., Bhattacharya, S.M, and Chakroborty. D., Arsenic in ground water in six districts of West Bengal, India : The biggest arsenic calamity in the World.Analyst, 119, 168 N - 1970 N. 1994.

Chatterjee. A; Das, D; Mandal, B.K; Chowdhary. T.R; Samanta, G. and Chakraborty, D; Arsenic in ground water in six districts of West Bengal, india : The biggest arsenic calamity in the world. Part -1. Arsenic species in drinking water and urine of the affected people, Analyst, 120, 643 - 650, 1995.

Bhattacharya, P; Chatterjee, D. and Jacks, G; Safeguarding ground water from arseniferous aquifers.Proc. 22nd WEDC Conference on Reaching the Unreached :Challanges for the 21st Century, New Delhi, India, PP 220- 223,1996.

Bhattacharya, P; Chatterjee, D. and Jacks, G; Occurrence of arsenic contaminated groundwater in alluvial aquifers from Delta Plains, Eastern india : Options for safe drinking water supply, Water Resour. Dev. 13(1),79-92, 1997.

Chakraborty, A.K; Banerjee, D; Ghoshal, S and Barman, P; Arsenical dermatitis from tubwell water in West Bengal, India J, Med. Res; 85, 326 - 334. 1987.

Robertson, F.N; Arsenic in ground water under oxidizing conditions, South - west United state, Environmental Geochemistry and Health, 11, 171- 185, 1989.

Forst, F; Frank, D; Pierson, K; Woodruff, L; Raasina, B; Davis, R and Davies, J; A seasonal Stydy of arsenic in groundwater, Snohomish Country, Washington, USA, Environmental Geochemistry and Health, 15,209-214, 1993

Bromssen Von Mattias; Genesis of high arsenic groundwater in the Bengal Delta Plains, West.Bengal and Bengladesh; Royal Institute of Technology, Stockholm, Sweden, Thesis Report Series 1999 : 18.

Roy, A. Kumar; Chemistry of arsenic and arsenic minerals relvent to contamination of groundwater and Soil from Subterranean source. Everyman's science, Vol XXXIV No. -1, P 15-21, 1999

Hering, J.G. and Elimelech, M; International prospective on arsenic in groundwater: Problems and Treatment strategies, Proc. American Water works Association, Annual Conference, June 18 -22, 1995.

Jain, C.K and Ali, I; Arsenic : Occurrence, toxicity and speciation techniques, Water Res; 34(17), 4304-4312,2000.

Pandey, A.K; Pandey, G.C; Neraliya, S; Evaluation of the toxicity of Thiramand Ziram fungicides to freshwater fish, *Heteropneustis fossilis*. Proc. of 1st. Annual Conf. of ISVPT, pp. 62 - 66.

Reitman, S. and Frankel, D. A Calorimetric method for the determination of Serum oxaloacetic and glutamic pyruvate transaminase. An. J. clin Pathol 28 : 53 - 56.

Snedecor, G and Cochran. W.S.1967. Statistical methods Oxford and IBH. Company, New Delhi: pp 593.

Flora, S- J. Arsenic induced oxidative stress and its reversibility following combined administration of N - acetylcysteine and meso 2,3 - dimercaptosuccinic acid in rats. Clin - Exp - Pharmacol - Physiol. 26 (11)865-9, Nov 1999.

Flora - S -J -S; Kumar - P; Biochemical and immunotoxicological alterations following repeated gallium arsenide exposure and their recoveries by meso-2-3-dimercaptosuccinic acid and -2, 3-dimercaptopropane 1- Sulfonate administration in rat. Environmental Toxicology and pharmacology 2 (4) : 315 -320. 1996.

Biswas - U (a); Sarkar - S; Bhowmik - M - K; Roy - S; Clinicopathological Profile of induced Chronic arsenic toxicity in goats. Indian journal of Animal Science, April, 1998; 68 (4) 320 - 323.

Kornberg, H.L., "Anaplerotic Sequence and their role in Metabolism" in P.N. Campbell and G.D. Greville (eds), Essays in Biochemistry, Vol -2, Academic Press inc; New York, 1966. The amphibolic nature of the Krebs cycle and the replenishment of cycle intermediates.

CHAPTER 41

EFFECT OF TEMPERATURE ON THE VIRULENCY OF *FLAVOBACTERIUM* SPP. ISOLATED FORM INDIAN CATFISH (*CLARIAS BATRACHUS*) OF HIMALAYAN AND SUB-HIMALAYAN REGIONS

Prasad*, Yogendra and Verma, Vinay

**Aquatic Biotechnology and Fish Pathology Lab.*
Department of Animal Science, M.J.P. Rohilkhand University,
BAREILLY- 243006

ABSTRACT

During one year period (Oct.-2000 to Sept.-2001) about 2100 live catfishes were examined at three sampling stations, namely: Nanak sagar, Sharda reservoir and Nakatia river (Sub-Himalayan region) and Nainital (Himalayan region). First two sampling stations Nanak sagar and Sharda reservoir are fed with Sharda river, originated from Himalayan region. Fish were sampled with the help of fishermen by using nets and hooks. Only alive fish were examined for being symptomatic. They were placed in plastic bags filled with hand pipe water and transported to the laboratory. Out of 2100 fish examined only 110 fish were showing symptoms like the presence of shallow ulcerations on the general surface of the body and rottening at the base of the fins. Inocula were taken from the ulcered, rotten regions and the different organs like liver, kidney etc. of the symptomatic fish and inoculated on to Cytophaga agar. Twenty isolates ($F1$-F_{20}) *of Flavobacterium* were isolated from symptomatic fish and pure cultures of the strains were obtained onto the same medium. Virulence nature of the isolates was determined by direct challenge. Experimentally infected fishes showed the symptoms of haemorrhages at the site of injection which developed in to a shallow ulcer after 21 to 30 days of incubation. Reisolated strains from artificially infected fish were similar in their morphological and biochemical characteristics. Impact of temperature on the virulency of *Flavobacterium* is presented in this paper.

INTRODUCTION

Bacterial fish pathogens are the major constraints of aquaculture. They adversely affect the fish production and cause great economic loss. Among bacterial fish pathogens *Flavobacterium* is important pathogen because it causes various diseases in freshwater (Farkas and Olah, 1986 and Holt *et al.*, 1993) and marine (Wakabayashi *et al.* 1984 and Bernardet *et al.* 1990) fishes. The eutrophic water bodies provide a natural habitat for the growth and proliferation of bacteria due to the availability of organic and inorganic substances in that water bodies. Different strains of *Flavobacterium* viz. *F. columnare, F. branchiophilum* and *F. psychrophilum* are ubiquious in aquatic ecosystem and cause columnaris disease in *Clarias batrachus* (Prasad & Qureshi, 1995); gill disease in *Oncorhynchus mykiss* (Decostere *et al.*, 1999) and cold water disease in *O. mykiss* (Madsen & Dalsgaard, 1999), respectively. The peduncle disease in fish is also reported to be caused by *F. psychrophilum* (Van Duijn, (1973); Kimura *et al.* (1978); Wakabayashi, (1989); Ekman *et al.*, (1999) and Wiklund *et al.*, (2000). In fact *Flavobacterium* is known by various synonames i.e. *Cytophaga, Flexibacter* but the present name is given by Bernardet *et al.* (1996) which has been finally accepted. Unfortunately they have not been placed in the approved list of bacterial names till today.

In columnaris disease fish infected regions get highly necrotised and small flax of the tissue are observed dangling in the water. Occurrence of columnaris disease has been reported from both tropical and temperate regions and stressors such as water temperature, pH, Alkalinity and stocking density play a significant role in the incidences of this disease (Wakabayashi, 1991). However, this disease is mostly studied in aquarium fish "cotton wool" disease particularly in black and red mollies *(Poecilia sphenops)* and guppies *(Poecilia reticulata)* (Decostere and Haesebrouck, 1999). The main significance of this paper is to evaluate the impact of water temperature on the virulency of *Flavobacterium* spp.

MATERIALS AND METHODS

2100 fishes have been sampled from different sampling stations, namely: Nanak sagar, Sarda dam, Nakatia river (Sub-Himalayan region) and Nainital (Himalayan region). Out of which only 110 fishes were found to be symptomatic. Live and moribund fish were used in this study to obtain the bacterial inocula. Moribund fishes were sacrificed at the sampling sites and their organs like liver, kidney, heart etc. were collected in a saline solution (0.3% of NaCl) where as alive symptomatic fishes were placed in plastic bags/bucket filled with hand pipe/bore well water and were carried to the laboratory where they were kept in 500L plastic pools ("3 x 2.5") filled with unchlorinated static water. Few symptomatic fishes were sacrificed and their organs were collected in saline solution. Sampled organs were homogenized in 0.3% NaCl solution and homogenate was centrifuged at 6000 rpm for 15 minutes and supernatant was used for the isolation of bacteria on Cytophaga agar / broth (Anacker & Ordal, 1959) containing 0.05% tryptone, 0.05% yeast extract, 0.02% sodium acetate, 0.02% beef extract and 9% of Agar / liter. The pH was adjusted to 7.2 - 7.4. All the inoculated plates were incubated at 37±1°C for 24 hrs.

All the isolated strains of bacteria have been identified by following the different biochemical tests such as Gram staining, shape, presence of flexirubin type pigmentation, aerobic growth, oxidase reaction, indole production, catalase production, gelatin degradation and H_2S production, etc.

LD_{50}

Isolated bacteria were enumerated by using spread plate count (SPC) and Pour plate count methods. For this purpose 24 hrs. old colonies were used and they have been quantified by plating them on Cytophaga agar. LD_{50} was determined by inoculating 0.2 ml suspension containing inocula of $2x10^3$, $2x10^4$ and $2x10^5$ cfu/ml and the temperatures were maintained at 15 to 35°C (15, 20, 25, 30, 35°C) by using thermostat. Mortalities were recorded after 24 hrs of incubation for about 30 days. They wercr confirmed by reisolating the bacterium from the sample of liver/kidney collected from freshly dead experimental fish.

EXPERIMENTAL INFECTION TRIAL

10 *C. batrachus* per concentration were injected intramuscularly with *Flavobacterium* spp. @$2x10^3$, $2x10^4$ and $2x10^5$ cfu/ml and were kept in 500L plastic pools filled with clean bore well static water. The water temperatures were maintained between 15 to 35°C for 30 days.

RESULTS

During one year of survey about 2100 fishes have been screened from different water bodies / sampling stations. The symptomatic fishes were mostly recorded during summer months when the water temperature was ranging between 15 - 25°C in Himalayan region and 25 - 35°C in Sub-Himalayan region (Tables - 41.1a & 41.1b). The symptomatic fishes showed the clinical signs of ulceration, rottening, erosin and necrotization of the skin on the darso-latral side of the body.

Table- 41.1a
Showing number of *Flavobacterium* isolates in catfish sampled from different water bodies of Himalayan region.

Case No.	Year	Water bodies	Fish	Water temp. °C	No. of strains isolated
1.	2000-2001	Nainital (Uttranchal) Sampling station (A)	Catfish (*Clarias batrachus*)	15-20	2
2.	2000-2001	Nainital (Uttranchal) Sampling station (B)	-do-	15-20	3

Table- 41.1 b

Showing number of *Flavobacterium* isolates in catfish sampled from different water bodies of Sub-Himalayan region.

Case No.	Year	Water bodies	Fish	Water temp. °C	No. of strains isolated
1.	2000-2001	Nakatia river Bareilly, U.P.	Catfish (*Calrias batrachus*)	20-30	2
2.	2000-2001	Sarda dam UdhamSingh Nagar (Uttranchal)	-do-	20-30	5
3.	2000-2001	Nanaksagar UdhamSingh Nagar (Uttranchal)	-do-	20-30	8

Bacteriological investigations revealed the presence of long, slender, Gram negative rods, Oxidase positive and showed the gliding movement on Cytophaga medium. More or less similar colonies have also been isolated from different organs like liver, kidney, heart, spleen and other body organs of the symptomatic fishes. Some biochemical characteristics of the strains isolated have been presented in Table-41.2. All isolates were incubated at temperature of 37°C and colonies appeared were found to be flat and yellow - green (Flexirubin type pigmentation) within 24 hrs. of incubation. LD_{50} tests showed that 0.2 ml suspension containing 2x10, 2x10 and 2x10 cfu/ml were significant because that dilutions gave 60-80% mortalities at 25, 30 and 35°C, respectively.

Table - 41.2

Some Bio-chemical characteristics *of Flavobacterium* spp. isolated from *Clarias batrachus*.

S. No.	Characters	Bacteria (*Flavobacterium* = F)			
		F_{1-5}	F_{6-10}	F_{11-15}	F_{16-20}
1.	Gram reaction/shape	-/R	-/R	-/R	-/R
2.	Motility	+	+	+	+
3.	Colour of colony	Y-O	O	Y-O	Y-G
4.	Growth at different temperature	37°C	25°C	35°C	37°C
5.	Growth in different salinity				
	0.5%	+	+	+	+
	1.0%	-	-	-	-
6.	Oxidative(O)/Fermentative(F)	O	O	O	F
7.	Oxidase reaction	+	+	+	+
8.	Indol production	-	-	-	-
9.	H_2S production	+	+	+	+
10.	Catalase production	+	+	+	+
11.	Orinithin decarboxylase	-	-	-	-
12.	Phosphatase	+	+	+	+
13.	Gelatin degradation	+	+	+	+

Experimental infection trials reveled that challenged fishes exhibited the symptoms of haemorrhages and necrotization at the site of injection. In case of fishes inoculated with 0.2 ml suspention of *Flavobacterium* spp. containing $2x10^3$, $2x10^4$ and $2x10^5$ cfu/ml and were kept in 500 L plastic pools at the water temperature 15, 20 and 25°C exhibited the mild symptoms of the disease on the onset of 30, 26, 23 and 28, 25, 21 and 27, 24 and 20 days, respectively (Table-41.3a). Similarly in case fishes inoculated with same doses of *Flavobacterium* spp. and kept at water temperature 25,30 and 35°C exhibited the symptoms of the disease on the onset of 26, 23, 20 and 24, 22, 19 and 23, 20 and 18 days respectively (Table-41.3b).

Table- 41.3a:
Showing experimental infection trials at the different temperatures and inocula of *Flavobacterium* in *C. batrachus.*

Viable cell/ml.		$2x10^3$			$2x10^4$			$2x10^5$	
Temp (°C)	15	20	25	15	20	25	15	20	25
No. of fish used	10	10	10	10	10	10	10	10	10
Mean wt. (gm.)	90	90	90	90	90	90	90	90	90
Dose (ml.)	0.2	0.2	0.2	0.2	0.2	0.2	0.2	0.2	0.2
No. of days	30	28	27	26	25	24	23	21	20
Lesions		Haemorrhage			Haemorrhage			Haemorrhage	
Mortality		10/0			10/1			10/2	
Isolation of Bacteria		10/0			10/0			10/0	

Table- 41.3b:
Showing experimental infection trials at the different temperatures and inocula of *Flavobacterium* in *C. batrachus.*

Viable cell/ml.		$2x10^3$			$2x10^4$			$2x10^5$	
Temp (°C)	25	30	35	25	30	35	25	30	35
No. of fish used	10	10	10	10	10	10	10	10	10
Mean wt. (gm.)	90	90	90	90	90	90	90	90	90
Dose (ml.)	0.2	0.2	0.2	0.2	0.2	0.2	0.2	0.2	0.2
No. of days	26	24	23	23	22	20	20	19	18
Lesions		Haemorrhage			Haemorrhage			Haemorrhage	
Mortality		10/2			10/6			10/8	
Isolation of Bacteria		10/2+			10/6+			10/8+	

DISCUSSION

Occurrence of columaris disease in *Clarias batrachns* is recorded in 5.65% of fishes examined. It suggests that incidences of this disease is significant in the Nanak sagar and Sarda dam. It is interestig that though the 5.65% of fishes were showing the symptoms of columnaris disease but *Flavobacterium* spp. were found persistantly only in 0.95% of the fish sampled. It is evident from *in vitro* culture that the isolates of this bacterium prefer 35°C+ for proper growth. Simultaneously, occurrence of this disease was recorded mostly during the summer months when the water temperatures were above 25°C. It forces us to predict that catfishes in the Nanak sagar and the Sarda reservoir are prone to *Flavobacterium* infection above 25°C. Which coincide with the view of Holt (1975), Decostere and Haesebrouchk (1999), and Prasad and Qureshi (1995).

Persistant occurrence of the strains were recorded from the different organs as well as necrotized region of infected fish. It suggests that this bacterium is not only infecting the external body surface or gills but it also gets into the internal organs which is contradicting with the findings of Snieszko (1981) who reported that this bacterium infects the skin and gills while other organs remain uninfected. The symptoms such as necrotization, haemorrhages and erosion of skin, exhibited by the symptomatic and artificially infected fish are similar to those described by Bullock *et al.* (1971), Ekman *et al.* (1999). Moreover, the columnaris disease is mostly studied in salmo trout (Holt *et al*, 1989), Eel (Chen *et al.*, 1982) and Carps (Bootsma and Clerx, 1976) and minimum attention has been paid on the Indian catfish.

The bacterium is long, slender, Gr-ve rods, oxidase +ve and shows gliding movement on Cytophaga agar. Appearance yellow green colonies indicate the presence of Flexirubin type pigmentation that is the characteristic feature *of Flavobacterium.* Similar view has been expressed by Bernardet *et al.*, (1996). Experimental infection trials suggest that this organism is highly pathogenic in nature. Temperatures play a significant role on the pathogenesis of this bacterium because when the same doses (0.2ml) of $2x10^3$, $2x10^4$ and $2x10^5$ concentration were enoculated in the same length and weight groups of fishes and were kept at different temperature they showed clinical sign in minimum dose on the onset of 18 days at 35° and they exhibited the clinical sign on the onset of 30 days when kept at 15°C. This finding is similar to Morrison *et al.* (1981) who reported that outbreaks of columnaris disease in atlantic salmon is being more sever at higher end of the range. Holt *et al.*, (1975) found that mortality in experimental *F. columnaris* infection increases with increasing temperature. On contrary *F. psychrophilum* causes severe outbreaks of bacterial gill disease (rottening and fusion of gill filaments) when water temperature is below the 10°C (Hilger *et al.*, 1991 and Holt *et al.* 1993).

Result also indicates that 25-35°C more appropriate for the growth of *Flovaobacterium.* Same results have been given by Soltani and Burke (1994) and Prasad and Qureshi (2000) who reported the optimum growth of this micro organism is ranging between 20-30°C. It is evident from the challenge tests that appearance of symptoms of coloumnaris disease was more or less similar to that of natural infection. Thus it can be conclused that *F. columare* grows more rapidly at temperature ranging between 25-35°C.

ACKNOWLEDGMENTS

The authors are thankful to Indian Council of Agricultural Research, New Delhi for granting research project. Thanks are also due to Dr. T. A. Qureshi for his constructive criticism and healthy suggestion rendered in the preparation of this manuscript.

REFERENCES

Anacker, R. L.and Ordal, E. J. (1959). Studies on the myxobacterium *Chondrococcus columnaris.* 1. Serological typing. J. Bacteriol 78:25-32.

Bernadet, J.F.; Campbell, A.C. and Buswell J.A. (1990). *Flexibacter maritimus* is the agent of black patch necrosis' in Dover sole in Scotland. Dis. Aquat. Org. 8 : 233-237.

Bernardet, J. F.; Segers, P.; Vancanneyet, M; Berthe, M.; Kersters, K. and Vandamme, P. (1996). Cutting a Gordian Knot. Emended classification and description of the genus *Flavobacterium*. Emended description of the family *Flavobacteriaceae* and proposal of *Flavobacterium hydatis* nom. nov. (basonym *Cytophaga aquatilis* strohl and Taist 1978). Int. J. Syst. Bacteriol. 46:128-148.

Bootsma, R., Clerx J. P.M. (1976). Columnaris disease of cultural carp. *Cyprinus carpio* L. characteristics of the causative agent. Aqua. 7,371-384.

Bullock, G. L.; Conroy, D.A. and Snieszko, S.F. (1971). Bacterial diseases of fishes. In: Disease of Fishes, Book 2A(Ed), Snieszko, S.F. and Axelrod H.R., Neptune, T.F.H. Publishers, N.J. p 151.

Chen, C.R.L.; Chung, Y.Y.; Kuo, G.H. (1982). Studies on the pathogenicity of *Flexibacter columnaris*. 1. Effect of dissolved oxygen and ammonia of the pathogenicity of *Flexibacter columnaris* to eel *(Anguilla japanica.)* CAPD Fisheries series No. 8, Reports on fish Dis. Res. 4 : 57-67.

Decostere, A. and Haesebrouck, F. (1999). Outbreak of colmnnaris disease in tropical aquarium. Vet record 144:23-24.

Decostere, A.; Haesebrouck, F.; Charlier, G. and Ducatelle, R. (1999). The association of *Flavobacterium columnare* strains of high and low virulence with gill tissue of black mollies *(Poecilia sphenops)*. Vet. Microbiol 67:287-298.

Ekman, E.; Borjeson, H. and Johansson, N. (1999). *Flavobacterium psychrophilum* in Baltic salmo of salmo brood fish and their offspring. Dis. Aquat. Org. 37 (3) : 159-163.

Farks J. and Olh J. (1986) Gill necrosis - a complex disease of carp. Aquaculture 58 :17-26.

Hilger, I.; Ullrich, S. and Anders, K. (1991). A few ulcerative flexibacteriosis like disease (Yellow pest) affecting young Atlantic cod *Gadus morhua* from the German Wadden Sea. Dis. Aquat. Org. 11 : 19-29.

Holt, R.A.; Amandi, A.; Rohovec, J.S. and Fryer, J.L. (1989). Relation of water temperature to bacterial cold water disease in Coho salmon, Chinook salmon and rainbow trout. J. Aquat. Anim. Health. 1 : 94-101.

Holt, R. A.; Rohovec, J. S. and Fryer, J. L. (1993). Bacterial cold water disease. In. V. Inglis R. J. Roberts and N. R. Bromage (Editors). Bacterial disease of Fish. Blackwell Scientific Publications, Oxford pp 3-22.

Holt, R.A.; Sanders, J.E.; Zinn, J.L.; Freyer, J.L. and Pilcher, K.S. (1975). Relation of water temperature to *Flexibactor columnaris* infected in steel head trout *(Salmo gairneri)* Coho *(Oncorhynchus kisutch)* and Chinook *(O. teshawytscha)* Salmon. J. Fish. Res. Board Can. 32:1553-1559.

Kimura, T.; Wakabayashi, H. and Kudo, S. (1978a). Studies on bacterial gill disease in salmonids. 1. Selection on bacterium transmitting gill disease. Fish pathol. 12 : 233-242.

Madsen, L. and Dalsgard, I. (1999). Reproducible method for experimental infection with *Flovobacterium psychrophilum* in rainbow trout *(Oncorhynchus mykiss)* Dis. Aquat. Org. 36(3) : 169-176.

Morrison, C.; Cornick, J.; Shum G. and Zwicker, B. (1981). Microbiology and histopathology of Saddleback disease of underyearling Atlantic salmon *Salmo solar* L. J. Fish Dis. 4 :243-258.

Prasad, Y. and Qureshi, T. A. (1995). Incidences of columnaris disease in *Glorias batrachus*. Aquaculture for 2000 A.D. : 275-282.

Snieszko, S.F. (1981). Bacterial gill disease of freshwater fishes. Fish Disease leaflet No. 62, Fish and Wildlife service, Washington, DC, 11pp.

Soltani, M. and Burke, C.M. (1994). Responses of fish-pathogenic *Cytophaga/Flexibacter* like bacteria (CFLB) to environmental conditions. Bull. Eur. Assco. Fish Pathol. 4:185-187.

Van Duijn, C. and C. Jr. (1973). Disease of fish (e.d.), Thomas, C. C. springfield gllinois, 372 pp.

Wakabayashi, H. (1991). Effect of environmental conditions on the infectivity of *Flexibacter columnaris* to fish J. Fish Diseases. 14:279-290.

Wakabayashi, H.; Egusa S. and Fryer, J. L. (1980). Characteristics of filamentous bacteria isolated from a gill disease of salmonids. Can. J. Fish. Aquat. Sci. 37 : 1499-1505.

Wakabayashi, H.; Huh, G. J. and Kimura, N. (1989). *Flavobacterium branchiophilia* sp. nov., a causative agent of bacterial gill disease of fresh water fishes. International J. Sys. Bacteriol 39 : 213-216.

Wiklund, T.; Madsen, L.; Bruun, M.S. and Dalsgaard, T. (2000). Detection *of Flavobacterium branchiophilia* from fish tissue and water samples by PCR amplification. J. App. Microbiol. 88(2) : 299-307.

CHAPTER 42

Evaluation of Biocleaning Efficiency of Few Cyanobacterial Strains in Fish Farm Effluents

S. Chidambaram Pillai

Lecturer (S.G) in Botany,
P.G. Department of Botany, V.O.Chidambaram College,
Tuticorin - 628 008.

ABSTRACT

Fresh water cyanobacterial strains *of Anabaena variabilis, Phormidiwn valderianum* and *Nostoc calcicola* were screened in fish farm effluent in laboratory bioassays to compare their bioclea-ning efficiency. A.variabilis and P.valderianum responded well compared with *Nostoc calcicola* and the results were interpreted and probable reasons were discussed.

Key words: *Fish farm effluent treatment, Biocleaning, Cyanobacterial strains, Growth parameters.*

INTRODUCTION

Rejection of used waters in the natural water bodies raises increasing concern on water pollution. Unfortunately aquaculture has assumed the dimension from farming to industry which generate and discharge assimilable pollutant waste rich in nitrates, phosphates and ammonia which magnify the eutrophication problems. As microalgae especially cyanobacteria have been considered ideal waste removers, an attempt has been made to use few strains to compare their biocleaning efficiency of fish farm effluents. The effect of this effluent on the growth of these algae has also been undertaken.

MATERIALS AND METHODS

Freshwater fish farm effluent was collected from the Manju fish farm, Karambai, Kallidaikurichi Tirunelveli District, Tamimadu. Cyanobacterial strains of *Anabaena variabilis, Phormidium valderianum* and *Nostoc calcicola* were used from the culture collections of Department of Botany,

V.O. Chidambaram College, Tuticorin and have been maintained in the BG11 medium under continuous illumination of light at 1400lux and 27±2°C. The effluent was filtered through cotton to remove unwanted debris etc and used as such without dilution.

Experiments were conducted in duplicates. To study the efficiency of these cyanobacteria in biocleaning the effluent, the treatments were undertaken for a total duration of 20 days under laboratory conditions. One ml of uniform suspension of *Anabaena, Phormidium* and *Nostoc* species was inoculated as initial inoculum in all the flasks. The effluent samples were periodically (every 10th day) analysed for various physico-chemical parameters using APHA 1980. Growth of these cyanobacterial strains was measured in terms of chlorphyll-a *carotenoids* and protein as the biomass component. (Subramanian et. al., 1994)

RESULTS AND DISCUSSION

Characteristics of fish farm effluent has been depicted in Table 42.1. Table 42.2 represents the chemical characteristics of the effluent under different treatments using the cyanobacterial strains and effluent influenced change in few biochemical constituents of these strains are depicted in Table 42.3.

Inoculated cyanobacteria are known to grow fairly efficiently in fish farm effluents and their growth was measured in terms of chlorophyll, carotenoids and protein changes.

In the fresh effluent the pH value was slightly alkaline and in the cyanobacteria screened effluent a marked increase in pH. The increased pH is related to the depletion of inorganic carbon during photosynthesis by these algae and this typical capacity of cyanobacteria to bring about

Table 42.1 Characteristics of fish farm effluent (mg/l) except pH

Parameters	Day (0)
pH	7.4
DO	6.2
CO_3	0.0
HCO_3	3.6
TS	175
BOD	4.2
COD	32
Sulphide	2.8
Nitrate	3.80
Nitrite	BDL
Phosphate	4.6
Ammonia	13.74

BDL - Below Detectable Limits

Table 42.2 Chemical Characteristics of fish farm effluent before and after cyanobacterial treatments (mg/l) except pH

Parameters	Day	Anabaena. sp 10th day	Anabaena. sp 20th day	Phormidium.sp 10th day	Phormidium.sp 20th day	Nostoc sp. 10th day	Nostoc sp. 20th day
pH	7.4	9.95	9.82	9.60	9.58	8.52	9.20
DO	6.2	7.7	7.6	8.7	8.4	7.2	7.8
CO_3	0	48	46	32	32	44	42
HCO_3	36	-	-	-	-	-	-
TS	175	145	140	132	129	158	151
BOD	4.2	2.8	2.2	2.0	1.6	2.6	2.1
COD	32	28	22	24	20	22	20
Sulphide	2.8	1.6	1.2	2.6	2.4	2.8	2.2
Nitrate	3.8	3.52	2.98	1.98	1.18	3.24	3.04
Nitrite	BDL	0.112	BDL	0.207	BDL	0.132	BDL
Phosphate	4.6	2.4	1.8	3.2	BDL	3.8	2.6
Ammonia	13.74	4.74	2.82	2.36	1.48	5.68	4.64

BDL - Below Detectable Limits

Table 42.3 Chlorophyll-a, Carotenoids and Protein content of three cyanobacterial strains before and after effluent treatment

Organisms	Initial inoculum 'o' day	Control 20th day	Treatment Days 10th day	Treatment Days 20th day
CHLOROPHYLL.a				
(rag ml $^{-1}$)				
Anabaena sp.	0.94±0.18	2.28±0.16	2.44±0.17	2.45±0.30
Phormidium sp.	0.72±0.16	1.78±0.14	2.04±0.15	2.28±0.16
Nostoc sp.	1.68±0.24	2.57±0.30	1.51±0.23	0.76±0.16
CAROTENOIDS				
(mg ml^{-1})				
Anabaena sp.	0.041 3±0.020	0.0463±0.020	0.0634±0.020	0.0682±0.020
Phormidium sp.	0.0148±0.013	0.0185±0.015	0.0226±0.016	0.0244±0.017
Nostoc sp.	0.020±0.028	0.026±0.027	0.0178±0.075	0.0128±0.021
PROTEIN				
(Mg ml^{-1})				
Anabaena sp.	166.07±0.040	170.66±0.81	133.66±0.72	100.98±0.62
Phormidium sp.	126.44±0.700	136.53±0.73	145.26±0.75	139.40±0.7
Nostoc sp.	86.12±1.780	92.55±1.85	77.69±1.69	60.37±1.49

changes in pH is evident in this study also. Increased pH often favours the growth of cyanobacteria (Tarn and Wong, 1989; Manoharan and Subramanian 1993).

Cyanobacteria screened effluent showed a pronounced increase in dissolved oxygen also. This increment is due to algal photosynthetic activity and also the solubility of oxygen from the atmosphere under laboratory condition in which they are incubated. High dissolved Oxygen in batch cultures enhance the nutrient removal which subsequently reduce the oxygen demands of the effluents (Kankal et. al., 1987).

The fresh effluent showed only bicarbonate ions. These algae by metabolising the bicarbonate ions, increase the pH value. HCO_3^- utilization by algae is an adaptive phenomenon (Senthil and Goyal, 1996). The gradual withdrawl of CO_2 from HCO_3^- resulted in the liberation of OH^- ions which has made the effluent water more alkaline.

Cyanobacterial treatments in fish farm effluent showed considerable reduction in total solids. This reduction observed might be due to the uptake of dissolved solids by these organisms. Similar observation of solids removal by *Oscillatoria* and *Halobacterium* sp. combination in effluent waters had already been documented (Uma and Subramanian, 1990).

Cyanobacteria treated fish farm effluent showed reduction in both BOD and COD levels. BOD reduction in waste waters using algal cultures had been reported (Pouliot et. al., 1989; Tripathi et.al., 1991). Similar observation of COD reduction by *Oscillatoria chlorina* screened effluent had also been documented earlier (Rana and Palria, 1987). It is clear from this study that these algae not only retrieve inorganic nutrients, but also helps in reducing BOD and COD level.

Sulphide is generally toxic with lower pH values. This toxicity can be overcome with increased pH by the algae during their active growth in the effluent (Grilles Laliberte et. al., 1990). In this study, *Nostoc* and *Phormidium* species screened effluent showed no remarkable removal in the sulphide levels when compared with *Anabaena* treated effluent, which may be due to poor utilization. The utilization of sulphide during photosynthesis by *Oscillatoria* species obtained from the sulphide enriched waters can not also be ruled out here (Cohen et. al., 1975).

Nitrate serves as nitrogen source for the growth of algae. High level of nitrogenous compounds in waste waters can be effectively removed by algae in general (Tam and Wong, 1989), cyanobacteria in particular (Proulx and dela Noue 1988). Even after a prolonged exposure in the effluent, *Anabaena* and *Nostoc* species showed poor utilization of nitrate nitrogen which might be due to the fact that these algae during their growth might differentiate heterocysts, the site of nitrogen fixation, which contribute poor removal of nitrate in the effluent. Similar observation was made by Haselkon (1978). Also the ammonical nitrogen utilization instead of nitrate nitrogen by these algae leaving substantial nitrate level in the effluent. On the other hand, *Phormidium* screened effluent showed apparent assimilation of all the nitrogen sources from the effluent. Varma and Wilcomb (1964) contend that nitrate nitrogen promotes better algal growth than ammonia nitrogen in waste waters.

Nitrite at low concentration serves as nitrogen source for the growth of cyanobacteria. In the fresh effluent the nitrite level was well below the detectable limits. But after cyanobacterial treatment, within 10 days its level has a marked increase in all the cases which may be due to the reduction of nitrate into nitrite which in turn is removed completely with in another 10 days. Similar initial level increase and subsequent depletions of nitrite in paper mill effluent treatment using the cyanobacterium *Oscillatoria pseudogeminata* has been made earlier (Manoharan and Subramanian, 1992).

While *Anabaena* and *Nostoc sp* showed poor utilization of phosphates, *Phormidium* screened effluent showed its level below detectable limits. The capacity of cyanobacteria to remove large amount of phosphorous from waste water had been documented earlier (Chan et. al., 1979, Tam and Wong, 1989). Filamentous cyanobacteria are not only able to retain these pollutants for a longer period, but can be suitably harvested easily (Kumar and Sharma, 197S). The removal of phosphate may be attributed with their uptake during the active growth and this observation of inverse relationship between increased algal growth and decreased phosphate concentration in wastewaters Charles Neos and Varma (1966) can not also be ruled out.

The effluent after treatment with these three strains showed considerable reduction of ammonia within 20 days. This reduction of ammonia level may be due to effective uptake by these microalgae to maintain their metabolism which is in accordance with the observation made by (Shelef et. al., 1978). As ammonia did not seem to interfere with these algal growth, they penetrate the cell membrane by diffusion which could explain its removal from the effluent.

Influence of effluent on the biochemistry of three cyanobacterial strains

Due to varied physiological adaptation of these algae in the effluent, there is a significant change on me biochemical composition of the organisms. The growth of these organisms in terms of chlorophyll-a, carotenoids and proteins showed considerable variations.

During the treatment period, the two strains *Anabaena* and *Phormidium* sp showed remarkable increase in chlorophyll-a and carotenoid pigments than the *Nostoc* sp which showed gradual reduction in the above pigment composition indicating their poor adaptation in the effluent.

As far as the total protein level is concerned, the fish farm effluent had brought down the total protein content of *Anabaena* and *Nostoc* species during the treatment period. Contrary to this in *Phormidium* sp, a initial level increase and subsequent reduction in protein level was noticed. The decrease in protein proportion can be expected when nitrogen and phosphorus have been taken up by these algae during their active growth phase (Talbot and de la None, 1993). Similar observation of protein level reduction in *Oscillatoria* sp. was made while treating a paper mill effluent (Manoharan and Subramanian 1992). Also the environmental stress induced modification in the protein level while they grow in the effluent can not also be ruled out (Kimpel and Key, 1985; Bhagwat and Apte, 1989).

Cyanobacteria are not a metabolically sluggish group of organisms as they play a major role in removing various pollutants as their nutrients from the effluents. Based on the discussions of the above results, it can be concluded that of the three strains employed in the above study, *Anabaena variabilis* and *Phormidium valderianum* were found the better choice candidates which were minimally influenced by the adverse condition of the effluent, but helped in removing / reducing the fish farm pollutants maximally.

REFERENCES

APHA 1980. Standard methods for Examination water and waste water. 15th Ed. American Public Health Association, New York.

Bhagwat, A.A. and Apte, S.K. 1989. Comparative analysis of protein induced by heat shock, salinity and osmotic stress in the nitrogen fixing cyanobacterium *Anabaena* sp. Strain L.31 Journal of Bacteriology. 171: 5187-5189.

Chan, K.Y., Wong, K.H. and Wong P.K. 1979 Nitrogen and Phosphorus removal from sewage effluent with high salinity by *Chlorella salina*. Environ. Pollu. 18 : 139-146.

Charles Neos and Varma MM. 1966. The removal of phosphate by algae. Water and sewage works p. 456-459.

Cohen, Y., Padan, E & Shilo, M. 1975. Facultative anoxygenic photolosyn thesis in the cyanobacterium *Oscillatoria limnetica.* Journal of Bacteriology 123 : 855 - 861.

Gilles Laliberte, Daniel Proulx, Niels De pauw and Joel de la Noue 1990 Algal Technology in waste water Treatment p. 283 - 297.

Haselkorn, R., 1978. Heterocysts, Ann. Rev. Plant Physiol. 29, 319-344.

Kankal, N.C., Nema, P., Gokhe, B.H and Mehata C.G. 1987. Dissolved oxygen and detention time as process parameters in sewage treatment by aerated lagoon. IACPW Tech. Annual. 14: 61-68.

Kimpel, J.A. and Key, J.L. 1985. Heat shock in plants. Trends, Biochem. Sci. 85 : 353 - 357.

Kumar, H.D. and Sharma, V. 1975. Experimental studies on the uptake of mineral nutrients by algae isolated from polluted habitats Biochem Physiol Flanzen (BPP) 166 : 221 - 232.

Manoharan, C. and Subramanian, G. 1992. Interaction between paper mill effluent and the cyanobacterium *Oscillatoria pseudogeminata var. unigranulata.* Pollution Research 11(2): 73 - 84.

Manoharan, C. and Subramanian, G. 1993. Feasibility studies on using cyanobacteria in effluent treatment. Indian Journal of Environmental Health, 35(2): 88 - 96.

Proulx and de la Noue 1988. Removal of macronutrients from wastewaters by immobilized microalgae. In Bioreactor immobilized enzymes and cell fundamental and applications. Ed.M.Moo - Young. Elsevier Applied Science, pp. 301 - 310.

Pouliot, Y., Buelna, G., Racine, C. and de la Noue 1989. Culture of cyanobacteria for tertiary wastewater treatment and biomass production. Biological wastes 29, 81 - 91.

Rana, B.C, Palria S. 1987. Pollution abatement of certain waste waters using biological and chemical methods. Indian Journal of Ecology 14 (1): 1-6.

Senthil, C. and Goyal, S.K. 1996. Inorganic carbon assimilation in algae and its response to elevated levels of carbondioxide. Phykos 35 (1 & 2): 1-12.

Shelef, G., Moraine, R., and ORON, G. 1978. Photosynthestic biomass production from sewage. In Ergebnisse der Limnologie, Heft II: Microalgae for food and feed. Edited by C.J. Soeder and R. Binsack. Stuttgart, Germany, pp. 3 -14.

Subramanian, G., Uma, T., Thajuddin, N., Sekar, S., Sundararaman, M., and Sophia Rajini 1994. Laboratory manual, STORMCUP, NFMC, Bharathidasan University, Trichy.

Talbot, P. and J. de la Noue., 1993. Tertiary treatment of waste water with *Phormidium bohneri* (Schmidle) under various light and temperature conditions. Water Research 27(1): 153 -159.

Tam, N.F.Y. and Wong, Y.S. 1989. Waste water nutrient removal by *Chlorella pyrenoidosa* and *Scendesmus* sp. Environ. Pollut. 58:19-84.

Tripathi, B.D. and Shukla, S.C. 1991. Biological treatment of waste water by selected Aquatic plants. Environmental Pollution 69, 69 - 78.

Uma, L. and Subramanian, G. 1990. Effect of cyanobacteria in effluent treatment. (Proceedings of National symposium in cyanobacterial Nitrogen fixation). Indian Agricultural Research Institute, New Delhi, pp. 437 - 444.

Varma, Man M. & Wilcomb, Maxwell J. 1964. Effect of Light Intensity on Photosynthesis. Water and Sewage works, 110, 191.

CHAPTER 43

USE OF RANDOM AMPLIFIED POLYMORPHIC DNA (RAPD) FOR DETECTION OF SOME GENETIC CHANGES IN *LABEO ROHITA* (HAM.) OF MIDNAPORE, INDIA

Partha Bandyopadhyay *, Arup Tewary *, Biplab Sarkar and Bidhan C. Patra***

* *Aquaculture Research Unit*
Department of Zoology, Vidyasagar University
Midnapore- 721 102, West Bengal, India

** *Central Institute of Freshwater Aquaculture Kausalyaganga,*
Bhubaneswar - 751 002, Orissa, India

ABSTRACT

Regular use of pituitary extract for induced breeding makes some changes in genetic level. Random Amplified Polymorphic DNA (RAPD) is very much useful for the detection of genetic changes. A study has been carried out by collecting *Labeo rohita (Ham.)* from three different sources within the district of West Midnapore. Among them two were from flowing rivers and one from a induced breeding cum cultured farm. A single arbitrary nucleotide primer (OPC - 02) was randomly used for assay the genetic changes. Result of RAPD analysis shows that the genetic variation occurs in the induced breed farm raised fish as compared to the riverine fishes of the same species.

Key words: *RAPD. OPC - 02 primer, induced breeding, polymorphism. Labeo rohiia*

INTRODUCTION:

Recent advances in techniques for DNA analysis and subsequent data analysis have greatly

Correspondent author
Telephone Office : (91) 03222-276554 Ext: 455 Residence : (91) 03222-266035
Fax No. : (91) 03222-275329. e-mail : bcpatra@yahoo.com

increased the ability to understand the genetic relationships among species at the molecular level. Among the techniques, the Random Amplified Polymorphic DNA (RAPD) has emerged for studying genetic variation at the population and species level. Polymorphic Chain Reaction (PCR) short primers homologues are using to random target sites in the genome bases RAPD markers on amplification of genomic DNA. Polymorphism can simply be identified at the presence or absence of amplification product on ethidium bromide strained agerose gel (Bardakci & Skibinski, 1993; Rao *et al.*, 1996; Larson *et al.*, 1996: Williams *et al.*. 1998). It provides a relatively easy and rapid assessment of the deferences in the genetic composition of related individuals and species. Polymorphism generated by the RAPD techniques has been used for assessment of variation in population genetic analysis, genetic mapping, species identification of breed of specific species (Williams *et al.*. 1990. Welsh & Mcclelland. 1990). By using the usefulness of RAPD to determine the breed of specific species; we were able to measure the impact of induced breeding, due to crossbreeding and inbreeding. Generally crossbreeding and inbreeding occurred during the induced breeding. As a result some of the hatchlings showed chromosomal variation, which can easily be determined by the application of RAPD. An experiment was conducted by using the Indian Major Carp *(Labeo rohita* Ham.) from three different sources among them two were from rivers (Kanshabati and Rupnarayan in Midnapore) and one from a carp culture farm of Midnapore town.

Our main aim of this investigation is to evaluate the genetic variation between the wild sources and captive sources of fishes through RAPD technique.

MATERIALS AND METHODS:

Blood samples were collected in 1.5 ml sterile tubes containing 0. 5 ml of preservation mixture (4 M Guanidium - thiocyanate; 2% *N* - lauroyl - sarcosine ; 50mM Tris - HCL, pH 7.4 ; 50 mM Ethylene Diamine Tetraacetic Acid (EDTA) and 0.1% B-mercaptoethanoi) (Laulier *et al.*, 1995). A 1.0 ml blood sample of *Labeo rohita* was used to produce a genomic DNA stock for the PCR assays. The samples were incubated for 30 minutes at room temperature, extracted with buffered (pH 7.0) phenol, followed by chloroform isoamyl alcohol (19 : 1) extraction. Nucleic acids were precipitated with 1 : 10 volume of 3M sodium acetate, 2.5 volumes of ethanol and by incubating for 1 hour at - 80 ° C. The DNA was pelleted, washed twice with 70% ethanol, dried and suspended in sterile water. The DNA quality was checked by electrophoresis on agarose gels.

RAPD conditions

PCR was performed using 10 - 50 ng of total DNA from each sample tested. The DNA was dissolved in 10 mM Tris - HC1, pH 9; 2.5 mM MgCl2; 50 mM KC1; 0.1% Triton x 100; 150 µM each of dATP, dCTP, dGTP, dTTP; 0.1 µM oligonucleotide in a total volume of 50µl, and added 2 units of Taq DNA polymerase (Apligene). PCR was carried out in a microprocessor linked programmable thermal controller (Perkin - Elmer) for 45 cycles. Each cycle consisted of 1 minute denaturation at 94 °C, 1 minute annealing at 35° C (5° C above the T_m of the used primer) and 2 minute extension at 72° C. After the last cycle, there was a 5 minute extension at 72° C.

RAPD analysis

The random primers are selected to flank the DNA region to be amplified. Primers are oligonucleotides that are single stranded fragments of DNA, complementary to the 5 end of the target DNA to be amplified. There must be nearly complete homology between the target DNA and the primer nucleotide sequence for primer annealing, specifically at the sites of the DNA region to be amplified. Typically primers are 15 - 30 nucleotide long and it is necessary that the 5 - 6 bases at the 3 end of the primers exhibit precise base pairing with the target DNA. All primers for target amplification should have the same melting temperature (Temperature of annealing) and G + C content is about 60 - 70 %. OPC - 02 - GTGAGGCGTC, a commercial primer from Operon Technologies Inc. (Alameda, CA, USA) was used with following 5 - 3 sequence.

Analysis of RAPD product through agarose gel electrophoresis

The PCR amplified samples were analysed through 1.8 % agarose gel eletrophoresis. Agarose gel (1.8 g) was taken in control flask and boiled in 100 ml of TBE buffer (Tris 50 nM, EDTA 100 mM, Boric acid 20 ml, pH - 8.0). Gel tray was prepared by pasting adhesive tape on both side and was placed the comb. Then boil agarose poured in gel tray and allowed to solidify. After solidification the comb was removed gently and tray was placed in gel tank. Then 1 x TBE buffer was poured in gel tank till cast gel submerges. Amplified DNA samples (6-8 μl), gel loading dye (2 μl) and distilled water (3-4 μl) were mixed together and loaded in wells. Known molecular marker (OPC - 02) (250 bp) was also loaded along with the samples. The power pack was connected to the gel tray set at 100 V. The power pack was stopped when blue dye reached to the end of the gel and gel was transported to the glass tray for staining. Staining was done in glass tray using ethidium bromide for about 1015 minutes. Then the gel was removed and observed under UV transilluminator. High molecular DNA remains nearer to well and smear of DNA, which indicates degradation of DNA. Photographs were taken onto thermal paper using a video camera and video copy processor (Mitsubishi, Japan).

RESULTS

In the proposed experiment, *Labeo rohita* was used as an experimental animal, which was collected from three different sources and after that DNA fragments are isolated from the fish. These DNA fragments are used for RAPD technique with a single arbitrary primer OPC - 02. The primer is rich in G+C content is about 60-70 %. The sequence of the primer from 5 - 3 direction is GTGAGGCGTC (Williams *et al.*, 1990). In the figure 43.1 A, B. & C band indicates the genomic structure *of Labeo rohita*, collected from the river Kanshabati and Rupnarayan (Wild population) and a captive culture farm respectively. The figure 43.1 clearly shows that lane A and B make similar kind of band pattern with the marker protein but lane C showed different pattern. This differential band separates the fish genetically from the other fishes of *Labeo* sp. Using RAPD got several amplified DNA fragments. This amplified DNA fragments are analysed by agarose gel electrophoresis. As a result DNA segments are amplified and showed polymorphism. The molecular weight of the fragments of C band is largely differing from the other two bands.

DISCUSSION

This study demonstrates the intraspecific genetic variation within a population of Indian major carp *Labeo rohita*. In general with other molecular techniques, RAPD analysis has been successful in identifying the intraspecific genetic variation (Hadrys *et* al., 1992; Hedrik, 1992; Riedy *et* al., 1992; Scota *et al.*, 1992; Powell *et al.*, 1995). RAPD analysis also identified DNA marker which can distinguishes the specific variation of *Labeo rohita* and in this respect RAPD has been more successful than alozyme analysed (Seyoum, 1990, Seyoum and Kornfield, 1992). The fish is collected from the culture farm showed the different band pattern (C) as compared to other two lane (A & B) with the marker DNA. Due to rapid induced breeding and the use of synthetic hormone as a inducing

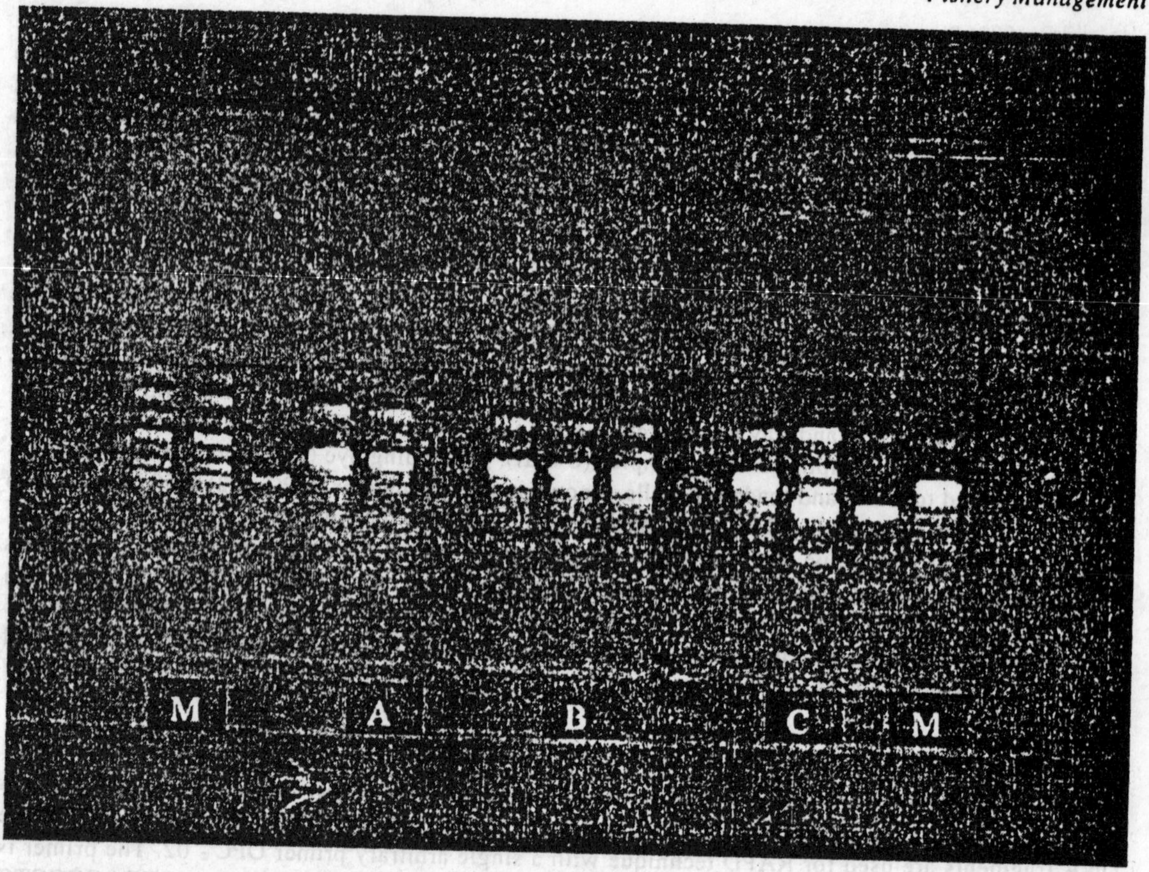

Figure 43.1

The different bands of the fragment produced by the RAPD using with a single primer (OPC-02)

agent, sometimes become cause of the change of the genetic variation in the next generation. From the above result, we can conclude that the use of synthetic hormone and the rapid induced breeding causes some kind of species heterogenecity as well as influence in formation of subspecies. In this aspect this is the preliminary study and the first report in that area, but it needs further study.

ACKNOWLEDGEMENTS

The authors are highly thankful to the Indian Council of Agricultural Research and the Head, Department of Zoology, Vidyasagar University, India for necessary facilities and the research scholars of the Aquaculture Research Unit for their valuable co-operation during preparation of this manuscript.

REFERENCES

Bardakci. F and Skibinski, F.O.D. 1994 Application of the RAPD technique in tilapia fish: species and sub-species identification. The genetical Society of Great Britain., 73: 117-123.

Hadrys, H., Balick, M, and Schierwater, B. 1992. Application of random amplified polymorphic DNA (RAPD) in molecular Ecology. Mol. Eco. 1 : 55 - 63

Hedrik, P. 1992. Shooting the RAPDs. Nature. 335: 679 - 680

Larson, E. S., Anderson, L. P., Miller, N. A., Cousin, E. C., Richards, S. C., Lewis, A. F. and Knight M. 1996. Use of RAPD - PCR to differentiate genetically defined lines of an intermeadete host *Schistosoma mansonii, Biomphalaria glabrata.* J. Parasitology. 82 (2) : 227 - 244

Powell, W., Orozco-Castilo, C., Chalmers, K. J., Provan, J. and Waugh R. 1995. Polymerase Chain reaction based assays for the characterization of plant genetic resources. Electophoresis. 16 : 1726- 1730

Rao, A. C. B. K., Bhatt, V. K. and Totey M. S. 1996. Detection of species specific genetic markers in farm animals through random amplified polymorphic DNA (RAPD). Genetic Analysis: Biomolecular Engineering. 13 : 135 - 138

Riedy, M. F., Hamilton, W. J. and Aquadro, C. F. 1992. Excess of non-parential .bands in offspring from known pedigress analysed using RAPD PCR. Nucl. Acid res. 20 : 918

Scott, M. P., Haymes, K. M. and Williams, S. M. 1992. Parentage analysis using RAPD-PCR. Nucl. Acid Res. 20: 5493

Seyoum, S. 1990. Allozyme variation in subspecies of Oreochromis noloticus. Isozyme Bull., 23, 97

Seyoum, S. and Kornfield, I. 1992. Identification of the subspecies of *O. niloticus.* Using restriction endonuclease analysis of mitochondrial DNA. Aquaculture. 102 : 29 - 42

Welsh, J. and Maclelland, M. 1990. Fingerprinting genomes using PCR with arbitrary primers. Nucl. Acid Res. 18 : 7213 - 7218

Williams, K. G. J., Kubelik, R. A., Kenneth, J. L., Rafalski A. J. and Tingey V. S. 1990. DNA polymorphisms amplified by arbitrary primers are useful as genetic markers. Nucleic Acids Research. 18(22) : 531 - 535

Williams, J. D., Kazianis S. and Walter B. R. 1998. Use of random amplified polymorphic DNA (RAPD) for identification of Largemouth Bass subspecies and their intergrades. Trans. American Fis. Soc. 127 : 825 - 832

Hedrik, P. 1992. Shooting the RAPDs. Nature. 135: 679 - 680.

Larson, E. S., Anderson, L. P., Miller, N. A., Cousin, R. C., Richards, S. C., Evans, A. E. and Knight, M. 1990. Use of RAPD - PCR to differentiate genetically defined lines of an intermediate host *Schistosoma mansoni*, *Biomphalaria glabrata*. J. Parasitology, 82 (2): 237 - 244.

Powell, W., Orozco-Castillo, C., Chalmers, K. J., Provan, J. and Waugh, R. 1995. Polymerase Chain reaction based assays for the Characterisation of plant genetic resources. Electrophoresis. 16 : 1726 - 1730.

Rao, A. C. B., K., Bhat, V. K. and Totey, S. C. 1996. Detection of species specific genetic marker in farm animals through random amplified polymorphic DNA (RAPD). Genetic Analysis: Biomolecular Engineering. 13 : 135 - 138.

Riedy, M. F., Hamilton, W. J. and Aquadro, C. F. 1992. Excess of non-parental bands in offspring from known pedigrees analysed using RAPD PCR. Nucl. Acid Res. 20: 918.

Scott, M. P., Haymes, K. M. and Williams, S. M. 1992. Parentage analysis using RAPD PCR. Nucl. Acid Res. 20: 5493.

Seyoum, S. 1990. Allozyme variation in subspecies of *Oreochromis niloticus*. Isozyme Bull., 23: 97.

Seyoum, S. and Kornfield, I. 1992. Identification of the subspecies of *O. niloticus*. Using restriction endonuclease analysis of mitochondrial DNA. Aquaculture. 102 : 29 - 42.

Welsh, J. and Mcclelland, M. 1990. Fingerprinting genomes using PCR with arbitrary primers. Nucl. Acid Res. 18 : 7213 - 7218.

Williams, K. G. J., Kubelik, R. A., Kenneth, J. L., Rafalski, A. J. and Tingey, V. S. 1990. DNA polymorphisms amplified by arbitrary primers are useful as genetic markers. Nucleic Acids Research. 18(22): 6531 - 6535.

Williams, J. D., Kazianis, S. and Walter, G. R. 1998. Use of random amplified polymorphic DNA (RAPD) for identification of Largemouth Bass subspecies and their intergrades. Trans. American Fis. Soc. 127: 825 - 832.

CHAPTER 44

STATUS AND CONSERVATION OF BRACKISHWATER FISHERIES OF NORTH-EAST COAST OF INDIA (SUNDERBAN & CHILKA LAKE)

Sudhir K. Das

Dept. of Fishery Biology & Resources Management, West Bengal University of Animal and Fishery Science, Mohanpur, Nadia, Pin - 741252

ABSTRACT

The north east coast of India includes 480 Km coast line of Orissa and 64 km coastline of West Bengal. Chilka lake and Sunderban deltaic belt are the prominent brackishwater fishery resource of the north east coast. Sunderban region harbours about 92 species of fishes of 41 families while that of Chilka lake is 225 species. Threatened fishes of Sunderban area are presented. Specific and general conservation measures for fisheries of the region are discussed.

The north east coast of India comprises 480 Km coastline of Orissa and 64 km. coastline of West Bengal. It also includes world's biggest deltaic region, i.e. Sunderban area of Hooghly-Mattah estuary, Asia's biggest brackishwater lake i.e. Chilka lake as well as several estuaries among which Mahanadi estuary and Bhitar Kanika mangrove belt are very important. The fish production form Hooghly Matlah estuary was about 51,000 tonnes in the year 1996-97 while that of Chilka lake was about 5000 tonnes in the year 2000—2001. Coastal capture fisheries provide livelihood to several poor people residing around the water bodies. It also contributes considerable proportion to marine product exports. Last but not least, it provide cheap animal protein to local rural people and nearby urbanites and thus plays important role in preventing undernurrition of people.

SUNDARBAN DELTAIC AREA

Central Inland Capture Fisheries Research Institute (CICFRI), Barrackpore has made considerable works on ecology and fisheries of Hooghly Matlah estuary (Mitra *et al*, 1997). The river Ganga has extensive estuarine system covering southern part of W. B. and Bangladesh. The principal component is Hooghly river and it receives water from its way to sea from river Rupnarayan, Saptamukhi, Thakusan, Matlah, Gosaba, Bidya, Harinbhanga, Ichamati and Raimangal forming an estuarine complex. The lower portion of this estuarine system i.e. the main channel with several distributaries, creeks etc. form the Sunderban delta. This delta sustains multispecies commercial fisheries exploited by multigears.

The estuarine system has an area of about 8029 km^2 of which Sunderban area is about 2340 km^2. It is very rich in flora and fauna including fisheries. It provide rich breeding and nursery grounds for innumerable marine and freshwater finfish and shell fish. After commissioning of Farakka barrage in 1975, the main Hooghly estuary is fed directly by the Ganga through feeder canal and Bhagirathi. The additional discharge of freshwater into the system has changed ecology of estuary (De, 1999).

Fisheries : There is occurrence of 92 species of fishes of 41 families in Sunderban area. The important species in Sunderban area of Hooghly Matlah estuary are as follows :

Tenualosa ilisha, Harpodon nehereus, Trichiurus spp., *Setipinna* spp., *Pama pama, Polynemus indicus, Polynemus paradiseus, Eleutheronema tetradactylum, Lates calcarifer, Coilia* spp., *Stromateus cinereus, Arius sona, A. sagor, Ilisha elongata, Anchoviella commersoni, Polydactylus* spp., *Otolithoides biauritus, Liza parsia, Liza tade, Chirocentrus dorab, Osteogeniosus militaris, Raconda russeliana, Plotosuus canius, Scatophagus argus, Etroplus suratensis, Therapon jarbua, Cynoglossus* spp., *Strogylura strogylura, Synbranchus bengalensis* etc.

Important shrimps inhabiting in the area are *Penaeus monodon, Penaeus indicus, P. semisulcatus, Parapenaeopsis sculptilis, P. stylifera, Metapenaeus brevicornis, M. monoceros, Expalaemon stylifera* etc.

Sunderban estuarine system harbours good quantity of seeds of shrimps and fishes. The magnitude of seed had decreased after construction of Farakka barrage.

Gears : Important gears used for fisheries of Sunderban area are as follows : (1) Trawl nets (2) Set gill nets (3) Drift gill nets (4) Lift nets (5) Bag nets (stationary) (6) Set barrier nets (7) Cast nets (8) Large and small seine nets (9) Hook and lines (10) Traps.

Bag nets and drift gill nets are the most common gears.

Fishing in the region is mainly carried out throughout the year. But fish catch shows an increasing trend from August to December-January due to calm weather. The commercial fishing in Sunderban is restricted from February-March to June due to strong southwinds blowing in the area. The faunal structure of the area has been changing due to degradation of mangrove, rapid industrialisation, biotic pressure, siltation of river bed, intensive use of adjacent land for crop raising etc. (Pandit *et al*, 1999). A declining trend in availability of some spp. of fishes observed due to change in ecological conditions. Menon (1989), Jhingran (1989) and Pandit *et al* (1991) emphasised the need of conservation of natural fish resources and advocated for proper protective and preventive measure to conserve vast, varied and valuable fish fauna of Sunderbans.

Pandit *et al* (1994) reported endangered, vulnerable and rare species of Sunderban as follows.

A. Endangered fishes :

(1) *Mugil cephalus*

(2) *Plotosus canius*

(3) *Osteogeniosus militaris*

(4) *Chanos chanos*

(5) *Hemirhamphus* spp.

(6) *Cynoglossus* sp.

(7) *Elcotris fusca*

(8) *Echineis* sp.

(9) *Etroplus maculatus*

B. Vulnerable species :

(1) *Pangasius pangasius*

(2) *Liza tade*

(3) *Rhinomugil corsula*

(4) *Megalops cyprinoides*

(5) *Tachysums jella*

(6) *Sciaena biauritus*

(7) *Coilia* sp.

(8) *C. borneensis*

(9) C. *ramcarti*

C. Rare species :

(1) *Lates calcarifer*

(2) *Mystus gulio*

(3) *Elops saurus*

(4) *Sillago penigus*

CONSERVATION MEASURES :

(1) Mesh size to be restricted to protect juvenile fishes by bag net fishing.

(2) Exploitation of brood fishes during breeding season should be banned.

(3) Development of artificial propagation of important fishes and raching.

(4) Awareness to be created among local people to protect fisheries.

CHILKA LAKE

Chilka lagoon is situated in north east coast of India in state of Orissa (lat. 19° 28' to 19° 54' N and long. 85° 05' to 85° 38 E). Sea water is exchanged through two openings i.e. natural feeder channel near Arkhakuda village and artificial dug out canal joining Rushikulya estuary known as Palur canal. The lagoon has catchment area of 3200 sq. km. Approximately 2 to 2.75 lakh cusecs monsoon runoff is poured in the lake through about 35 water ways including Daya, Nun and Bhargovi, the distributaries of river Mahanadi. The monsoon discharge contributes 10 to 13 million tonnes of silt every year to the lagoon. The drifted silt gets deposited irregularly and makes the lagoon bed uneven (Anon, 1998).

Chilka lake has waterspread area of 906 sq. km. in summer and about 1165 sq. km in monsoon season.

Mohanty *et al.* (2001) reported occurrence of 225 species of fish and prawn in the lake. Among them, 28 species are marine, 38 species are freshwater and 159 species are brackishwater species. But eleven groups of fishes are commercially important. They are mullets, clupeids, perches, threadfins, catfishes, fishes of Beloniformes, sciaenids, cichlids, triacanthus, prawns etc. The commonly available fishes of the lake are as follows.

1. Mugil cephalus 2. Liza troscheli 3. Mystus gulio 4. Eleutheronema tetradactylum 5. Tenualosa ilisha 6. Lates calcarifer 7. Pseudosiaena coibor 8. *Nematolosa nasus 9. Tachysurus arius 10. Gerres setifer 11. Etroplus suratensis 12. Sparus sarba 13. Plotossus canius 14. Strongylura strongylema 15. Liza tade 16. Elops saurus 17. Chanos chanos 18. Megalops cyprinoides 19. Anadontostoma chakunda 20. Anchovilla indica 21. Therapon jarbua 22. Leognathidae* sp. *23. Lutjanus* sp. *24. Polynemus indicus 25. Plotosus canius 26. Tricanthus brevirostris 27. Sciaena* spp. etc.

Shrimps *1. Penaeus monodon 2. P. indicus 3. Metapenaeus monoceros 4. M. dobsoni 5. M. affinis.*

Crabs *1. Scylla serrata* 2. *Portunus pelagicus*

Gears : Different gears used for fisheries in Chilka lake are as follows:

1. Fixed gill net 2. Khonda (disconet) a fixed net 3. Gill net

4. Drag net 5. Cast net 6. Scoop net 7. Seive net etc.

There are 132 fishermen villages having population of 1,22,339 fisher folk. (22,039 families) inhabiting around Chilka lake. There are 86 primary fisherman cooperative societies functioning in the area. Total number of boats (2000-2001) operating in the lake is 56 (4 motorised and rest traditional) with 872 gears (Anon, 2002). The threatened status of fishes of Chilka lake is not studied well. There is need of deep and analytical study in this aspect.

Govt. of Orissa had made legislations for conservation of fisheries in Chilka lake vide order no. 1508 dt. 27.1.88 of Dept. of Forest, Fisheries & A. H. It is meant to regulate fishing in Palur canal and Chilka lake in the following manner (Anon. 2002).

(1) Fishing by any method is prohibited in Palur canal throughout the year.

(2) Capture of Khainga, Kabla, Bhekti below 150 mm size and prawns like Bagda and Chapra below 100 mm size by any means is prohibited throughout the year in Chilka lake.

(3) Fishing is completely prohibited in the outer channel of Chilka lake during months of December and January.

(4) Prawns which are not growing above 2 inches as specified below cannot be caught in the outer channel of Chilka lake during the months of February and March.
(a) *Metapenaeus monoceros* (b) *M dobsonii and* (c) *M offinis.*

(5) No fishing by means of net shall be allowed in the outer channel throughout the years. Fishing however, by other means like Khanda (disconet), Thata and Baja can be undertaken in the outer channel. But no Khanda shall enter into the deeper zone of the outer channel.

Areas : Outer channel = 25 km (Satpara to Chilka mouth)

Palur canal =15 km (Rambha to Rushikulya mouth).

GENERAL CONSERVATION MEASURES FOR FISHERIES OF NORTH EAST COAST:

1. Determination of threatened status (extinct, endangered, vulnerable, rare) of different individual estuarine of brackishwater fishes in each region.

2. Steps should be taken to establish new populations for endangered and extinct fishes. Isolated population of such fishes to be safeguarded.

3 There is damage to many fish habitats and the situation is not often easy to reverse.

   Habitat restoration of selected areas can be done in long term basis. Siltation and sedimentation is a problem in some waterbodies (example Chilka lake). Dredging at regular intervals will be helpful for restoration of habitat.

4. When a particular fish is under severe threat in a particular environment, then stock transfer of matured fishes can be made to another congenial environment.

5. Hatchery should be set up in Sunderban and Chilka lake areas. Brood fishes of threatened category can be raised through proper farm management. Seed should be produced through induced spawning and released to the aqua environment as a measure of ranching the depleted population.

6. Cryopreservation of gametes of fishes at low temperature should be standardised. After freezing for many years and then thawing, the material is still viable. This technique is more successful in sperms and have problems in eggs (Maitland, 1993). Steps are needed to standardise cryopreservation of eggs through R & D efforts.

7. Threatened fishes should be monitored regularly both in time and space. There is need of regular biological comparisons (growth, fecundity, population, feeding, ecology etc.) of such species in each environment even though it is expensive.

8. The estuarine environment should be monitored and damage to young stocks need to be quantified. The biocommunities should be assessed both qualitative and quantitative manner.

9. Local fishermen collect seed of tiger shrimps from estuarine environment using different gears and damage lot of seeds of other fishes. This practice should be stopped.

10. Considerable awareness should be created among local people in order to conserve the threatened fishes. It can be done through motivation, training, awareness camps etc. with participation of voluntary organisations, panchayat workers, primary school teachers, village leaders, youths etc. (Das, 2002):

REFERENCES

Anon (1998) : Chilka lake - Present and Past. CICFRI, Barrackpore, Bulletin No. 80.

Anon (2002) : Handbook on Fisheries statistics, Orissa, 2002 - 2001. Directorate of Fisheries, Orissa, Cuttack.

Das, S. K. (2002) : Fish Production through combined effort of Panchayat and Non Govt. Organisations (in Oriya language). Krushak Sambad, 14 (5) : p. 6.

De, D. K. (1999) : Ecology and Fisheries of Hooghly Matlah estuarine system. Summer School on Ecology, Fisheries & Fish Stock Assessment in Indian Rivers, CICFRI (July 14 - Aug. 12, 1999), Barrackpore.

Maitland, P. S. (1993): Conservation of freshwater fish in India. *In* : Advances in Fish Research (Ed. B. R. Singh), Narendra Publishing House, Delhi.

Menon, A. G. K. (1989) : Conservation of ichthyofauna of India. *Conservation and Management of Inland Capture Fisheries Resources of India,* Ed. A. G. Jhingran and V. V. Sugunan.

Mitra, P. M., Karmakar, H. C., Sinha, M., Ghosh, A. and Saigal, B. N. (1997) : Fisheries of the Hooghly-Matlah estuarine system - An appraisal. CICFRI, Barrackpore, Bulletin No. 67.

Mohanty, R. K., Mohanty, S. K. and Badapanda, H. S. (2001) : Fishery dynamics and management of Chilka lagoon. *Fishing chimes,* 21 (9) : 7 - 10.

Pandit, P. K., Bhaumik, U. and Chatterjee, J. G. (1994). Threatened Fishes of Sunderbans, West Bengal. *Threatened Fishes of India,* Nation Publication, 4 : 253 - 259.

CHAPTER 45

A Morphological Study of *Gambusia affinis* (Baird & Girard, 1853).

Vishal Tiwari, K.K.Gaur and Rita Rani
Department of Zoology, R.B.S. College, Agra-282002.

ABSTRACT

The present study aimed to notice the morphology of the most popular, exotic, freshwater, larvicidal, livebẽarer-aquarium fish, *Gambisia affinis* (Baird & Girard, 1853). Morphological features including body size, colouration, pigmentation patterns along with the presence of gonopodium in male and gravid spot in female fish were some of the remarkable morphological features noticed during the investigation. These were noticed as promiscuous dimorphic features, thereby, facilitating their recognition and distinction for mating.

Key words: *Pisces; Poeciliidae; Gambusia affinis; Morpholical features.*

INTRODUCTION

Gambusia affinis (Mosquitofish) belonging to the family *Poeciliidae* is the most popular, exotic, freshwater, larvicidal, livebearer-aquarium, fish. It plays an important role in the development along with the spread of the aquarium. It is the most widespread freshwater fish in the world having been introduced as a biological control over malaria.

A morphological study of this will certainly enlighten the adaptive features, which are of prime importance governing the reproductive biology to a great extent, as in fish, the relations of the sexes are usually very promiscuous during mating period. Some of the important contributions on the said field of studies are from Okada and Yamashita (1944); Marr (1955); Rosen and Bailey (1963); Wourms (1981); and that of Batty and Lim (1999).

MATERIALS AND METHODS

A set of 30 healthy adult male and female of *Gambusia affinis* species of the same age group in the ratio 2:1 respectively were put in a glass aquarium of 24 X 12 X 12 inches filled with potable water. The signs, to look for, was to identify healty fish, their solid looking body, bright eyes, no wounds, high level of swimming activity and clearly defined colour patterns. The aquarium used was equipped with adequate illumination, meeting the requirement of the plants, filter, a thermostat, an aerator pump and the accessories. The optimum physico-chemical parameters of water were monitored periodically. Aquarium water was changed at regular intervals. A reaymade Hitachi food along with some live food was given to them. Anti-bacterial and anti-fungal medicines were also added in the aquarium water at some regular interval.

Morphometrical observations on the experimental fish were made with the help of Vernier Calipers.

RESULTS AND DISCUSSION

The observations on the morphological features of male and female *Gambusia affinis* revealed following results. (Table 45.1)

Table 45.1
Morphological features of Gambusia affinis

Body shape	Body size	Colouration	Gonopodium (modified anal fin)	Gravid Spot	Dorsal fins	Caudal fin	Pelvic Fin
Male							
Normal	3-4 cm (1.2 to 1.6 inches).	Translucent grey, belly silvery.	Developed, rod like, functional 8-10 and fin rays and half to the length of anal fin.	Absent	Round; begins behind the anal fin, colourful (orange to red), 7-9 dorsal fin rays	Have gross variations inshapes.	Designed to aid in stablishing and directing the swing of the anal fin, grey in colour
Female							
Pot-bellied	5-7 cm (2 to 2.8 inches).	Grey coloured	Normal shapped, 8-10 anal fin rays.	Present in pregnant females.	Round, begins behind the anal fin, have black spots, 7-9 dorsal fin rays.	Have black spots	Normal light orange in colour.

Body form, shape and size

It acts as one of the important secondary sex-determining characteristic, which had always played a key role in facilitating mutual distinction and recognition. Body short, cylindrical, compressed, abdomen rounded, head short, snout pointed, mouth oblique.

Head and body with cycloid scales. Lateral line scales 27-28. Terminal and upward pointed mouth is adapted for feeding at the water's surface. Eyes larger and situated in between dorsal and ventral profiles of head. The males were noted normal, while the females were noted much more pot-bellied, with round and tight stomach, which in late pregnancy become cuboidal in shape. Males were noticed considerably smaller than the females. Males were noted 3-4 cm. (1.2 to 1.6 inches), while females 5-7 cm (2 to 2.8 inches). Females need to be large,because she literally caries the precious cargo of dozen or more babies.

Colouration

It serves as a mark of sexual distinction and recognition. This phenomenon is termed as dichromatism. Both the sexes of *Gambusia affinis* has a black vertical line through the eye. It is a modestly coloured fish. The body colour is mainly grey, dusted with black spots and blotches. Males were translucent grey, with a bluish sheen on the flanks, black olive-brown, belly silvery. Female fish were similarly grey coloured but exhibit dull, olivaceous pigments. Females can also be recognized with light orange pelvic fins or a yellow-orange spot at vent. The bri colour of males may be used for advertising his health, virility and presence to females, while the drabness of the females observed might possible offer them a considerable degree of protection and acts as an important factor for camouflage, especially when they are pregnant.

Anal fin

It is the most significant external discriminatory feature between male and female, separating livebearers from all other types of fish. It thus acts as a valuable index in finding relationship between the groups of fishes.

In order to fertilize the female, the anal fin of male *Gambusia affinis* developed into a functional structure to insert the packets of sperms into the female fish. This structure is known as "gonopodium". The female anal fin was noted girthy, large and normal shaped. Gonopodium is supported internally by bony structures and can be moved backward, forward and sideways. 8-10 anal fin rays were noticed in both the sexes. In males the length of anal fin rays is less than half the length of anal fin (gonopodium). Spines at tip of third anal fin ray of male fish is one to three times longer than wide.

Caudal fin, Dorsal fin and Pelvic fin

Caudal fin of males exhibit gross variations in shapes. The variations in shapes and colours of caudal fin are supposed to catch the attention of females towards the males for mating. A male has a colourful (orange to dark red) dorsal fin. A female has black spots on the caudal and dorsal fin. 7-9 dorsal fin rays were noted in both the sexes. Dorsal fin can be distinguished from guppies *(Poecilia reticulata)* by position of the dorsal fin in relation to the anal fin. In it, the dorsal fin begins behind the anal fin, whereas in guppies the dorsal fin is opposite to the anal fin.

Pelvic fin of male fish is noticed that it is designed in such a way that they aid in stabilizing and directing the swing of the gonopodium, while that of a female fish is normal.

Gravid Spot

It acts as a valuable index in discriminating males and females *of Gambusia affinis*. In the pregnant females, at the posterior end of the gut near to the urinogenital vent, there appears a characteristic black spot or gravid spot. It may be considered equivalent to the womb, but unlike mammals, the eggs contain an embryo well furnished with nutritive elements on which it feeds during development. The eyes of the developing fry were visible through the thin walls of the gravid spot. It was quite fascinating to watch the developing babies through this gravid spot, as it is an area where her scales are transparent. They may have this transparent arc in order to orient the unborn babies to their watery world. It may also be presumed that the two gravid spots (each on lateral side) provide a "bull eye" for the males to aim at the vent, where insemination and birth occurs.

ACKNOWLEDGEMENTS

The authors are thankful to the Principal, R.B.S. college, Agra for providing the essential research facilities during this investigation.

REFERENCES

Batty, J. and Lim, R. 1999. Morphological and reproductive characteristic of male mosquitofish, *Gambusia affinis holbrooki* inhabiting sewage contaminated waters in new South Wales, Australia. *Arch. Environ. Conta. Toxicol.* 36(3): 301-307.

Marr, J. C. 1955. The use of morphometric data in systematic racial and relative growth studies in fishes. *Copeia.* 1: 23-31.

Okada, Y. K. and Yamashita, H. 1944. Experimental investigation of the sexual characters of *Poeciliid fish. J. Fac. Sci. Imp., Univ. Tokyo.* 6: 589-633.

Rosen, D. E. and Bailey, R. M. 1963. The *Poeciliid* fishes (Cyprmodontiformes), their structure, zoogeography and systematics. *Bull. Amer. Mus. Nat. Hist.* 126: 1-176.

Wilhm, J. L. and T. C. Dorrid. 1968. Biological parameters for water quality. *Biosci.* 18: 477-481.

Wourms, J.P. 1981. Viviparity: The maternal fetal relationship in fishes. *Amer. Zool.* 21: 473-515.

CHAPTER 46

PESTICIDE AND FISH: A WORKHOUSE FOR THE DETECTION, EVALUATION AND ABATEMENT OF WATER POLLUTION

Biplab Sarkar*, Partha Bandyopadhyay, S. Adhikari*, S. Ayyappan*** and Bidhan C. Patra****

**Central Institute of Freshwater Aquaculture, Kausalyaganga, Bhubaneswar -751 002, Orissa, India*

***Aquaculture Research Unit, Department of Zoology, Vidyasagar University, Midnapore-721 102, West Bengal, India*

****Deputy Director General (Fisheries), Indian Council of Agricultural Research Krishi Anusandhan Bhavan, Pusa, New Delhi - 110001, India*

ABSTRACT

The aquatic bodies to the earth have been treated as a natural place to discard unwanted waste. In India with occurrence of blue revolution the demand of pesticides were increasing day by day. Now-a-days carbamate and pyrethoid insecticides are getting more preference over traditional organophosphates because of their special characterision even cypermethrin using to control *Argulas*. Behavioral or morphological changes through pesticide may make the fish more conspicuous in the environment and more susceptible to predation or parasitization, thereby reducing the ability of the population to survive and reproduce. It has been well said that in order to eliminate a species, a pollutant need not be lethal. Though it is the ultimate aim of pollution study, no such antidote or mechanism for controlling pesticide toxicity of fish is developed. But it needs much more studies to come out successfully with some antidotes against these pesticides.

Key words: *Pesticide toxicity, fish, antidote*

Correspondent author

Telephone Office : (91) 03222-62297 Residence : (91) 03222-66035

Fax No. : (91) 03222-62329 e-mail : bcpatra@yahoo.com

INTRODUCTION

Before First World War, toxicity testing was in very infantile stage and sporadic. This testing (One fish-in-a-jar) was done to confirm the causative agents of any fish killings. First major break through came when Power (1917) published his research paper "The gold fish as test animal, in the study of toxicity". Being interested by this, Ellis (1937), Hart *et.al.* (1945) started their works and later peoples accepted several species of fishes also, which sometimes proved more sensitive than gold fish. Subsequently an epidemic case of the "gold fish-allergy-syndrome" developed and for therapeutic reasons one began to test everything which will be fitted best into a jar. (Mount 1977, Macek, 1980).

Thus, since 1950 the acute toxicity testing has become the 'work house' for the detection, evaluation and abatement of water pollution (Buikema *et.al.*, 1982). The 96 hour exposure is taken as most common time interval because the effects of test material occur rapidly and are well defined in these time period (Reish and Oshida, 1987). For the assessment of hazard of chemicals in view of legislative measures, sequential and tiered testing programme were developed (EPA, 1978, EC 1979, OECD, 1979) starting with short term testing and ending with long term test. Later several sophisticated semi chronic and chronic tests were designed taking into account sublethal criteria like growth and reproduction, histology, haematology, toxico kinetics and. studying physiological and biochemical effects. (Macek, 1980). Assessment became more complicated by the introduction of lots of hazardous chemicals and their mixtures. Gradually ecological significance were established and the weakness of proposed testing and evaluation methods were realised (Cairns, 1981).

At the last of 20th Century, toxicitogical analysis of different hazardous chemicals and their degraded forms, has been done by new techniques and their residual concentration has been quantified - which gives a new dimension to toxicological research (Murty, 1988).

ACUTE TOXICITY

Acute toxicity denotes relatively short term lethal or other effect; usually defined as occurring within 4 day for fish and macro invertebrates and shorter times (2 day) for organisms with shorter life spans. (APHA, 1997). It has two fold application in environmental risk analysis i.e. toxicological screening and acute toxicity determination. Acute toxicity is better expressed in the terms of LC_{50} or lethal concentration 50. It is the concentration of a substance that is lethal to 50% of the test organisms in 96 hours. The 96 hour bioassay is advantageous because of minimal laboratory requirements and rapid results (FAO 1987). A major demerit is that it does not give any information about the site of toxic action. Several factors like environmental parameters, seasonal variation, species multiplicity, age and more over physiological conditions of the organism take great role for the differences of LC_{50} valuations (Singh and Narain, 1982). Especially, in short term experiments, the relatively hydrophobic chemicals, steady state is not achieved and the toxicity may be severely underestimated (Van Hougen, 1985).

Though many reports have been published describing detailed methods and procedures of use in acute toxicity test with aquatic biota (Jones 1940, Katz, 1961), recommended procedures are provided by the U.S. Environmental Protection Agency (1978), The American Public Health Association (1985). The Organization of European Community Development (OECD, 1981).

A large number of papers had been published on the acute toxicity of pesticides to fishes (Mukhopadhay *et.al,* 1980, Wijngaarden *et.al,* 1993; Fernandez, *et.al,* 1996; Domitrovic 1997; Abdelghani *et. al,* 1997, Farrell *et.al,* 1998; Pathiratne *et.al,* 1998; Dalvi *et.al,* 1998; Rebach, 1999)

TEST ANIMAL

According to United States Environmental Protection Agency (USEPA, 1979) the selection of test organisms is to be based on four criteria:

(i) The organism must be a representative of an ecologically important group.

(ii) The organism must occupy a position within a food chain leading to man.

(iii) The organism must be widely available, amendable to laboratory tests, easily maintained and genetically uniform.

(iv) There must be adequate background data on the organism *(i.e.* on physiology, genetics, taxonomy etc.)

Rosenberger *et. al.* (1978) added other criteria which included economic importance, type of test, sensitivity to the toxicant and consistency in response to the toxicant.

Many researchers has studied several kinds of fish species for some of their merits. Lloyd (1963) worked on the toxicity *of Salmo gairdneri;* Vittozzi and Angelies (1991) on *Lepomis macrochirus;* Gomez *et.al.* (1998) on tench *(Tinca tinca* L.); Pathiratne and George (1998) on Nile tilapia *(Oreochromis niloticus);* Dalvi and Davis (1998) on Channel catfish; Soengas *et.al* (1997) on rainbow trout, Tsuda *et.al.* (1997) on killifish *(Oryzias latipes)* etc.

Newsome (1980) examined the suitability of several species of fish for use in multigeneration toxicity test and rejected American flag fish *(Jordanella floridae),* gold fish, Zebra danio *(Brachydanio rario)* but considered convict cichlid *(Cyclosoma nigro fasciatum)* to be the most suitable.

In India, test species selected by Bhattacharya *et. al.* (1988) was *Channa punctatus,* by Mukhopadhay *et.al.* (1984) was *Clarias batrachus,* by Dutta (1980) was *Colisa faciata,* by Devi (1984) *Puntius ticto,* by Fernandez, *et.al.* (1996) were *Etroplus maculatus,* by Thakur and Bais (1999) on *Heteropneustes fossilis* (Bl.), by Santhakumar and Balaji (2000) were *Anabas testudiueus* (bl.), etc.

For determining toxicity of a wide varieties of toxic substances, species such as G. giuris, *Clarias batachus,* and *Sacrobranchus fossilis* are not true sensitive fish species. They were included in most cases due to local availability and further easy maintenance in the laboratory (Das, 1996).

Jhingran (1982) has made a literature survey of Indian scientists and published a variable acute toxicity chart of different fish species and analysed their prospective results. Fish size, sex, age, environmental condition, and more over volume of test solution may alter the result.

CHRONIC TOXICITY AND SAFE CONCENTRATION

The ultimate aim of chronic toxicity testing is to determine the significant adverse effects on aquatic ecosystem. The various types of chronic tests are 30 day tests, multigeneration and whole life cycle tests, partial life cycle and embryo-larval tests etc. The life stages of fishes most sensitive to toxicant are generally early life stages (Mckim, 1985; Van leeuwen *et.al.,* 1985). Mckim (1985) compared 56 fish life cycle tests involving 34 chemicals and 4 species of fish and showed that the embryo-larval tests offer the same information with comparable dependability as the long-term chronic tests or from full life cycle toxicity test. Fishes survival and activity will be longer when concentration of test chemical decreased. Tenfold reduction of lethal concentration gives a tenfold increase in survival time (Jones, 1964).

Safe concentration is the maximum concentration of a toxicant that has no observable harmful effects after long-term exposure over one or more generations. Sprague (1971) emphasized that the term "Safe level" is not an entirely satisfactory one, but in the absence of a better term, it has gained wide acceptance. Lots of scientist and many organization has given formula and suggestions for the estimation of safe concentration. Hart *et.al.* (1948) proposed the formula which is already established and popular. European Fisheries Advisory Commission (EFAC, 1979) suggested that a concentration less than four orders of magnitude lower than the 96h LC_{50} is an acceptable or safe concentration. Another approach to arrive at safe level is to use the one tenth of the 48h LC_{50} value.

Mount and Stephan (1969) proposed to use Laboratory Fish Production Index (LFPI), which incorporates data generated over at least one generation of the test species, regarding effects on survival, growth, reproduction etc. to calculate the Maximum Acceptable Toxicant Concentration (MATC). It may be recalled that MATC is a range of concentration, the lower end of which does not produce any adverse effect and upper end is the lowest concentration that would cause adverse effect on some life stages. Later Mount and Stephan proposed that the AF (application factor) could be calculated by dividing the MATC by 48 or 96h LC_{50} value. The use of AF for calculating safe levels for other species was explained and discussed in detail by Eaton (1973).

Johnsoñ and Julin (1980) considered that MATC is acceptable more than other methods because it is less time consuming, less expensive to generate and are widely accepted.

PESTICIDES

Carbamates

Carbamates are moderately toxic to fish, but highly toxic to invertebrates, and introduced into the late 60's for its modified characters over organophosphates. Among them, Carbofuran and Carbaryl are commonly used (Johnson *et.al.*, 1980) Singh, *et. al.* (1990) worked on the pesticides (carbamate) induced changes in circulating thyroid hormones in the freshwater catfish, *Clarias batrachus*. Belisle *et.al.* (1988) did specific analysis of Carbamate Pesticides in sediments using column extraction and gas chromatography. Bhattacharya *et.al.* (1992) and Bocquene *et. al.* (1993) studied on the effect of carbamate on acetyl cholinesterase system of fish. Hota *et.al.* (1993) analysed the metabolic effects of kilex carbaryl on a freshwater teleost, *channa punctatus* (bloch.).

Few researchers has also concentrated their studies on the impact of Carbofuran to several species of fishes. But no or very few of them has emphasised on Indian Major Carps. Verma *et. al.* (1982) worked on the determination of LC_{50} value of Carbofuran, on fish species like *Mystus vittatus, Sacrobranchus fossilis* etc. Gill (1980) concluded from experiments that Carbofuran was metabolised by the microsomal-NADPH system in fish, *Trichogaster pectoralis,* oxidative N-hydroxylation was the major route of metabolism of carbofuran. Ram (1988) studied on the Carbofuran induced histophysiological changes in thyroid of teleost fish, *Channa punctatus* (bloch.). Hohrciter *et.al.* (1991) worked on the effects of insectisides carbofuran on adenylate parameters (ATP concentration adenylate energy change etc.) in bluegill sunfish *(Lepomis macrochirus)*.

Carbofuran alone or in combination with other carbamate insecticides influences the level of reproductive and metabolic hormones such as thyroxine and corticosterone, and results in impairment of endocrine, immune and behavioural functions (Yoon J. Y. *et al.* 2001).

Synthetic Pyrethroid

Among synthetic pyrethroids, important pesticides are Cypermethrin, Deltamethrin, Fenvalerate, Esfenvalerate etc. Although Pyrethroids are extremely toxic under laboratory conditions, in the field their toxicity is significantly reduced, presumably by adsorption of the toxicant to suspended solids (NRCC, 1986). Though previously, very little information was available for Pyrethroid research, in last decade of twentieth century, it got a little bit momentum. Symonik *et.al.* (1989) studied the effect of fenvalerate on metabolic ion dynamics in fathead minnow *(Pimephales promelas)* and blue gill *(Lepomis macrochirus)*. In India Malla-Reddy *et.al.* (1992) studied on fenvalerate toxicity to the system of Acetylcholinesterase (ACHE) in *Cyprinus carpio* Ohtsuka (1993) worked on synthesis and insecticidal activity of silicon-containing pyrethroid with low fish toxicity, Barry *et.al.* (1995) analysed the effect of age and weight specific respiration rate on toxicity of esfenvalerate to the Australian Crimson spotted rainbow. fish *(Melanotaenia fluviatilis)*. Bhatnagar (1995) described Pyrethroid induced toxicity to Phosphatase in *Clarias batrachus* (Linn.)

Cypermethrin has been developed in England by Elliot *et.al.*, 1975. There is very little information available regarding the toxicity of Cypermethrin to fish. Only some early life stage tests and detection has been performed. The acute toxicity of Cypermethrin to Brown trout, Rainbow trout, Tilapia was studied by Stephenson (1982). Singh *et.al.* (1993) worked on the effect of Cypermethrin on lactate and succinate dehydrogenase and cytochrome oxidase of fish and Reddy *et.al.* (1995) studied on sublethal concentration of Cypermethrin to the protein metabolism level in selected tissues of *Cyprinus carpio*. Davies (1993) described on sublethal effects on brown trout, Salmo trouts, caused by Cypermethrin spraying on Tasmanian stream and Cagauan (1992) worked on rice cum fish culture and the possible use of Cypermethrin on it. Very little or almost no works has been done on the Cypermethrin toxicity to Indian Major Carps and its different sublethal effects. So, by considering the up rising importance of Cypermethrin in aquaculture, it is a new and challenging avenue of research. Ernst *et al.* (2001) worked on the dispersion and toxicity to non-target aquatic organisms of Cypermethrin used to treat sea lice on salmon in net pen enclosures and analysed that Cypermethrin was having the potential to cause toxic effects over areas of hectares from a single release.

ENVIRONMENTAL PARAMETERS

Environmental parameters have a great role in effecting the toxicity of different pesticides. The major parameters are temperature, dissolved oxygen, pH, hardness, alkalinity etc.

Among all the environmental parameters, the effect of temperature on the toxicity of pesticides has been studied a lot and it is very important because it directly influence factors like enzyme activity, metabolic rate, oxygen uptake etc. Generally, toxicity increases with the high temperature. Macek (1969) studied the toxicity of 10 pesticides to the rainbow trout and 11 pesticides to the bluegills at different temperatures and found that the toxicity increased with increasing temperature. Same kind of results has been reported by Singh *et.al.* (1982) on endosulfan to *Heteropneustes fossilis*, by Schoettger (1970) on endosulfan to rainbow trout, by Duangsawasdi *et.al.* (1979) on fenitrothion and acephate to rainbow trout. But a completely opposite results has been shown by DDT and some pyrethroids where toxicity is less at higher temperature. Ogilvie and Anderson (1965) showed that

DDT exposed to Atlantic salmon differs from control. Mauck *et.al.* (1976) described that natural pyrethrum and some synthetic pyrethroids (dimethrin, RU-11679, resmethrin) was higher at 12°C than 17°C and dimethrin were more toxic on 17°C than 22°C. Cairns (1975) noted that though vast literature available on effects of temperature on aquatic organisms, yet when one tries to apply this information to specific interactions, often very little evidence is available.

pH has an excellent role in influencing various physiochemical properties of pesticides like hydrolysis, volatilization and in balancing the dissociated and undissociated forms (Weber, 1972). With few exceptions, the toxicity of organochlorine (OC) and organophophate (OP) compounds are not influenced by pH. The toxic effects of 2, 4-D were reduced when the pH was raised by the addition of sodium chloride (Holcombe *et al.*, 1980). Malathion is toxic below pH 7, and loses its toxicity in alkaline pH (Bender, 1969). In the case of carbamate compounds, generally toxicity was reported to increase with pH. Mexacarbate was 38 times more toxic to bluegills at pH 9.5 than at 7.5 (Mauck *et al.*, 1977). The toxicity of picloram to culthroat trout and lake trout increased with a rise in pH from 6.5 to 8.5. (Marking et.al., 1981). Generally, the activity of synthetic Pyrethroids was not altered by a change in the pH, the biological activity of natural pyrethrums was influenced by pH, its toxicity being higher at lower pH (Mauck *et.al.*, 1976).

Hardness of water does not seem to influence the toxicity of OC and OP Compounds (Pickering *et.al.* 1962). The toxicity of three formulations of 2, 4 D, three formulations of endothall, fenoprop., PCP and dichlrobenil was not affected by hardness. (Inglis *et.al.*, 1972). Lioyd and Jordan (1964) found that survival time of Rainbow trout in rapidly lethal pH value shortened with decrease of calcium carbonate content of water. Hardness of water had no effect on toxicity of maxacarbate (Mauck *et.al.* 1976) and Carbaryl (Johnson *et.al.* 1980). While water hardness had no effect on the toxicity of synthetic Pyrethroids, natural pyrethrums was more toxic in hard water (Mauck *et.al.* 1976).

Not much work seems to have been carried out on the influence of low oxygen on the toxicity of pesticides to fish. Davies (1975) attempted to formulate the criteria for minimum dissolved oxygen requirement of fish. His approaches was on examining the threshold levels of dissolved oxygen that cause changes in some physiological parameters. Oxygen deficiency leads to histological lesions. Channel Catfish exposed for 72h to an oxygen content of 1.5 ppm showed anomalies in gill, liver, kidney and spleen (Scott and Rogers, 1980).

BEHAVIOURAL STUDY

Behaviour (or activities) of an organism represents the final integrated result of a diversity of biochemical and physiological processes and behavioural patterns are known to be highly sensitive to changes in the steady state of an organism (Warner *et al.*, 1966). Lots of scientist expressed the patterns of behavioural changes which varies according to fish species, concentration of chemical, physico-chemical conditions and on rate of excretion of fish species. Poisons with a resorptive action, the symptoms are manifested primarily through their action on the nervous system, cause a rapid loss of equilibrium in fish (Metlev *et al.*, 1971). Symtoms of Furadon and malathion toxicity were reported by Konar and Ghosh (1982). Within 24 hours of exposure to 0.5 ppm furadon, fish had shown slow opercular movement, lethargy and occassional jumping. After 48 hours, fish showed loss of balance and bending of body while malathion shown loss of balance, lethargy and eratic swimming behaviour. Heath *et al.* (1993) observed that high concentration of carbofuran caused an impairment of swimming performances in Japanese medaka *(Oryzias latipes)*. Sagilo *et al.* (1996) worked on the effects of concentration (1, 10, 100 μg/L) and duration (4, 8, 12 h.) of exposure to carbofuran were assessed on the swimming activity, social interactions and behavioural responses of goldfish to a flow of water and observed a rate and concentration dependent increase in behavioural pattern and after 12 hour, it was less apparent.

Shin *et al.* (2001) worked on the effect of diazion on the behaviour of Japanese medaka *(Oryzias latipes)* and gene expression of tyrosine hydroxylase (TH) as a biomarker and concluded that a suppression of TH activity would be one of the causes for the abnormal behaviour of the medaka that could be quantitatively monitored using an image processing system. This study provides molecular and neurobehavioural bases of a biomonitoring system for toxic chemicals using a model organism such as fish.

Bretaud *et al.* (2002) described about a correlation between biochemical and behavioural effects of carbofuran in gold fish, *(Carassius auratus)*. They expressed that behavioural endpoints related to swimming pattern or social interactions had a direct relation to dose and duration of carbofuran exposed biochemical parameters like brain catecholamines, plasma glucose, hepatic glycogen content etc. and relative sensitivities of these different types of responses has been explained in the light of data on neurotoxic effects of carbamate insecticide in fish.

HISTOPATHOLOGY

Now-a-days, Histopathological studies of different tissues exposed to toxicants has got enormous attention. Tissues changes due to sublethal doses of toxicants not only giving an idea about the nature of toxicant but also predicting upon its mode of action or mechanism of toxicity, about the probable physiological impairment and assist greatly in the diagnosis of nature of toxicant (Nagler, 1984). Histology of fish and aquatic invertebrates is still a less studied discipline compared to counter parts in mammals (Mayers and Hendricks, 1985).

Liver

The liver is an organ concerned with several vital functions such as absorption of digested food stuff, detoxification, excretion of harmful substances, synthesis of several components of blood plasma, storage of glycogen, release of glucose and control over general metabolism. Hence its study in fish serves as a suitable index for toxicity of polluted water (Sahai and Singh, 1989).

Because of its physiological importance, already some literatures are available on liver. Couch (1975) reviewed the histopathological effects of pesticides and related chemicals on the livers of fish and concluded that many of them were nonspecific. Bhattacharya *et. al.* (1975) observed liver cord necrosis and cytoplasmic and nuclear disintegration of the hepatopancreas in the teleost *Clarias batrachus* due to endrin. Amminikutty and Rege (1977) reported that chronic exposure to thiodon and agallol caused rapid degeneration and vacuolation of hepatocytes in liver fish. Dubale and Shah (1979) observed coagulation of blood in sinusoids, necrosis and cytoplasmic disintegration in the hepatopancreas of *Channa punctatus* exposed to malathion. Altamirano-Reynoso (1984) analysed that lindane caused hydropic degeneration and the tubular epithelium necrosis were the most frequent microscopic injuries observed in liver of *Tilapia hornorum.* Sahai and Singh (1984) described vacuolation of hepatocytes of *Rasbora dandricus* while exposed to malathion toxicity.

Urdaneta (1989) reported hepatic necrosis due to effect of 12 pesticides in some species of cultivable fishes. Narayan and Singh (1991) observed extensive degeneration of cytoplasm, pyknosis of nuclei and loss of glycogen of *Heteropneustes fossilis* while subjected to acute thiodon. John *et.al.* (1993) described the impact of endosulfan on the liver of *Cyprinus Carpio* and expressed that extensive vacuolation of the cells, indistinct cell boundaries, loss of polygonal shape of the cells and degenerative necrosis were the histopathological changes observed in the liver.

Our study on the impact of Carbofuran and Cypermethrin on the hepatocytes of I.M.C. *(Labeo rohita)* is a important and new avenue of research.

Ovary

Some literatures are available on the ovarian damage due to pesticide toxicity but very few or no researches has done on the histopathological changes of ovarian cycle on Indian Major Carps. In India, most of the ovarian research has been concentrated on air-breathing fishes or cat-fishes and in OC and OP pesticides. Kulshrestha, *et al.* (1984) worked on the exposure of endosulfan and carbaryl on the ovaries of *Channa straitus* (bloch.) and observed reduction in the number of oocytes, increased number of damaged oocytes, development of inter follicular spaces, reduction in gonado-somatic index etc. Shukla *et.al.* (1984) have noted decreased ovarian activity and atretic oocytes in *Saratherodon mossambica* exposed to Malathion and Ghosh *et.al.* (1985) has described on damage in ovary, viz., degeneration of follicular, wall, ooplasm and connective tissues due to malathion toxicity on *Heteropneustes fossilis* and Ansari *et.al.* (1987) worked on malathion toxicity on the ovary of zebra fish *(Brachydanio rerio)*. Singh *et.al.* (1991, 1992, 1993) studied extensively on the impact of gamma-BHC to the steroid hormonal profiles of *Heteropneustes fossilis* and correlated it to ovarian damage, and shown that pesticide suppressed the levels of these hormones, viz. testosterone (T), estradiol-17 beta. 17 alpha-OHPg etc and described that Ovine lutenizing hormone releasing hormone and mystus gonadotropin has modulatory role on it. Pandey (1988) worked on the impact of endosulfan (thiodon) EC 35 on dynamics of oocyte development in the teleostean fish, *Colisa (Trichogaster) fasciatus* and described that ovarian activity was retarded greatly, the diameter of oogonia and stage I oocytes was greatly reduced and ovarian wall become thicken. Hazarika (1998) worked on the toxicological impact of different sublethal doses of BHC on the ovary of the air-breathing catfish *Heteropneustes fossilis* (bloch.) and investigated detailed histopathological changes.

Very few works has been done on the pesticidal effect on ovarian dynamics. Nagler (1984) investigated sublethal effect of pentachlorophenol (PCP) on ovarian development in rainbow trout *(Salmo gairdneri)*. Kling (1986) worked vividly on the toxic effects of sublethal doses of Lebaycid upon the ovaries in *Sarotherodon leucostictus* and observed that exposure caused total atresia, inhibition of vitellogenesis in the gonads and decreased gonado-somatic index. Peczely (1998) described effects of a fungicide (VITAVAX 200 EF) on the reproductive-endocrine functions in mallards and observed ovarian changes and changing pattern of steroid levels.

Reproduction

Reproductive toxicity indicates changes on the pattern of breeding response, on fecundity, on fertilization rate, hatchability of larvae and above all survivability of larvae, fry etc. Though, it is a prime subject of research because it directly relates to productivity of fishes. Very less researches has been conducted in this field. Very less or no research has been done on the effect of carbamate and synthetic Pyrethroid on fishes specially to Indian Major Carps.

ON FECUNDITY, FERTILIZATION ETC.

At concentration above the threshold toxicity, many organic compounds caused large pregnant female of *Gambussia affinis* to abort (Boyd, 1963) Following the feeding of DDT to brook trout at different doses for 22 weeks, the fishes fed with lower doses showed significantly higher number of ova than those of higher dose (Macek, 1968). In experimental ponds, 2, 4-D butylester at the highest concentration caused a delay in the spawning of blue gills (Cope, *et al.*, 1970). Oviposition of *Oryzias latipes* was suppressed heavily by the 96 hour exposure of chlordecone (Curtis, 1978). Fenitrothion, at sublethal concentration affected egg production, whereas temephos at sublethal concentration affected normal birth (Yasuno *et.al.* 1980). Fecundity and fertility of female sheepshead minnows exposed to different doses of chlordecone shown marked reduction (Goodman *et.al.*, 1982). Several doses of mirex also effected adversely to the spawning and fecundity of fathead minnows. Bhattacharya (1989) worked on histopathological changes of ovary due to pesticides and correlated that ovarian damages also interfering with the breeding performances of fishes, Petersen *et.al.* (1997) analysed that xenobiotics like pesticide may influence the viability of eggs and larvae. Oern *et al.* (1998) did extensive research of the impact on reproduction of an orally administered mixture of selected PCBs in zebra fish *(Danio rerio)*. He treated the fishes with different level of PCB concentration for 13 weeks period and observed that egg production was reduced in all three groups exposed and intermediate and higher groups contained a reduced number of mature oocytes.

ON HATCHING

Holden (1972) reported that brown trout eggs containing aldrin failed to hatched. Hatching of blue gill eggs exposed to a different doses of a formulation of fenoprop (Kuron) was not affected, but all the fry hatching from eggs exposed to the highest concentration died (Wilbur *et.al.*, 1973). On the contrary, hatchability of coho salmon eggs, exposed to PCBs, was much reduced. (Halter, 1974). Eggs of DDT exposed adults showed abnormal gastrulation, and 39% of them had vertebral deformities upon hatching. The percent of deformed larvae was dose-dependent. (Smith *et.al.*, 1974). Embryos of Atlantic silverside was treated singly with P.P-DDT, malathion or Carbaryl and was observed that they reduced the survival time of the embryoes. These effects were observed even at concentration that may occur temporarily in the environment (Weis *et.al.* 1976). Embryos, sacfry, and larvae of Caranx exposed to lindane had advanced hatching and larval characters. Yolk utilization and sac absorption were also impaired. (Venugopalan *et.al.*, 1979). Hatching success of fish eggs exposed to low concentration of mirex increased and hatching of fathead minnows eggs exposed to different concentration of chlordecone was significantly lower than that of the controls. (Buckler *et.al.*, 1981). No differences in hatching frequency or median hatching time were recorded in zebra fish *(Danio rerio)* which were orally exposed to a mixture of 20 PCBs in three different dose levels (Oern *et.al.* 1998).

HAEMATOLOGICAL PARAMETERS

Though physiological effects of toxicant or to diagnose the disease are well established in human and veterinary medicine, yet in fish these clinico-pathological tests are yet to be standardised. In this background haematological studies present variation in blood parameters, as reliable indices of fish health.

These parameters are influenced and modified by many factors such as handling (Houston *et.al.* 1971), metabolism (Booke, 1964), age (Tugarina and Ryzhova, 1970), maturity (Qureshi *et.al.*,

1971), sex (Steveke and Atherton, 1965), diet (Smith, 1968), stress (Bouch and Ball, 1966) and pollutants (Spraque, 1971). Some of researches has already been conducted on haematological parameters by several types of pesticides on different fish species by the scientists of India and abroad. Witt *et.al.* (1966) found a good correlation between DDT levels in blood and those in adipose tissue. Moss *et.al.* (1966) shown that entering the circulatory system, pesticides rapidly bind to the blood proteins. A sublethal concentration of malathion induced decreased erythrocyte and increased leukocyte count and increased serum-free amino acid in a freshwater catfish (Chliomovitch *et.al.* 1977). *Channa punctatus* expoosed to sublethal level of quinalphos for 15 days had lowered haemoglobin content (Mebride *et.al.* 1981). *Clarias batrachus* were treated with dichlorvos and shown decreased values of blood parameter like Hemoglobin, red blood cell counts, haematocrit, MCV and MCH values (Rajeswari *et.al.*, 1989). Gill *et.al.* (1991) worked extensively on the haematological impact created by different pesticides (endosulfan, phosphamidon, aldicarb etc.) to several fish species *(Puntius conchonius, Bar bus conchonius* etc.). Their selected parameters were mainly differential count, or biochemical analysis of blood etc. When *Heteropneustes fossilis* were treated with carbomate pesticide, Propoxur at different sublethal dose, a decreased result of TLC, TEC, Haemoglobin and increased values of ESR, clotting time was observed. Same sorts of trend was also seen in *Oreochromis mossambicus*, Ekalux (Sampath *et.al.* 1993), in *channa punctatus* by malathion, in *Cyprinus carpio* by endosulfan (Chandrsekar *et.al.* 1993), in *Heteropneustes fossilis* by lindane (Nath *et.al,* 1996) in *Heteropneustes fossilis* by Thiodan (Ramamjam *et.al.*, 1997), in *Cyprinion watsoni* by malathion (Khattak *et.al.*, 1996). A opposite data were also presented in *Channa punctatus* with Monocil (Mishra *et.al.*, 1994).

Blood glucose and hepatic glycogen can be utilized as a parameter of stress response (Chavin, 1973). Relatively innocuous stimuli such as aquarium transfer (Chavin and Young, 1970), thermal shock (Wede and Meyer, 1973) and putting them in a small shaking cages (Skeda *et.al.* 1975) elicited hyper glycemic response in fish.

Holmberg *et.al.* (1972) worked on the impact of Pentachlorophenol (PCP) to eel *Anguilla anguilla* and analysed that hyperglycemia and increased blood glucose was evident, which persisted for 55 days after transfer to clean water. Blood glucose and hepatic glycogen profile in *Channa punctatus* exposed to non-lethal concentration of single and mixture pollutants respond diversely to two types of treatments (Bhattacharya *et.al.*, 1987). According to her, blood glucose profile can be utilized as indicator of pollution stress in fish. Generally blood glucose value were elevated after pesticide exposure to fish. These sorts of result has been observed in *Auguilla anguilla* by the application of endosulfan (Gimeno *et.al.*, 1995), in *Cyprinus carpio* by deltamethrin (Balint *et.al,* 1995), and (Chandrasekar *et.al,* 1993), in *Orechronis mossambicus* by quinolphos (Ramakrishnan *et.al.* 1993) etc. But a very exceptional and opposite case was also recorded in *Channa punctatus* by Sastry *et.al.*, (1982). Gimeno *et.al.* (1994) analysed the correlation profile of glucose content between blood and liver of *Anguilla anguilla* and with endosulfan intoxicants commented that liver glycogen content decreased during exposure period and at the same time glucose content increased in blood.

Though lots of works has already been done on haematological profiles in fish in relation to pesticide, no work has been done on the impact of carbofuran and cypermethrin to Indian Major Carps, to especially to *Labeo rohita.*

ANTIDOTE STUDY

To counter the pesticidal effects of water bodies and to increase the survivability level of fishes and other living organisms in water, is the ultimate aim of Pollution research and burning work area for contemporary scientists. Almost no references are available in this field except some basic approaches.

Zamfir (1979) worked on the possibilities for the removal of Pesticide Polluted water in treatment stations and described some methodologies, i.e. flocculation and filtration can remove partially-DOT, 2, 4, 5-T, edrin, parathLon and lindane, Chlorine oxidation can remove parathion, diuron etc., ozone and potassium permanganate appear to exert effects similar to those of chlorimation. Activated charcoal has positive effects in the removal of absorption of most pesticide and U-V ray also can remove a certain amount of pesticides.

Some indirect approaches has also been taken by some scientists and their methods were environmental or nutritional manipulation. Sado *et.al.* (1992) reported that increased temperature and optimum level of dissolved oxygen (by aerator) can decrease the pesticidal action and Das (1996) described for the application of lime to increase the pH for counteraction. Ghazahy (1994) and Mukherjee (1996) commented for the efficacy of ascorbic acid (Vitamin-C) for the intoxication of different pollutants including pesticides. But application of different herbal extracts for this purposes are a very purposeful and new avenue of research.

CONCLUSION

Most of the pesticides acts on the respiratory process and cholinergic nervous system and hampers the cell metabolism. It is necessary to find a suitable antagonist or antidote to check the poisoning caused due to pesticides/pollutant and to restablish the life processes. Though it is the ultimate aim of pollution study, no such antidote or mechanism for controlling pesticide toxicity to fish is developed. Considering all the above points some plant extracts and organic matters has been applied as antidote to counteract pesticidal action. But it need much more studied to come out successful sustainable aquaculture with some antidotes against these pesticides for proper environmental impact assessment.

ACKNOWLEDGEMENTS

The authors are thankful to the Director, Central Institute of Freshwater Aquaculture, Kausalyaganga, Bhubaneswar -751 002, Orissa, India for carrying out this work and the Research Scholars of Soil and Water Chemistry Laboratory, Central Institute of Freshwater Aquaculture, Kausalyaganga, Bhubaneswar -751 002, Orissa, India for their valuable co-operation to prepare this manuscript.

REFERENCES

Abdelghani, A. A., Tehounwou, P.B., Anderson, A. C., Sujono, H., Heyer, L.R., Monkiedje, A. 1997. Toxicity evaluation of single and chemical mixtures of round up, Garlon-3A, 2, 4-D and syndets surfactant to channel catfish *(Ictalurus punctatus)*, bluegill sunfish *(Lepomis microchirus)* and craw fish *(Procambarus* sp.). Environ. Toxicol. Water-qual 12(3): 237-244.

Altamirano-Reynoso, R. 1985. Microscopic liver and kidney injuries in *Tilapia hornorum* caused by an acute intoxication with a commercial product that had lindane Thesis or Dissertation Universidad Nac. Autonoma. de Mexico (Mexico). Fac. De Med. Vet. Zootec.

APHA 1997. Standard methods for the examination of water, sewage and industrial waters. Am. Publ. Health, org. (17th Edn.) Newyork PP: 522.

Balint, T., Szegletes, T., Szegletes, Z., Halasy, K. and Namesok, J. 1995. Biochemical and subcellular changes in Carp exposed to the organophophorus methidathion and the pyrethroid deltamethrin. Aquat. Toxicol. 33 (3-4): 279-295.

Barry M.J., Logan D.C., Van-Dam, R.A., Ahokas, J.T. and Holdway, D.A. 1985. Effect of age and weight specific respiration on toxicity of esfenvalerate pulse-exposure to the Australian Crimson-spotted rainbow fish *(Melanotaenia fluviatilis)*. Aquat. Toxicol. 32(2-3): 115-126.

Belisle, A. A. and Swineford, D. M. 1988 Simple, specific analysis of organophosphorus and carbamate pesticides in sediments using column extraction and gas chromatography. Environ. Toxicol. Chem. 7(9):749-752.

Bender, M.E. 1969. The toxicity of the hydrolysis break down products of malathion to the fathead minow *(Pimephales promelas* Raf). Wat. Res. 3 : 571-582.

Bhatnagar, M.C., Tyagi, M, and Tamata, S. 1995. Pyrethroid induced toxicity to phosphatase in *clarias batrachus* (Linn.) J. Environ. Biol. 16(1): 11-14.

Bhattacharya, S., Mukherjee, S. and Bhattacharya, S. 1975. Toxic effects of endrin on hepatopancreas of the teleost fish. *Clarias batrachus* (L). Ind. J. Experiment Biol., 13 : 185-186.

Bhattacharya, T., Roy, A.K. and Bhattacharya, S.I 987. Histopathology of the kidney of *Channa punctatus* (Bloch) exposed to sublethal and chronic levels of Industrial effluents.II.Zool. Jb. Anat. 115: 263-272.

Bocquene, G., Galgani, F., Burgeot, T., Le-dean, L. and Truquet, P. 1993. Acetylcholinesterase levels in marine organisms along French Coasts. Mar. Pollut. Bull. 26(2): 101-106.

Booke, H.E. 1964. Blood serum protein and calcium levels in yearling brook trout. Prog. Fish. Cult. 26:107-110.

Bouck, G.R. and Ball, R.C. 1966. Influence of capture methods on blood character and mortality of rainbow trout *(Salmo gairdeneri)*. Trans. Am. Fish. Soc. 95(2):170-176.

Boyd, C.E. 1963. Insecticides cause mosquito fish to abort, Prog. Fish. Culturist, 26 : 138.

Bretaud, S., Saglio, P., Saligaut, C., Auperin, B. 2002. Biochemical and behavioral effects of carbofuran in gold fish *(Carassius auratus)*. Environ. Toxicol. Chem. 21(1): 175-81.

Brown, V.M., 1968. The Calculation of the acute toxicity of mixtures of poisons to rainbow trout, water Res. 2 : 723.

Brungs, W.A. and Mount, D.I. 1978. Introduction to a discussion of the use of aquatic toxicity tests for the evaluation of the effects of toxic substances, in Estimating the Hazard of chemical substances to Aquatic Life. ASTM STP 657, Cairns, J., Jr., Dickson, K.L. and Maki A.W. Eds. American Society for Testing and Materials, Philadelphia, 15.

Buckler D.R., Witt. A. Jr., Mayer, F.L. and Huckins, J.N.1981. Acute and Chronic effects of Kepone and mirex on the fathead minnow, Trans. Am. Fish. Soc. 110:270.

Buikema Jr., A. L. Nider-Lehner and Cairns; J. Jr. 1982. Biological monitoring Part IV-Toxicity testing. Ibid. 16:239-262.

Cagauan, A.G. and Arce R.G., 1992. Overview of pesticide use in rice fish farming in South-east Asia. Iclarm. Conf. Proc. no. 24.

Cairns J. Jr. 1981. The effects of the temperature and hardness of water upon the toxicity of zinc to pond water snail *Physa heterosstreplie*. Notulac Naturae. 308:1-11.

Cairns, J., Jr., Heath, A. G. and Parker, B.C. 1975. The effects of temperature upon the toxicity of chemicals to aquatic organisms. Hydrobiologia, 47 : 135.

Carlson, A.R. 1971. Effect of long term exposure of Carbaryl on survival, growth and reproduction of the fat head minnow *pimephales promelas.* J. Fish. Res. Bd. Canada, 29 : 583-587.

Carlson, A.R. 1971. Effects of long term exposure to Carbaryl (Sevin) on minnow (*Pimephales promelas*). J. Fish Res. Board. Can. 29 : 583.

Chandrasekar, S. and Jayabalan, N. 1993. Haematological responses of the common carp., *Cyprinus carpio* L. exposed to the pesticide endosulfan. Asian Fish. Sci. 6 (3): 331-340.

Chavin, W. 1973. In response of fish to invivo-mental changes. Eds. W.C. Chavin. Spring field. Llinoisis. PP.199.

Chavin, W. and Young, J.E. 1970. Factors in the determination of normal serum glucose levels of gold fish. Carassius auratus J. Comp. Biochem. Physiol. 33:629-653.

Chilamovitch, Y. P. and Kuhn, C. 1977. Behavioural, haematological and histological studies on acute toxicity of bis (tri-n-butyltin) oxide on *salmo gairdneri* Richardoon and *Tilapia rendalle* Boulenger. J. Fish. Biol. 10 : 1575.

Cope, O.B., Wood, E.M., and Watten, G. H. 1970. Some chronic effect of 2, 4 D on the bluegill (*Lepomis* macrochirus). Trans. Am. Fish. Soc., 99 : 1.

Couch, J.A. 1975. Histopathological effects of pesticides and related chemicals on the liver of fishes. In:Pathology of fishes. Ribelin and Migaki (eds). The University of Wisconsin Press. PP.559-584.

Curtis, L. R. and Beyers, R. J. 1978. Inhibition of oviposition in the telost *Oryzias latipes*, induced by subacute kepone exposure, Comp. Biochem. Physio. 61:15.

Dalvi, R.R. and Davis, S.W.1 998. Role of beta naphthoflavone in the acute toxicity of chlorpyrifos in channel catfish. Bull. Environ. Contam. Toxicol. 60(2) : 335-339.

Das B. 1996. Phd. Thesis. Orissa University of Agriculture and Technology, Bhubaneswar, Orissa.

Davies, J.S. 1975. Minimal dissolved oxygen requirements of aquatic life with emphasis on Canadian species: a review. J. Fish. Res. Bd. Can. 32(12) : 2295-2332.

Davies, P.E. and Cook, L.S.J. 1993. Catastrophic macroinvertebrate drift and sublethal effects on brown trout, *Salmo trutta,* caused by cypermemrin spraying on a Tasmanian Stream. Aquat. Toxicol. 27(3-4): 201-224.

Days, K.E. 1991. Effects of dissolved organic carbon on accumulation and acute toxicity of fenvalcrate, deltamethrin and cyhalothrin to *Daphnia magna* (Straus). Environ. Toxicol. Chem. 10:91-101.

Decic, S.I.V. and Lewis, S.M. 1991. Practical Haematology (VII Edn). J. and A Churchill Ltd., Livingston London, Melborne and New York.

Domitrovic, H.A. 1997. The use of autochthonous fishes in toxicity tests. Evaluation of the species Acquidens portalegrensis. Rev. Ictiol. 5 (1-2): 37-42.

Donaldson, E.M. 1975. Physiological and Physiocochemical factors associated with maturation and spawning. EIFAC Technical paper, 25: 53-71

Duangswawasdi, M. and Klaverkamp, J.F. 1979. Acephate and fenitrothion toxicity in rainbow trout : effects of temperature strem and investigations on the sites of action. Aquat. Toxicol. 35.

Duke, T.W. and Wilson, A.J. 1971. Chlorinated hydrocarbons in liver of fishes from the northeastern pacific ocean. Pestic. Monit. J.5: 228.

Eaton, J.G. 1973. Recent developments in the use of laboratory bioassays to determine "SAFE" levels of toxicants for fish, in Bioassay techniques and Environmental Chemistry. Ann Arbor, Science, Mirch : 107.

EC, 1979. Sixth ammendment directive 67/548. Council of the European communities. Off. J. Eur. Comm. L.259/ 24.

Elliot, M, Farham, A.W. Janes, N.F., Netham, P.H. and Pearson, B.C. 1975. Insecticidal activity of the pyrethrins and related compounds. VII.Insecticidal dihalo vinyl analogues of Cis and tras Chrysanthemates. Pest. Sci. 6 : 537.

Ellis, M.M. 1937. Detection and measurement of stream pollution. Bull U.S. Bur. Fish 48 : 365-437.

Ellis, M.M. 1937. Detection and Measurement of Stream pollution. U.S. Bur. Fish Bull. 48 (22): 365-437

EPA. 1978. Criteria and rational for decision making in aquatic hazard evaluation. Am. Inst. Biol. Sc. Arlington, Virginia 22209;pp.46.

Ernst W., Jackman P., Doe K., Page F., Julien G., Mackay K., Suthenland, T. 2001. Dispersion and toxicity to non-target aquatic organisms of pesticides used to treat sea lice on salmon in net pen enclosures. 42(6): 433-44.

Farrell, A. P., Stockner, E. and Kennedy, C. J. 1998. A study of the lethal and sublethal toxicity of polyphase P-100 an antisapstain fungicide containing 3-iodo-2 propynyl butyl carbamate (IPBC), on fish and aquatic invertebrates. Arch. Environ. Contam. Toxicol. 35(3): 472-478.

Fernandez, T.V. , Thomas, G. Vardhanan, Y. S., Kumar, R.M. 1996. An assessment of the toxicity levels of BHC on four aquatic organisms. J. Environ. Biol. 17(1): 21-24.

Ghazaly, K.S. 1994. Efficacy of Ascorbic acid (Vitamin C) on experimental copper intoxication in *Tilapia zilli.* Bull. Natl. Inst. Oceanogr. Fish, Egypt. 20(2) : 249-257.

Ghosh, R. and Dutta, J.1985. Malathion toxicity in air-breathing catfish singhi *(Heteropneustes fossilis).* Bulletin Ind. J. Fish. 32:447-452

Gill, T.S. Pande, J. & Tewari, H.1991. Individual and Combined toxicity of common pesticides to teleost *P. Conchonius. Ind.* J. Expt. Biol. 29:145-148.

Gill, T.S., Pande, J., Tewari H. 1991. Effects of endosulfan and phosphamidon poisoning on the peripheral blood of fish *(Barbus conchonius,* Hamilton). J. Environ. Sci. Health. A 26(2): 249-255.

Gills, S.S. 1980. In vitro metabolism of carbofuran by liver microsomes of the paddy field fish, *Trichogaster pectoralis,* Bull. Environ. Contam. Toxicol. 25 : 697.

Gimeno, L., Ferrando, M.D., Sanchez, S. and Andreu, E. 1996. Endosulfan effects on liver and blood of the eel, *Anguilla anguilla.* Comp. Biochem. Physiol. 108(3): 343-348.

Gomez, L., Masot, J., Martinez, S., Duran, E,, soler, F., Roncero, V. 1998. Acute 24-D poisoning in tench *(Tinca tinca* L.): Lesions in the hematopoietic portion of the kideny. Arch. Environ. Contam. Toxicol 35(3): 479-483.

Goodman, L.R., Hansen, D.J., Manning, C.S. and Favas, L.F. 1982. Effects of kepone on the sheepshead minnow in an entire life cycle toxicity test, Arch. Environ. Contam. Toxicol., 11 : 335.

Grohmann, A. and Sobhani, P. 1979. Beeinflusung der Giftwirkung von stoffen auf Fische durch Ionengleichfewiche (Harle des wassers and Toxicitat) Hydrochim. Hydrogeol. 3:147.

Haldane A.V. 1966. Organochlorine insecticide residues in salmonoid fish. J. Appl. Ecol. 3 : 45.

Halter, M.T. and Johnson, H.E. 1974. Acute toxicities of a Polychlorinated biphenyl (PCB) and DDT alone and in combination to early life stages of Coho salmon *(Oncorynchus Kisutch).*

Hanke, W., Gluth, G., Buhel, H. and Muller, R. 1983. Physiological changes in carps induced by pollution. Ecotoxicol. Environ. Saf. 7 : 229.

Hart, W.B. DoudRoff, P. and Greenback, J. 1945. The evaluation of Industrial wastes, chemicals and other substrial wastes, Chemicals and other substanes to freshwater fish. Wastes Contr. Lab. All. Co. 317.

Hart, W.B., Weston, R.F. and Demann, J.G. 1948. An apparatus for oxygenating test solution in which fish are used as test animals for evaluating toxicity. Trans. Am. Fish. Soc. 75 : 228.

Hassal, K.A. 1990. The biochemistry and uses of pesticides. II ed., English Language Book Society MacMillan Press Ltd., Hongkong.

Hazarika, R. and Das, M. 1998. Toxicological impact of BHC on the ovary of the air-breathing catfish *Heteropneustes fossilis* (Bloch). Bull. Environ. Contam. and Toxicol. 60(1): 16-21.

Heath, A. G. 1990. Water Pollution and Fish Physiology. CRC Press, USA. P.1-235.

Heath, A.G., Cech, J.J. Jr., Brink, L., Moberg, P., Zinkl, J. G. 1997. Physiological responses of fathead minnow larvae to rice pesticides. Ecotoxicol. Environ. Saf. 37(3): 280-8.

Heath, A.G., Cech, JJJr., zinkl, J.G., Steele, M.D. 1993. Sublethal effects of three pesticides on Japanese medaka. Arch. Environ. Contam. Toxicol. 25(4) : 485-94.

Hinsen, D.J., Parrish, P.R. Loser, J.I., Wilson, A.J.Jr. and Wilson, P.D. 1971. Chronic toxicity uptake and retention of Aroclor 1254 in two estuarine fishes. Bull. Environ. Contam. Toxicol. 6:113.

Hohreiter, D.W., Reinert, R.E. and Bush, P.B. 1991. Effects of the insecticides carbofuran and fenvalerate on adenylate parameters in bluegill sunfish *(Lepomis macrochirus).* Arch. Environ. Contam. Toxicol. 21(3): 325-331.

Holcombe, G.W., Flandt, J.T. and Phipps, G.L. 1980. Effects of pH increases and sodium chloride addition on the acute toxicity of 2-4 dichlorophenol to the fathead minnow. Water Res. 14 : 1073.

Holden, A.K. 1972. The effects of pesticides on life in freshwater. Proc. R. Soc. London, Ser. B. 180:383.

Holmberg, B., Jensen, S., Larsson, A., Lewander, K. and Olsson, M. 1972. Metabolic effects of technical pentachlorophenol (PCP) on the eel *Anguilla anguilla* L. Comp. Biochem. Physiol. 438:171-183.

Hota A.K., Mishra D.K. and Tripathy, P.C. 1993. Metabolic effects of kilex carbaryl on a freshwater teleost, *channa punctatus* (Bloch.) Muzaffarnagar-India Society of bioscience. 335-342.

Inglis, A. and Davis, E.L. 1972. Effects of water hardness on the toxicity of several organic and inorganic herbicides to fish.

James, R., Sampath K. and Alagurathinam, S.1996. Effects of lead on respiratory enzyme activity, glycogen and blood sugar levels of the teleost, *Oreochromis mossambicus* (Peters) during accumulation and depuration. Asian Fish. Sci. 9 : 86-99.

Jhingran V.G. 1982. Fish and Fisheries of India, pp 894.

Johansen, K. 1970. Ais-breathing fishes. In: Fish Physiology. W.S.Hoar and D.T.Randall (Eds). Academic Press, New York and London, Vol. IV. P. 361-411.

John, K.R., Jayabalan, N. and George, M.R. 1993. Impact of sublethal concentrations of endosulfan on the histology of *cyprinus carpio* liver and kidney. Proceedings of the national seminar on a Aquaculture-development in India-problems and prospects. 27-29 Nov. 1990 pp. 179-182.

Johnson, W.W. and Julin, A.M. 1980. Acute toxicity of toxaphene to fathead minnows. EPA-600/3-80. 005. U.S. Environmental Protection Agency, Duluth:Minn.

Johnson, W.W. and Finley, M.T. 1980. Handbook of acute toxicity of chemicals to fish and aquatic invertebrates. Resource publ. 137, US Department of Interior Fish and Wildlife Service Washington, DC.

Johnson, W.W. and Finley, M.T. 1980. Handbook of Acute toxicity of chemicals to fish and Aquatic invertebrates. Resource publ. 137. Fish and Wild Life Dervice. U.S. Department of the Interior. Washington D.C.

Jones, J.R.E. \ 940. The toxicity of the double chlorides of mercury and sodium. I. Experiments with the minow *Phoxinus phoxinus* (L.) J. Experiment Biol. 17 : 325-330.

Jones, J.R.E. 1964. Insecticides and Herbicides. In:Fish and River pollution. Butterworth and Co. Ltd. PP. 128-140.

Kanazawa, J. 1981. Measurement of the bioconcentration factors of pesticides by freshwater fish and their correlation with physico-chemical properties or acute toxicities, Pestic. Sec. 12 : 417.

Kapur, K., Kaur, K. and Toor, H.S. 1978. The effect of fenitrothion on reproduction of a teleost fish, *Cyprinus carpio* (Linn.): A biochemical study. Bull. Environ. Contam. Toxicol., 20 : 438-442.

Katz, M. 1961. Acute toxicity of some organic insecticides of three species of salmonids and spike stickleback. Trans. Amer. Fish. Soc. 90 : 310.

Khattax, I.U.D. and Hafeez, M.A. 1996. Effect of malathion on blood parameters of the fish *Cyprinion Watsoni* Pak J. Zool. 28(1): 45-49.

Kling, D.B. 1986. Toxic effects of sublethal concentration of Lebaycid upon the ovaries were examined in *sarotherodon leucostictus*. Z. Angero. Zool. 73(1) : 75-92.

Konar, S.K. and Ghosh, T.K. 1982. Effects of Furadon on fish, plankton, worm and water quality. IAWPC Tech. Annual 9 : 128-134.

Kulshrestha, S. K. and Arora, N.1984. Impairments induced by sublethal doses of two pesticides in the ovaris of a freshwater teleost *Channel striatus* Bloch. Toxicol. Letl. 20(1):93-98.

Lioyd. R. (1960) and Jordan, D.H.M. 1964. Some factors affecting the resistance of rainbow trout *(Salmo gairdneri* Richardson) to acid waters. Ind. J. Air. Wat. Pollut. 8 : 393-403.

Lloyd, R. and Jordan, D.H.M. 1963. Predicted and observed toxicities of serveral sewage effluent to Rainbow trout. J. Proc. Inst. Sewage Purif. 2:16.

Macek, K.J. and Sanders, H. O. 1970. Biological variation in the susceptibility of fish and aquatic invertebrate to DDT, Trans. An. Fish. Soc. 99 : 89.

Macek, K.J. 1968. Reproduction in brook trout *(Salvelinus fontinalis)* fed sublethal concentration of DDT, J. Fish Res. Board Can. 25 : 1787.

Macek, K.J. 1980. Aquatic toxicology : Fact or fiction ? Envir. Health. Persp. 34 : 159-163.

Maeck, K.J., Hutchinson, C. and Cope, O.B. 1969. The effect of temperature on the susceptibility of blue gills and rainbow trout to selected pesticides. Bull. Environ. Contam. Toxicol, 4 : 174.

MacFarlane, R.B. and Livingston, R. J. 1983. Effects of acidified water on the locomotor Behaviour of the gulf killifish, *Funtulus grandis* : A time series approach. Arch. Environ. Contam. Toxicol. 2: 163-168.

Mahajan, C. L. and Juneja, C. J. 1979. Effect of aldrin on peripheral blood of fish, *Channa punctatus* (Bloch.). Indian J. Environ. Hlth. 21 : 162-172.

Mahapatra, B.C. and Rengarajan, K. 1995. A manual on Bioassays in the laboratory and their techniques. CMFRI special publication. No.64.

Malhotra, Y. R. and Sharma, K. K. 1986. Effect of malathion on some aspects of early developmental stages of *Puntius conchonius* (Ham.). Proc. Natl. Symp. Fish & Env., 1986, PP. 1-4.

Malla-Reddy, P., Philip, G.H., and Bashamohideeh, M. 1992. Regulation of AchE system of freshwater fish, *Cyprinus carpio,* under fenvalerate toxicity. Bull. Environ. Contam. Toxicol. 48(1): 18-22.

Mani, K. and Saxena, P.K. 1985. Effect of safe concentration of some pesticides on ovarian recrudesence in the freshwater murrel, *channa punctatus* (Bl.): A quantitative study. Ecotoxicol. Environ. Saf, 9 : 241-249

Marking, L.L. 1974. Toxicity of the synthetic pyrethroid SBP-1382 to fish. Prog. Fish culturist. 36:144

Mathur, D.S. 1965. Histological changes in the livers of certain fishes induced by dieldrin. Sci. Cult. 31 : 258-259.

Mauck, W.L., Olson, L.E. and. Marking, L.L 1976. Toxicity of natural pyrethrins and five pyrethroids to fish. Arch. Environ. Contam. Toxicol. 4:18.

Mauck, W.L., Olson, L.E., and Hogan, J.W. 1977. Effects of water quality on deactivation and toxicity of mexacarbate (zectran) on fish. Arch. Environ. Contam. Toxicol. 6 : 385.

Mayer F.L., Mehrle, P.M. and Cruteher, P. L. 1978. Interaction of toxaphone and Vitamin C in Channel catfish. Trans. Am. Fish. Soc. 107 : 326.

Mayers, T.R. and Hendricks, J.D. 1985. Histopathology. In:Fundamentals of Aquatic Toxicology. G.M. Rand and S.R. Petrocelle (eds). Hemisphere Publ. Corp. Washington pp.283-331.

Mcbride, J. R., Dye, H. M. and Donaldson, E. M. 1981. Stress response of juvenile sockeye salmon *(oncorhynchus nerker)* to the butoxyethanol ester or 2, 4-dichlorophenoxyacetic acid. Bull. Environ. Contam. Toxicol. 27 : 877.

McKim, J.M. 1985,. Early life stage toxicity tests. In:fundamentals of Aquatic toxicology, G.M. Rand and S.R. Petrocelli (eds.) Hemisphere Publ. Corp. Washington PP.58-95.

Metelev, V.V., Kanaev, A.L. and Diasokhva, N.G. 1971. Water toxicity, Amerind publishing Co. Pvt. Ltd. New Delhi PP.3-203.

Metz. J. and Branacin, M. 1975. Biochemical and biophysical aspect of salt excretion by chloride cell in teleost. In:Excretion (Wessing, A. Ed.)

Mishra, J. and Srivastava, A.K. 1982. Malathion induced haematological and biochemical changes in the Indian catfish, *Heteropneustes fossilis.* Environ. Res. 30 : 393-398.

Moss, J. A. and Hathway, D.E. 1964. Transport of organic compounds in the mammal: partition of dieldrin and telodrin between the cellular components and soluble proteins of blood. Biochem. J. 91: 384.

Mount, D.J. 1977. An Assessment of Application factors in Aquatic toxicology. US-EPA-600/3-77-085. U.S. Environmental protection Agency. Corvallisore.

Mount, D.J. and Stephan, C. E. 1967. A method for establishing acceptable toxicant limits for fish: Malathion and the butoxyethyl ester of 2, 4-D Trans. Amer. Fish. Soc. 96: 185-193.

Muir, D.C.G., Hobden, B.R. and Servos, M. R. 1994. Bioconcentration of pyrethroid insecticides and DDT by rainbow trout: uptake, depuration, and effect of dissolved organic carbon. Aquat. Toxicol. 29:223-240.

Mukherjee, M. (1996). Antidote to combat white spot disease in Culture Ponds. Fish. Chimes. 16(5): 30-31.

Mukherjee, S. and Bhattacharya, S. 1975, Histopathological lesions in the hepatopancreas of fishes exposed industrial pollutants. Indian J. Exp. Bio. 13(6): 571-573.

Mukhopadhyay, P. K. and Dehadrai, P.V. 1980a. Studies on air breathing catfish, *Clarias batrachus* (Linn) under sublethal malathion exposure. Ind. J. Experiment Biol. 18 : 400-404.

Mukhopadhyay, P.K. and Dehadrai, P.V. 1980b. Biochemical changes in the air breathing catfish *Clarias batrachus* (Linn.) exposed to malathion: Environ. Pollut. A22 : 149-158.

Mukhopadhyay, P.K. and Dehadrai, P.V. 1980. Biochemical changes in the air-breathing catfish, *Clarias batrachus* exposed to malathion. Environ. Res. 22 : 149-158.

Mukhopadhyay, P. K., Mukherjee, A. P. and Dehadrai, P.V. 1984. Phospholipid and fatty acids in liver of *Clarias batrachus* exposed to sublethal malathion. J. Environ. Biol. 5 : 221-229.

Mulcahy, M. F. 1975. Fish blood changes associated with disease. A haematological study of pike lymphoma and salmon ulcerative dermal necrosis. In : The Pathology of Fishes. W.E. Ribelin, C. Migaki Madison (Eds.) University of Wisconsis, pp. 925-944.

Murty, A.S. 1988: Toxicity of pesticides to Fish, 1 : 19.

Mushthaq, M. and Nazar, A.R. 1994. Effect of pesticide, Nuvan, on *Catla catla* J. Aqua. Biol. 9(1 & 2): 80-82.

Nagler, J.J., Aysola, P. and Ruby, S.M. 1984. The seasonal effect of sub-lethal pentachlorophenol exposure on ovarian development in rainbow trout (Salmo gairdneri). 11th annual aquatic toxicity workship, Richmond, B.C., November 13-15, 1984 (np.)

Nagler, J.J., Prasad, A. and Ruby, S.M.1986. Effect of sublethal pentachlorophenol on early oogenesis in maturing female rainbow trout *(Salmo gairdneri)*. Arch. Environ. Contam. Toxicol., 15 : 549-555.

Narayan, A.S. and Singh, B.B. 1991. Histopathological lesions in *H. fossilis* subjects to acute thiodan toxicity. Acta. Hydrochem. Hydrobiol. 19(2): 235-243.

Nath, R. and Jaipuryar, A.R.K. 1996. Effect of sublethal concentration of lindane on peripheral haemogram in the fish *Heteropneuster fossilis* (Bloch). Environ. Ecol. 14(1): 158-162.

Newsome, C.S. 1980. A multigeneration fish toxicity tests as an aid in the hazard evaluation of aquatic pollutants. Ecotoxicol. Environ. Saf. 4 : 362.

OECD, 1979. Report on the assessment of potential environmental effects of Chemical. I. OECD, pp.28.

Oern, S., Andersson, P.L., Foerlin, L., Tysklind, M. and Norrgren, L. 1998. The impact on reproduction of an orally administered mixture of selected PCBs in Zebrafish *(Danio rerio)*. Arch. Environ. Contain. Toxicol. 35(1):52-57.

Ogilvie, D.M., and Anderson, J.M. 1965. Effect of DDT on temperature selection by young Atlantic Salmon *Salmo Salar, L* Fish. Res. Board Can. 22 : 503.

Pandey, A.C. 1988. Impact of endosulfan (thiodon) EC 35 on behaviour and dynamics of oocyte development in the teleostean fish, colisa *(Trichogastter fasciatus)*. Ecotoxicol. Environ. Saf. 15(2):221-225.

Pathiratne, A. and George, S. G. 1998. Toxicity of malathion to Nile tilapia, *Oreochromis niloticus* and modulation by other environmental contaminants. Aquat. Toxicol. 43(4) : 261-271.

Peczely, P. and Kovacs, K. 1998. Effects of a fungicide (VITAVAX 200 FF) on the reproductive endocrime functions in mallards. Environ. Toxicol. Water. Qual. 13(2):133-141.

Peterson, R. H., Daye, P.G., Lacroix, G. L. and Garside, E.T. 1982. Reproduction in fish experiencing acid and metal stress, in Acid Rain/Fisheries, Johnson, R.E. (Ed.), American Fisheries Society, Bethesda, Md. P. 177.

Pickering, O. R., Henderson, C. and Lemke, A.E. 1962. The toxicity of organic phosphorus insecticides to different species of warm fishes. Trans. Am. Fish. Soc. 91 : 175.

Pickering, A.D., Potliger, T.G. and Christie, P. 1982. Recovery of brown trout, *salmo trutla,* L. from acute handling stress : a time course study, J. Fish Biol. 20 : 229.

Power, E.B. 1917. The gold fish as a test animal in the study of toxicity. Illinois. Monogr. 4 : 93-127.

Qureshi, M.D., Heldin, R.V., Vanstone, W.E. and Anastassiadis, P.A. 1971. Levels of constitutents of glycoproteins in the sera pacific salmon. J. Fish. Res. Bd. Canada. 28(8): 1173-1179.

Rajeswari, K., Reddy, S. J., Reddy, D.C. and Ramamurthi R. 1989. Effect of dichlorvos on certain blood parameters of fish *clarias batrachus.* Environ. Ecol. 7(4) : 933-934.

Ram, R. N. 1988. Carbofuran induced histophysiological changes in thyroid of the teleost fish, *Channa punctatus* (Bloch.). Ecotoxicol. Environ. Saf. 16(2) : 106-113.

Ramanujam, S.N. and Mohanty, G. 1997. Thiodon induced sublethal responses on haematological parameters of *H. fossilis.* Journal of life sciences-Royal Dublin-Society. 2 : 8-13.

Ramkrishnan, R. and Sivakumar, A. A. 1993. Effect of quinolphos on the fish *Oreochromis mossambicus*. J. Ecobiol. 5(1) : 45-50.

Rebach, S. 1999. Acute toxicity of permethrin/piperonyl Butoxide on Hybrid Striped Bass. Bull. Environ. Contam. Toxicol. 62(4) : 448-454.

Reddy, P.M. and Bashamohideen, M. 1995. Alterations in protein metabolism in selected tissues of fish. *Cyprinus carpio,* during sublethal concentration of cypermethrin Environ. Monit. Assess. 36(2) : 183-190.

Reish D.L. and Oshida, P.S. 1987. Short-term bioassay. In: Manual of methods in aquatic environment research. FAO Fish Tech. Pap. 247(10): 1-62.

Sado, E. K. and Ita, E.G. 1992. The effects of capture methods, temperature and oxygen on the survival of live freshwater clupeids during acclimation and transportation. J. Aquacult. Trop. 7(2) : 165-176.

Saglio, P., Trijasse, Azam, S., D.1996. Behavioural effects of waterborne carbofuran in goldfish. Arch. Environ. Contam. Toxicol. 31(2) : 232-8.

Sahai, S. and Singh, S. 1989. Hepatotoxic effects of some pesticides in *Puntius ticto.* Environment and Experimental Toxicology. Jai Research Foundation, Valvada-396108.

Sampath, K., Velammal, S., Kennedy, L. J. J. and James, R. 1993. Haematological changes in *Oricochromis mossambicus* as a function of exposure period and sublethal levels of Ekalux. Acta Hydrobiol. 35: 73-83.

Sastry, K.V. , Mallik, P.V. and Singh, S.K. 1982. Alteration in some biochemical and enzymological parameters in the snake head fish *Channa punctatus,* exposed clinicallyto quinalphos. chemosphere. 11:1211.

Scott, A.L. and Rogers, W.A. 1980. Histological effects of prolonged sublethal hypoxia of channel catfish *Ictalurus punctuatus.* J. Fish. Dis. 3 : 305.

Shin, S.W., Chung, N.I., Kim, J.S., Chon, T.S.. Kwon, O.S., Lee, S.K., Koh, S.C. 2001. Effect of diazinon on behaviour of Japanese medaka *(oryzias latipes)* and gene expression of tyrosine hydroxylase as a biomarker. J. Environ. Sci. Health. 36(6) : 783-95.

Shukla, L., Shrivastava, A., Merwani, D. and Pandey, A.K. 1984. Effect of sublethal malathion on ovarian histopathology in *Sarothrodon mossambicus,* Comp. Physiol. Eco., 9: 13-17.

Singh S. and Sahai, S. 1986. Accumulation of malathion in the gills, kidney and liver of *Puntius ticto* as assessed by thin layer chromatography. J. Environ. Biol.7 (2): 107-112.

Singh S. and Sahai, S. 1987. Pesticide metabolism in a freshwater teleost, *Puntius ticto.* In: Environment and Ecotoxicology (Eds: R.C. Dalela, Y.N. Sahai and Saroj Gupta). 211-218.

Singh, A., and Agarwal, R.A. 1993. Effect of cypermethrin on lactate and succinic delydrogenase and cytochrome oxidase of snail and fish. Bull. Environ. Contain. Toxicol. 51(3) : 445-452.

Singh, B. B. and Narain, A.S. 1982. Acute toxicity of thiodon to catfish *(Heteropneustes fossilis).* Bull. Environ. Contam. Toxicol. 28 : 122.

Singh, P.B., Kime, D.E. and Singh, T.P. 1993. Modulatory actions of Mystus gonadotropin on gamma-BHC induced histological changes, cholesterol and sex steroid levels in *Heteropneustes fossilis.* Ecotoxicol. Environ. Saf. 25(2) : 141-153.

Smith, C.E. 1968. Haematological changes in the cohosalmon fed a folic acid deficient diet. J. Fish. Res. Bd. Can. 25:151-156.

Smith, R.K. and Cole, C.F. 1973. Effects of egg contaminations of DDT and dieldrin on development in winter flounder *(Pseudopleuronectes americanus),* J. Fish. Res. Board Can. 30: 1894.

Soengas, J. L., Strong, E.F., Aldegunde, M. and Andres, M. D. 1997. Effects of an acute exposure to lindane (gamma-hexachlorocyclohexane) on brain and liver carbohydrate metabolism of rainbowtrout. Ecotoxicol. Environ. Saf. 38(2) : 99-107.

Spraque, J.B. 1971. Measurement of pollutant toxicity to fish III. Sublethal effects and "safe" concentration. Water Res. 5 : 245-266.

Stephenson, C. E. 1982. In analysis of established pyrethroid insecticides. Residues Review. 39 : 188.

Steveke, E.W. and Atherton, C.R. 1965. Use of micro haematocrit value to six largemouth. Bass. Prog. Fish. Cult. 27:87-90.

Thakur, N. and Sahai, S. 1987. Changes in blood serum proteins, and other haematological parameters in the teleost, *Channa punctatus* on exposure to organochlorine pesticide, proceedings of the Academy of Environmental Biology. Environ. Pestic. Toxicol. 263-271.

Tsuda, T., Kajima, M., Harada, H., Nakajima, A. and Aoki, S .1997. Acute toxicity, accumulation and excretion of organophosphorus insecticides and their oxidation products in Killifish. Chemosphere, 35(5) : 939-949.

Tugarina, P.Y. and Ryzhova, L.N. 1970. Age related feature of the blood of the black Baikal graying *Thymallus arcticos* baicalensis. J. Ichthyol. 10:384-357.

United States Environmental Protection Agency. 1979. Guidance for premanufacture testing : Discussion of policy issues, alternative approaches and test methods. Federal Register, 44:16240.

Urdaneta, H. 1989. Organochlorine compound in ponds of a fish culture centre Pacz Municipality, Zulia State Venezuela. Bolcent-mvest. Biol. Maracoitio 23 : 1-10.

Vaid, S. and Mishra I. M. 1997. Impact of Macraberin Forte against malathion poisoning in fish brain. Poll. Res. 16(4):275-276.

Van Hougen and Apperhuizen. 1985. In APHA, 1989 Newyork.

Van Leeuwen, C.J. 1985. Short-term toxicity testing. In:Manual of Aquatic Ecotoxocology, Editor P.K. Ray, P.N. Viswanathan, H.A.M. de Kurijf and D.de.Zwart. Hemisphere Publ. Corp., Wa^hington.

Venugopalan, V.K. and P. Sasibhushana Rao 1979. Pesticide induced impairment in incubation and post embryonic development of planktonic eggs of esturine fish of Porto NOVO (South India) Waters. Proc. Symp. Environ. Biol. Academy of Environmental Biology, Muzzaffarnagar, India, 397.

Verma, S.R., Bansal, S. K., Gupta, A.K., Pal. N., Tyagi, A.K., Bhatnagar, M.C., Kumar, V. and Dalela, R.C. 1982. Bioassay trials with twenty three pesticides to a freshwater teleost *Sacrobranchus fossilis*, Wat. Res., 16 : 525-529.

Vittozzi, L. and. Angelis, G.D. 1991. A critical review of comparative acute toxicity data on fresh water fish. Aquat. Toxicol. 19 :167-204.

Warner, R.E., Peterson, K.K.. and Borgman, L. 1966. Behaviour pathology in fish: a quantitative study of sublethal pesticide toxication. J. Appl. Ecol. 3 : 223.

Weber, J.B. 1972. Interaction of organic pesticides with particulate matter in aquatic and soil systems, in fate of organic pesticides in the Aquatic Environment (Advances in Chemistry Series III). Faust, S.D., Eds. American Chemical Society Washington D.C. PP.55.

Wedemeyer, G. 1973. Some physiological aspects of sublethal heat stress in the juvenile steelhead trout *(Salmo gairdneri)* and coho salmon *(Onchorhynchus kisutch). L* Fish. Res. Bd. Can. 30:831.

Weis, J.S. and Weis. P.S. 1976. Optical malformations induced by insecticides in embryoes of the Atlantic silverside, *Menidia menidia*. Fish. Bull. 74 : 208.

Wilbur, R.L. and Whitney, E.W. 1973. Toxicity of the herbicide Kuron (Silvex) to bluegill eggs and fry, Trans. Am. Fish Soc. 102 : 630.

Witt, J.M., Brown, W.H., Shaw, G.I., Maynard, L.S., Sullivan, L. M., Whiting F.M. and Stuli, J.W. 1966. Rate of transfer of DDT from the blood compartment. Bull. Environ. Contain. Toxicol. 1 : 187.

Yoon Jy, oh SH, Yoo SM, Lee S.J., Lee H.S., Choi S.J., Moon C.K., Lee, B.H. 2001. N-nitrosocarbofuran, but not carbofuran, induces apoptosis and cell cyle arrest in CHL Cells. Toxicology, 169(2): 153-61.

Zamfir, Gh. 1979. Water pollution by pest - controlling agents and the possibilities for their removal in surface water treatment stations. Igiena 28(2) : 97-103.

CHAPTER 47

Observation on Mixed Fish Farming of Indian and Exotic Carps in a Seasonal Pond of Jodhpur

Anita Jhajhria

Department of Zoology, Government P. G. College. Jalore - 334 001. Rajasthan.

ABSTRACT

Mixed fish farming in Shekhawat pond (0.736 ha) was carried out for a period of 8 months from August 1999 to March 2000 using *Catla catla, Labeo rohita, Cirrhinus mrigala and Cyprinus carpio.* No supplementary feed and fertilizers were supplied to the fishes reared. The growth of *Cyprinus carpio* (1173.50 gm) was best followed by *Catla catla* (578.30 gm), *Labeo rohita* (263.40 gm) and *Cirrhinus mrigala* (252.50 gm). The fish production (4127 kg/ ha/8 months) was achieved which is higher than the average fish production (100-400 kg/ha/ yr) in similar water bodies of this region.

Key words: *Mixed fish farming, Shekhawat pond, Indian and exotic carps, Production.*

INTRODUCTION

The goal of achieving productivity from pond ecosystem is to sustain high yield of fish rich in protein. In aquatic ecosystems energy culminates at its different trophic levels with considerable loss of energy at each link of food chain. For augmenting high fish productivity of carps, the knowledge of bioecology of fresh water ponds is utmost important as ecology plays vital role in regulating productivity of these fishes.

The basic principle of mixed fish farming or polyculture is that when compatible fishes of different habits are stocked together, they secure for themselves in the most efficient manner, all life requisites available in the pond for fish production without harming each other.

In North-weastern Rajasthan. there are 176 seasonal water bodies with water spread area of 5700 ha in which the fish production is as low as 100-400 kg/ha/yr (Pragati Prativedan, 1995). In Jodhpur township, as well as in its vicinity, there are many ponds of varied dimensions where knowledge on ecology and related fish productivity of fresh water pond ecosystems for utilization of appropriate fish fauna is still lacking. With this consideration, the studies where conducted for assessing the fish production of Shekhawat pond.

MATERIALS AND METHODS

The fies of three Indian major carps namely, *Catla catla. Labeo rohita* and *Cirrhinus mrigala* alongwith an exotic carp *Cvprinux carpio* were obtained from the induced spawning exercises using the synthetic hormone. These fries were then transported in oxygen packed system of polythene bags in August 1999 for stocking in the pond. The number of fries to be stocked in the pond was computed by adopting the standard formula (Jhingran. 1975).

The harvesting was done using cast and drag net in March. 2000. The pond water was collected monthly during the stocking period between 8:30 a.m. - 11.30 a.m. and was analysed using the methods of APHA (1985) and Welch (1948).

RESULTS AND DISCUSSION

The results of stocking and production are summerised in Table 47.1. The aboitic parameters analysed are shown in the Table 47.2. The production of 4127 kg/ha/8 months was achieved in the pond. The average weight of (1173.50 gm) *Cyprinus carpio* was highest followed by *Catla catla* (578.30 gm), *Labeo rohita* (263.40 gm) and *Cirrhinus mrigala* (252.50 gm). Valuable contributions in the field of fish productivity of ecologically different ponds with reference to carps has been done by various workers. Lakshman *et al.* (1971) and Singh *el al. (*1972) conducted year long experiments at the pond culture, division of CIFRI combining seven Indian and exotic species of fish and obtained a production varying from 2234-4210 kg/ha/yr Sukumarna (1976) obtained a production of 6191-7332 kg/ha/8 months at Karnal. Sharma et al. (1979) reported a production of 4323 kg/ha/yr in fish-cum-duck rearing without adding any fertilizer to the pond and giving any supplementary feed. Mathew *et al.* (1979) obtained productions with six species combined as 2199-2242 kg/ha/8 months at Poona.

Table 47.1.

Species-wise stocking and production of Indian and exotic carps in Shekhawat pond (0.076 ha)

Fish species	Fin Formula	Habitats	Feeding Habits	Stocking density of fries	Total weight (Kg.)	Average weight (gm.)	Average length (cm.)
Catla Catla (Catla)	Diii-iv 14-16.Aio.IM2O Vi8	Surface feeder	Planktonic	900	580	578.3	25.23
Labeo rohita (Rohu)	Dii-iv 12-14: Aii-n5: Pi 16-18:Vi8	Column feeder	Herbivorous	2000	450	263.4	19.69
Cirrhinus mrigla (mrigal)	Diii-iv 12-13:Aii5: Pi 17: Vi8	Bottom feeder	Omnivorous	3200	410	252.	20.65
Cyprimis carpio (Common carp)	Diii-iv 18-20: Aiii5. Pi 15. Vi8	Bottom feeder	Omnivorous	4000	1600	11735	40.23

Table 47.2

Physico-chemical properties of pondwater during the mixed culture period
Stocking Months

Abiotic parameters	August 1999	Sept 1999	Oct. 1999	Nov. 1999	Dec. 1999	Jan. 2000	Feb. 2000	Mar. 2000
Water temperature (°C)	24.1	24 5	23.6	20.1	16.5	15.5	15.9	204
pH	7.6	7.7	7.8	7.8	7.9	8.0	7.8	8.0
Transparency (cm)	14 0	20.0	22 0	24.0	19.0	17.0	26.0	30.0
DO (ml/1)	5.643	4.559	4.453	4.749	4.997	5.088	4.640	5.159
Free CO_2 (ppm)	9.0	8.0	7.0	6.0	4.0	3.0	5.0	4.0
CO_3^{-2} (ppm)				-	-	-		-
HCO_3^{-2} (ppm)	84.0	98.0	95.0	102.0	110.0	115.0	120.0	124.0
Total Alkalinity (ppm)	84.0	98.0	95.0	102.0	110.0	115.0	120.0	124.0
Total Hardness (ppm)	70.0	890	86.0	110.0	120.0	125.0	135.0	140.0
Chloride (ppt)	0 302	0.526	0496	0574	0.608	0.649	0 797	0 803
Salinity (ppt)	0054	0 950	0895	0.104	0 109	0.117	0.144	0.145
Phosphate (µg/1)	0.212	0.354	0.410	0287	0.279	0.250	0398	0.559
Nitrate (µg/1)	3. 3 3 6	3.685	4.528	4.104	3.910	3432	4.728	6.223

Kumar (2001) recorded maximum growth of Catla (9.3 kg) followed by common carp (3.1 kg), rohu (1.9 kg) and for mrigal (320 gm) in one year from deep ponds of Patna. Baghel and Saxena (2002) reported a production of 4965-97 kg/ha/9 months in Tarai of Uttaranchal.

In the present study, the fish production (4127 kg/ha/8 months) relied only on the natural food and the kitchen wastes of the army mess. At least 50 % of the total food of carps in the fish ponds comes from the pond itself (Schaeperclaus. 1952). Gaudet (1967) reported that in Israel out of 2000 kg/ha, probably 400 kg comes from the natural food.

Thus, it may be concluded that on the basis of ecobased planning, judicious management and by incorporating the application of supplementary feed and fertilizers a many fold enhancement in fish production of the ponds of this region is possible.

ACKNOWLEDGEMENTS

The author expresses sincere gratitude to Dr. Devendra Mohan. Associate Professor. Department of Zoology, Jai Narain Vyas University. Jodhpur for evincing keen interest and guidance in the research work.

REFERENCES

APHA, 1985. Standard methods for the examination of water and wastewater. AWWA. WPCE. New York. 16th edition.

Baghel, D. As. and Saxena. V. 2002. Composite fish culture in Tarai region of Uttaranchal. Cicohios. 29 : 101 - 104.

Gaudet, J. I.. 1967. The status of warm-water pond fish culture in Europe. FAO. Fish. Rep. pp. 174.

Jhingran, V. G. 1975. Fish and fisheries of India. 1st edition. Hindustan Publishing Corporation. New Delhi.

Kumar, D. 2001. Fish productivity of ecologically different ponds with reference to carps and air breathing fishes. .1. Inland. Fish. Soc. India. 33 : 8 - 16.

Lakshmanan, M. A. V.: Sukumaran. K. K.: Murthy. D. S.: Chakraborty. D. P. and Philipose. M. T. 1971. Preliminary observations of intensive fish farming in fresh water pond by the composite culture of Indian and exotic species. J. Inland. Fish Soc. India. 2:1-21.

Mathew, P. M.: Singh. B. K. and Chakraborty. D. P. 1979. Stocking of fry in composite fish culture. In : Symposium on Inland Aquaculture. CIFRI. Barrackpore. pp. 1 - 44.

Pragati, Prativedan (Raj). 1995. Inter State Fishery Planning Cell (1994-95). State Fisheries Department. Rajasthan. Jaipur.

Schaperclaus, W. 1952. Die Karpfenteichshwirtschaft in der Deutsche Demokratischen Republik. Sitz. Ber. VII (7). Derlin. Deotsche Ahad. Landwirtschaftwiss.

Sharma, B. K.: Kumar. D.: Das. S. R. and Chakraborty. D. P. 1979. Observations on swine-dung recycling in composite fish culture. In : Symposium on Inland Aquaculture (Abstract) CIFRI. Barrackpore. February 12-14. pp. 99.

Singh, S. B.: Sukumaran. K. K.: Chakrabarti. P. C. and Bagchi. M. M. 1972. Observations on composite culture of exotic carps. J. Inland Fish. Soc. India. 4 : 38 - 50.

Sukumaran, K. K. 1976. Progress report of the karnal subcentre of Haryana for the period November 1972 to August 1975. In : Third workshop on the All India Coordinated Research project on composite fish culture and fish seed production. February 26-27. CIFRI. Puri

Welch, P. S. 1948. Limnological methods. The Blakiston Co.. Philadelphia.

CHAPTER 48

THE ECONOMIC VIABILITY OF DUAL CULTURE (RICE + FISH) IN COASTAL SALINE ZONE OF WEST BENGAL

M. Ghosh, D. Maiti and A.K. Nandi.

Department of Agronomy, Bidhan Chandra Krishi Viswavidyalaya, Mohanpur- 741252, Nadia, West Bengal

ABSTRACT

Field experiments were conducted for three consecutive year during the wet seasons of 1996,1997 and 1998 at Regional Research Station, (Coastal Saline Zone) Kakdwip, West Bengal, India to examine the economic viability on the performance of rice-fish integrated farming system at Coastal Saline Zone. The cultivation of rice was carried out under water-logged condition during the low saline period, from July to November. Salt resistant tall indica rice variety NC492 (Sobita) was used for cultivation. On the other hand three fish species, bagda (Penaeus monodon), Javapunti (Puntius javanicus) and tilapia (Oreochromis massambicus) were reared alone and in association with rice and inorganic fertilizer (N:P:K= 60 : 30: 30 Kgha^{-1}). The maximum net returns (Rs 46,348.30 ha^{-1}) was recorded when bagda was reared in association with rice and fertilizer. The highest benefit - cost ratio (1.40 mean value) was recorded in rice + bagda + fertilizer treatment.

Key Words: *Fish Seedlings, Gross and net return, Benefit- cost ratio.*

INTRODUCTION

The coastal saline soils in West Bengal encompass an area of 0.8 mha, which forms a major part of the total 2.1 mha Coastal Saline soils in India (Chattopadhyay et al 1983). Almost the entire area is monocropped paddy being the only crop grown during monsoon season when the salinity levels of the soil are low. During rest of the period of year, the land is kept fallow due to rise in soil salinity as well as non-availability of good quality irrigation water (Natarajan, 1983) Saha (1979) reported fish production of 4221 Kgha-^{1}yr-^{1} through application of NPK fertilizer, which was about 4 times the production obtained in ponds on inharrent fertility only. Sinhababu et al (1996) observed that rice-fish system can annually produce 16t of food crops, 0.61 fish and prawn, besides 5.2 t of fodder (straw).

An attempt was, therefore, made to examine the economic viability of rice and fish in rice-fish farming system.

MATERIALS AND METHODS

The experiment was conducted at the Regional Research Station (Coastal Saline Zone) of Bidhan Chandra Krishi Viswavidyalaya, Kakdwip, South 24 Parganas, West Bengal during the wet seasons of 19%, 1997 and 1998. The station is about 35 km from the Bay of Bengal. A randomised block design with three replications and eleven treatments was set up with a plot size of 320 m². An internal trapezoidal perimeter trench with a width 1.5 m, depth 0.5 m. and height 0.75 m was constructed. The soil of the experimental site was clay loam in texure, saline in nature with pH 8.0, medium organic carbon content was 0.68% and available NPK content in soil was 220 Kgha^{-1}, 40 Kgha^{-1} and 400 Kgha^{-1} respectively. The treatments consisted of cultivation of rice alone, rice plus fish without and with fertilizers ((N:P:K = 60:30:30 Kgha^{-1}). Supplementary feeding with 10.08 Kg of powdered mustard oil cake and rice bran (1:1) was applied in each fish plot during September and October. Forty day old seedlings of a long duration rice variety NC492 (Sobita) were planted in July at a spacing of 20 cm x 15 cm using a uniform basal dose of 60 Kg N, 30 Kg P_2O_5 and 30KgK_2Oha^{-1}.

Fry of bagda (Penaeus monodon) javapunti (Punthis javanicus) and tilapia (Oreochromis mossambicus) were slocked at 952, 442 and 552 per treatment respectively after 30 days of planting. The grasspea (Lathyrus Sativus L) cv. Nirmal was teated as a rainfed paira crop after transplanted rice in three consecutive years. After harvest of rice during December, the rice plot was gradully allowed to dry up, the fishes collect to the perimeters trench as harvest starts. Observations oon grain yield and yield attributes were recorded following the harvest of paddy. The length and weight of fishes were also recorded at harvest by standard fish measuring scale and balance respectively.

RESULTS AND DISCUSSION

Cost of Production

Cost-of production of different treatments differed greatly according to the cost of seed material, cost of operations, labour requirements and cost of different fish fingerlings of individual treatments (Table 48.1). The highest cost of production (Rs. 34,054/ ha^{-1}, mean) was recorded with T_6 (rice + bagda + fertilizer) treatment during kharif season and grasspea was taken as a paira crop during rabi season in the same plot followed by T_3 (rice + bagda) and T_9 (only bagda) which showed Rs. 32,434.90/- ha^{-1} and Rs. 25,463.707 ha^{-1} cost of production including paira crops respectively. The cost of productions was generally higher in dual culture of rice and fish as to other treatments.

Gross Return

Gross monetaary return in thousand rupees ha^{-1} from main and by products of different crops in kharif and rabi seasons and different fish species under different treatments were presented in Table 48.2. The total maximum gross return (Rs, 80,403/- ha^{-1} mean) was recorded with T_6 (rice + bagda + fertilizer) treated plots during kharif season and khesari (grasspea) was grown in same treated plots as a paira crops during rabi season followed by T_3 (rice + bagda), Tg (rice + tilapia + fertilizer) and T_9 (only bagda) treated plots with grass pea as a paira crop during rabi season, which showed Rs. 56,602/- ha^{-1}, Rs. 41,656/- ha^{-1} and Rs. 40,935/- ha^{-1} in respective treatments. The bagda in

Table : 48.1 Average expenditure incurred in different treatments for rice-fish farming system in coastal-saline zone of West Bengal with reference to Kakdwip, 24 Parganas (South)

Item	Quantity	Rate (Rs.)	Treatments										
			T_1 Rs. ha^{-1}	T_2 Rs. ha^{-1}	T_3 Rs. ha^{-1}	T_4 Rs. ha^{-1}	T_5 Rs. ha^{-1}	T_6 Rs. ha^{-1}	T_7 Rs. ha^{-1}	T_8 Rs. ha^{-1}	T_9 Rs. ha^{-1}	T_{10} Rs. ha^{-1}	T_{11} Rs. ha^{-1}
1. Seedling raising			180.00	180.00	180.00	180.00	180.00	180.00	180.00	180.00			
(a) Land preparation													
(b) Cost of seed	60kg	@ Rs. 8.00/kg	480.00	480.00	480.00	480.00	480.00	480.00	480.00	480.00			
2 Transplanting in main field													
(a) land preparation	PT @ 3 times	@ Rs. 240 / ha	1800.00	1800.00	1800.00	1800.00	1800.00	1800.00	1800.00	1800.00			
(b) transplanting	52 Mandays/ha	@Rs. 54.1/MD	2813.20	2813.20	2813.20	2813.20	2813.20	2813.20	2813.20	2813.20			
3. Cost of fertilizer													
(a) Urea	132 kg/ha	@ Rs. 4.20/kg		554.40				554.40	554.40	554.40			
(b) S.S.P	187.5 Kg/ha	@ Rs. 3.20/Kg		600.00				600.00	600.00	600.00			
(c) M.O.P	49.80 kg/ha	@ Rs. 5.00/Kg		249.00				249.00	249.00	249.00			
4 Cost of fertilizer application	4 mandays	@ Rs. 54.1 /MD		216.40				216.40	216.40	216.40			
5. Harvesting, Binding. Carrying and Threshing	30 mandays	@Rs. 54.1/MD	1623.00	1623.00	1623.00	1623.00	1623.00	1623.00	1623.00	1623.00			
6. Land revenue			75.00	75.00	75.00	75.00	75.00	75.00	75.00	75.00			
7. Fish seed					20700.00	11049.00	8625.00	20700.00	11049.00	8625.00	20700.00	1 1049.00	8625.00
(a) Bagda	@ 17250 FL/ha	@Rs. 1.2/FL											
(b) Javapunti	@13812FL/ha	@ Rs. 0.80/FL											
(c) Tilapia	@ 17250 FL/ha	@ Rs. 0.50/FL											
8 Fish feed													
(a) Mustard oil cake	@ 360 Kg/ha	@ Rs. 4.40/kg			1584.00	1584.00	1584.00	1584.00	1584.00	1584.00	1584.00	1584.00	1584.00
(b) Rice bran	@ 360 kg/ha	@ Rs. 1.00/Kg			360.00	360.00	360.00	360.00	360.00	360.00	360.00	360.00	360.00
9. Labour charge	@ 60 Days/ha	@ Rs. 25.00/day			1500.00	1500.00	1500.00	1500.00	1500.00	1500.00	1500.00	1500.00	1500.00
10. Paira crop (Grasspea)													
(a) Seed	@ 25 Kg/ha	@ Rs. 16.00/Kg	400.00	400.00	400.00	400.00	400.00	400.00	400.00	400.00	400.00	400.00	40000
(b) blowing (Broadcasting)	@ 2 mandays/ha	@Rs. 54.10/MD	108.20	108.20	108.20	108.20	108.20	108.20	108.20	108.20	108.20	108.20	108.20
(c) Harvesting. Threshing etc.	15 mandays/ha	@ Rs. 54.01/MD	811.50	811.50	811.50	811.50	811.50	811.50	811.50	811.50	811.50	811.50	811.50
Total Expenditure			8290.90	9910.0	32434.90	22783.90	20399.90	34054.70	24403.70	21979.70	25463.70	15812.70	13388.70

Table : 48.2

Economics of different treatments under consideration of rice-fish farming system in coastal-saline zone (Kakdwip) of West Bengal

Treatments	Grain yield (t ha^{-1})	Selling price of rice (Rs.)	Straw yield (t ha^{-1})	Price of straw (Rs.)	Fish yield (qua^{-1})	Selling price (Rs.)	Yield of Khesari (qua^{-1})	Price of Khesari (Rs.)	Total Income (Rs.)	Total Cost (Rs.)	Profit/loss (Rs.)	Benefit/Cost	T_1 equiva-lent	Rice equiva-lent
T_1 (Rice alone)	2.0	9750.0	2.1	1278.0			5.8	8625.0	19653.0	8290.0	11362.1	1.4	1.0	1.8
T_2 (Rice+ fertilizer)	3.4	16800.0	3.2	1920.0			7.6	11415.0	30135.0	9910.7	20224.3	2.0	1.5	27
T_3 (Rice - Bagda)	2.0	9850.0	3.3	1962.0	1.3	39300.0	5.7	8490.0	59602.0	32434.9	27167.1	0.8	3.0	5.4
T_4 (Rice + Javapunti)	2.0	10000.0	3.4	2034.0	2.3	6930.0	6.1	9075.0	28039.0	22783.9	5255.1	0.2	1.4	2.5
T_5 (Rice + Tilapia)	1.9	9600.0	3.4	2064.0	3.4	8575.0	5.9	8910.0	29149.0	20359.9	8789.1	0.4	1.5	2.6
T_6 (Rice+Bagda+Fert)	3.2	16050.0	3.7	2208.0	1.7	49500.0	8.4	12645.0	80403.0	34054.7	46348.3	1.4	4.1	7.3
T_7 (Rice+Javapunli+Fert)	2.8	13850.0	3.3	1956.0	3.4	10200.0	9.1	13680.0	39686.0	24403.7	15282.3	0.6	2.0	3.6
T_8 (Rice+Tilapia+Fert)	2.9	14550.0	4.2	2496.0	4.0	10000.0	10.4	15660.0	42706.0	21979.7	20726.3	0.9	2.2	3.9
T_9 (Bagda)					1.0	30000.0	5.7	8535.0	38535.0	25463.7	13071.3	0.5	2.0	3.5
T_{10} (Javapunti)					1.8	5400.0	6.1	9090.0	14490.0	15812.1	- 1322.7	-0.1	0.7	1.3 1
T_{11} (Tilapia)					3.0	7500.0	6.2	9270.0	16770.0	13388.7	3381.3	0.3	0.9	1.5

Note : The prevailing selling price of rice, paddy, straw, and bagda, javapunti and tilapia were respectively Rs. 500/qt. Rs. 60/qt and Rs. 300/kg, Rs 30/kg and Rs. 25/kg. While that of Khesari was Rs. 1500/qt.

association with rice and only bagda showed better gross return, it might be due to high selling price (Rs. 300 Kgha^{-1}) of bagda

Net return

Net monetary return (mean of three years data) of the individual treatments are exhibited in Table 48.2. The data revealed that rice + bagda with application of fertilizer (T_6) gave maximum (Rs. 46,348.30/ ha^{-1}) net return followed by rice + bagda (T_3) and rice+tilapia with fertilizer (T_8) having Rs. 27,167.10/ ha^{-1} and Rs. 19,676.307 ha^{-1} net returns under T_3 and T_8 treatments respectively. However, rice with application of fertilizer (T_2) paid higher (Rs. 20,224 30/ ha^{-1}) net return. The lowest net return (Rs. 3706. 30) - ha^{-1}) was recorded in T_{11} (only tilapia) while net loss of Rs, 1322.70/ ha^{-1} was recorded when only javapunti was reared in rice field.

Benefit: Cost ratio (BCR)

Benefit-cost ratio explaning the net return per rupee invested under individual treatments were computed and presented in Table 48.2. Rice with fertilizer (T_2) recorded maximum benefit cost ratio of Rs. 2.00. Among the rice-fish dual culture system with or without fertilizer, the highest benefit-cost ratio (1.40 mean value) was recorded in T_6 (rice+ bagda+ fertilizer) treatment followed by T_8 (rice + tilapia + fertilizer) and T_3 (rice + bagda) treatments having 0.9 and 0.8 mean, the benefit: cost ratio respectively. The negative benefit cost ratio (- 0.1, mean) was observed in T_{10} (only javapunti) treatment. Das et al (19%) reported that the net returns as well as cost : benefit ratio were maximum in paddy fish and prawn culture.

Rice equivalent yield

The highest rice equivalent yields (7.3 t ha^{-1}) was recorded with T_6 (rice + bagda + fertilizer) treatment followed by T_3 (rice + bagda), T_8 (rice + tilapia + fertilizer), T_9 (only bagda) and T_7 (rice + javapunti + fertilizer) treatments having 5.4, 3.8, 3.7 and 3.6 t ha^{-1} respectively. However, the rice equivalent yields of javapunti and tilapia in association with and without rice and without application of fertilizer were generally lower than T_2 (rice + fertilizer) treatment (Table 48.2).

REFERENCES

Chattopadhyay G.N. Ghosh, Apurba and Sana P.K. (1983) : A note on the possibilities of salinization of soil under paddy-cum-brackish water aquaculture in some coastal saline soils, J. Indian Soc. Coastal Agric. Res. 1(1): 43-45.

Das, R.N. Laha, W.K. and Mazumdar, B (1996): Studies on technology on paddy-cum-fish and prawn farming in low lying coastal zone of West Bengal. J. Indian, Soc. Coastal Agric. Res. 14 (1 and 2): 229-232.

Natarajan, A.V. (1983) : Possibilities of Brackish water paddy-cum-fish Farming in Coastal Saline Soils, J. Ind. Soc. Coastal Agric Res. 1(1): 27-30.

Saha (1979) : Relative response of three nitrogenous fertilizers in different pond soil in relation to primary productivity, plantation and survival and growth of Labeo rohita Spwan. J. Inland fisheries Co. India. 7: 162-171.

Sinhababu, D.P. Jha, K.P. Mathur, K.C. and Ayappan, S. (1996) : Integreated farming system: Rice-fish for low land ecologies, Proc. of International Symposium on Rainfed Rice for sustainable Food Security, CRRI, Cuttack, Sep, 23-25.

CHAPTER 49

STUDIES ON THE REPRODUCTIVE BIOLOGY OF GUPPY, *POECILIA RETICULATA* (PETERS, 1860)

Vishal Tiwari

Department of Zoology, R.B.S. College, Agra-282002 (Uttar Pradesh), India.

ABSTRACT

The present study aimed to notice the reproductive biology of the most popular, exotic, freshwater, larvicidal, aquarium fish, *Poecilia reticulata* (Peters, 1860). Rapid sexual maturity, prolific breeding habit, sex recognition, courtship, fair formation, ova-viviparity, parental care and superfaetation were some of the most important aspects of the findings noticed during the work.

Keywords: *Fish, Poecilia reticulata, Reproductive biology*

INTRODUCTION

P. reticulata belonging to the *Poeciliidae* family is one of the renowned and cosmopolitan fish in terms of its unique beauty and larvicidal nature throughout the world. They are of great biological, ornamental and commercial value. They play an important role in the development along with the spread of the aquarium and has been firmly established among the top ten popular aquarium fishes for long most people care to remember. Owing to its brilliantly coloured patterns, it is also known as "rainbow fish" or "Peacock fish". They are supposed to be one of the most beautiful coloured fish of the live-bearing toothed carps. They were selected as an example of diversity of form in livebearers.

Perhaps this has triggered off many hobbyist, commercial breeders and exporters to systematically search for an attractive species and varieties by making a deep study of their reproductive biology. A good knowledge of the biology, feeding behaviour and ambient environment condition of the fish species is a pre-requisite for culture technique. Keeping this in view, the present study has been proposed to work out the detailed reproductive biology, which will bring out explorable bifold benefits in the fishery science. Some important contributions on the said field of studies are from Breeder and Coates (1935); Clark and Aronson (1951); Amoroso (1960); Bhatt (1968); Abrahams (1993) and that of Godin and Dugatkin (1995).

MATERIALS AND METHODS

A set of 30 healthy adult male and female of *P. reticulata* of the same age group in the ratio 2: 1 respectively were put in a glass aquarium of 24 X 12 X 12 inches (12 Gal or 55 ltr.) filled with potable water. It was equipped with adequate illumination, meeting the requirements of the plants, filter, a thermostat, an aerator pump and the accessories. Another small sized densely planted aquarium of 12 X 6 X 6 inches (6 Gal) was also prepared in the same manner having the same accessories except an aerator as a breeding tank. This was done in order to separate the females by putting them into it after successful mating. Optimum water quality parameters were maintained in both of these. The water of both was changed at regular intervals. A readymade food (Hitachi food) along with some live food (dried daphnia etc.) was given to the fishes. Anti-bacterial and anti-fungal medicines were also added in the form of drops in the aquarium water at some regular interval.

RESULTS AND DISCUSSION

The males of *P. reticulata* were noted comparatively smaller, more brighter, colourful and as a polygamous active members for mating preferences. In order to fertilize the female, the anal fin of male has developed into a functional structure to insert the packets of sperms into the female fish. This structure is known as "gonopodium" (Fig. 49.4). The bright spectacular colours of males may be used for advertising their health, virility and presence to the females. They were continuously observed as persistent divers, engaged in courtship display, always chasing the females. Males sooner or later were send darting around a female, his fins spread and quivering. Acceptance of a male's advances is often signaled by head shaking on the part of the female, followed by a dipping of the anterior part of her body (with a corresponding raising of the posterior half), which thus renders her genital aperture more accessible to the male once he adopts the correct mating position. The males were noted to court the female by spreading all his fins like a peacock flaunting his tail, and excitedly approaches the female, usually a little from the rear or parallel. Males bends their body into 'S' shape. This type of behaviour was repeated (Fig. 49.2). As the males wings the gonopodium forward to make contant, the female will often fold her anal fin out of the way helping him even further. When the males were in an advantageous position, they makes a quick thrust with the gonopodium which can be promptly moved at right angles, forward and side ways, and with uncanny precision inserts its tip in the egg duct of the female, delivering a charge of milt.

After successful meting the females transferred to the breeding tank showed their chest fills out almost upto the head and her middle gets quite distended, as the days of delivery approaches. Females about to give birth were recognized by their round tight stomach, which in late pregnancy become cuboidal in shape. A pregnant female were identified by the gravid spot (darkened area) behind her and fin just posterior to the belly (Fig. 49.1). When viewed from above, her strides appear swollen. Disturbing a pregnant female may result in premature deliveries.

About 3-5 weeks of gestation period, the females gives an average birth up to 30-50 young ones (ovo-viviparity) (Fig. 49.3). The fry come out of the mother with their tails first followed by their anterior part of the body. The young ones were measured about 1/8 inch (3 mm) long. At the age of 30-40 days, the sexes can be distinguished by the development of larger body of female and other sexual dimorphic features too. The fry were noticed gaining fast maturity and within average 3-5 months they get ready for taking part in a successful mating. Bearing live young among ***Poeciliids*** presents a further stage in the protection of the young (Parental care).

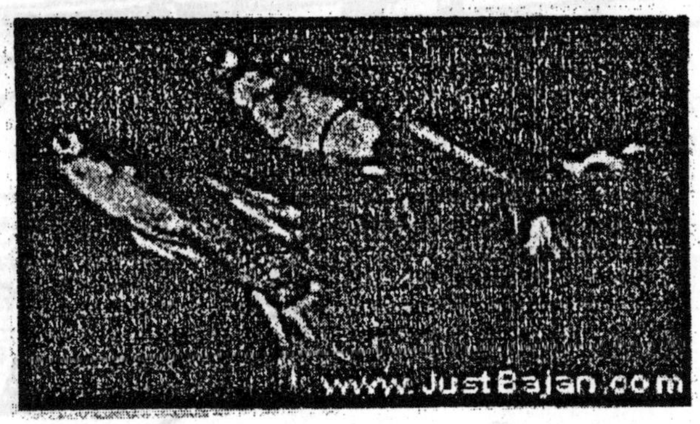

Fig. 49.1

Pregnant Female of *P. reticulata* (above male) with gravid spot on stomach.

Fig. 49.2

Courtship Dance during mating in *P. reticulata*

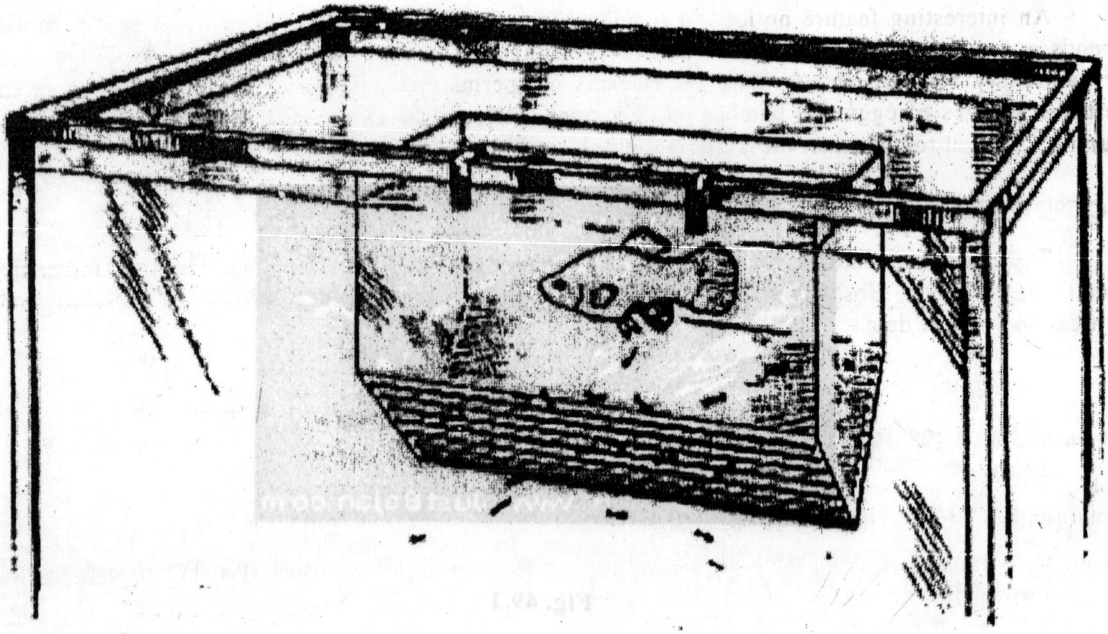

Fig. 49.3

Birth of Livebabies in *Poecilia Reticulata*

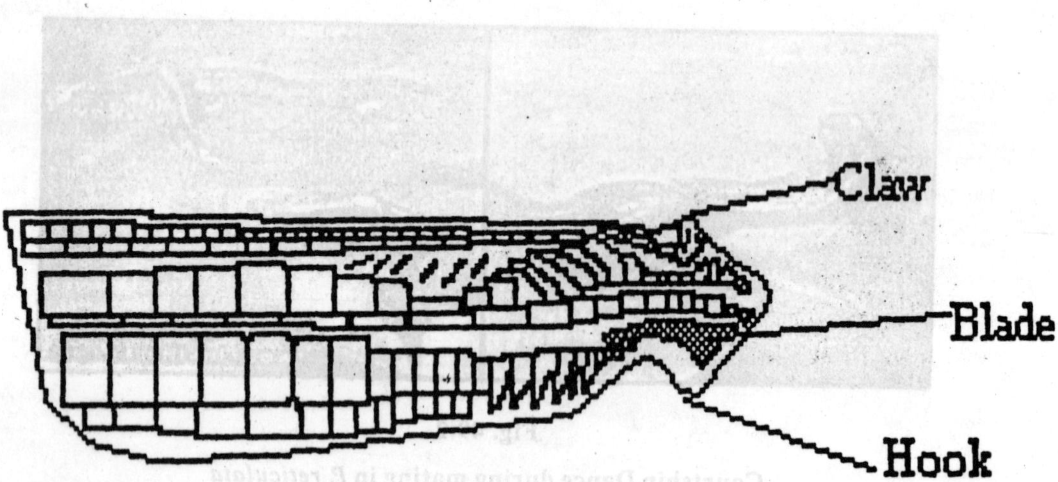

Fig. 49.4

Modified Anal Fin of *Poecilia Reticulata*

An interesting feature noticed in females that a fertile female have as many as half a dozen broods even if no males were present in the breeding tank. This behaviour may be attributed as a precautionary measure and to store the packets of sperms ejected by the male and uses them to fertilize a series of eggs and hatches as they ripen, following a single mating, females produces a number of fry at regular interval (superfaetation).

ACKNOWLEDGEMENTS

The author is thankful to Dr. A. K. Sharma, Scientist Incharge, C.I.F.E., ICAR. Deemed University Center Powarkheda, Hoshangabad (M.P.) for invaluable cooperation, procurement of literature and critical analysis of data.

REFERENCES

Abrahams, M. V. 1993. The tradeoff between foraging and courting in male guppies, *Poecilia reticulata. Anim. Behav.* 45: 673-681.

Amoroso, E. C. 1960. Viviparity in fishes. *Symposia. Zoological Soc. Loud.* 1: 153-181.

Bhatt, V. S. 1968. Studies on the reproductive biology of some freshwater fishes (Part IV) *Heteropneustes fossilis* (Bloch). *Ind. J. Fish.* 15: 99-155.

Breder, C. M. and Coates, C. W. 1935. Sex recognition in the guppy, *Lebistes reticulatus. Zoologica.* 19: 187-207.

Clark, E. and Aronson, L. R. 1951. Sexual behaviour in guppy, *Lebistes reticulatus* (Peters). *Zoologica.* 35: 49-66.

Godin, J. and Dugatkin, J. G. J. 1995. Variability and repeatability of female mating preference in *Poecilia reticulata. Anim. Behav.* 49: 1427-1433.

INDEX

O

P

R

O

P

R

X

Z